AF722166

II GRANADA LECTURES IN COMPUTATIONAL PHYSICS

II GRANADA LECTURES IN COMPUTATIONAL PHYSICS

Almuñécar (Granada), Spain 7 – 12 September 1992

edited by

Pedro L. Garrido
Joaquín Marro
University of Granada

World Scientific
Singapore • New Jersey • London • Hong Kong

Published by

World Scientific Publishing Co Pte Ltd
5 Toh Tuck Link, Singapore 596224
USA office: 27 Warren Street, Suite 401-402, Hackensack, NJ 07601
UK office: 57 Shelton Street, Covent Garden, London WC2H 9HE

British Library Cataloguing-in-Publication Data
A catalogue record for this book is available from the British Library.

II Granada Lectures on
COMPUTATIONAL PHYSICS

Copyright © 1993 by World Scientific Publishing Co Pte Ltd

All rights reserved. This book, or parts thereof, may not be reproduced in any form or by any means, electronic or mechanical, including photocopying, recording or any information storage and retrieval system now known or to be invented, without written permission from the publisher.

For photocopying of material in this volume, please paye a copying fee through the Copyright Clearance Center, Inc., 222 Rosewood Drive, Danvers, MA 01923, USA. In this case permission to photocopy is not required from the publisher.

ISBN-13 978-981-02-1163-9
ISBN-10 981-02-1163-5

II GRANADA SEMINAR on COMPUTATIONAL PHYSICS
Almuñécar (Granada, Spain), September 7-12, 1992

STEERING COMMITTEE

J. F. Fernández — *Centro de Física, Instituto Venezolano de Investigaciones Científicas, Caracas, Venezuela.*

P. L. Garrido, Scientific Coordinator — *Departamento de Física Moderna, Universidad de Granada, Spain.*

R. de la Llave — *Department of Mathematics, University of Texas, Austin, Texas, U.S.A.*

J. Marro, Director of Seminar — *Departamento de Física Aplicada, Universidad de Granada, Spain.*

PREFACE

The term *Computational Physics* refers to the creative use of computers in scientific research. The generality of this definition is intentional: computational methods developed to deal with problems in physics (and mathematics) have also proved most useful in many problems of biology, chemistry, and geology, for instance. This interdisciplinary nature gives Computational Physics an important educational value. Furthermore, this discipline is recognized as having a significant potential for fundamentally benefiting the education of physics students. Namely, it helps the latter develop a deeper understanding of key concepts, and enhances their problem-solving ability that is the hallmark of well-qualified physicists. Educators need to face the task of outlining an appropriate curriculum for the subject by trying to take advantage of such unique qualities.

A basic one-semester course on Computational Physics may be efficiently implemented by following one of the few existing books written for the purpose. Alternatively, one may follow any good, standard textbook on numerical analysis, supplemented with concepts, methods, and equations variously selected according to the preferences of the teacher. This second procedure is perfectly reasonable, since consensus on a canonical program does not easily come by in such a heterogeneous field.

The situation is more complicated when one tries to teach a more advanced, state-of-the-art course which either complements or replaces the basic one-semester course. It is not easy to find anything that fully serves as a reference with this aim. There is a variety of interesting topics to choose from, but they are only described in the specialized literature, and this does not pay much attention to the pedagogical aspects. This is probably due to the fact that Computational Physics is nowadays a rapidly evolving field. That is, advances in both numerical analysis and the efficient use of computers are constantly incorporating new problems to the field. In fact, the numerical treatment of some old, difficult problems of physics is now so specific and productive that these have become most relevant within the context of Computational Physics. I presume that the experts are so excited about the new problems falling into their hands, about the development of clever methods of computation and programming, and about the proper choice and design of hardware and computer architecture, that insufficient attention is still being paid to pedagogical and educational aspects in general.

This concern has moved us to publish the contents of the Seminar on Computational Physics which is organized in Granada every two years. The Seminar aims at bringing together small, selected groups of students and (bilingual) active researchers on different aspects of Computational Physics. It is part of the doctoral programme of the

Institute for Theoretical and Computational Physics which is at present being set up at the University of Granada. The 1990 edition was devoted to an introduction to the Monte Carlo and Molecular Dynamics methods that has not yet been published. The *II Granada Graduate Seminar on Computational Physics* (Almuñécar, Granada, Spain, 7-12 September 1992) focused on some ideas and techniques for dealing with Dynamical Systems which are of interest in many different fields. Some of the key concepts described were: dissipative systems, chaos and strange attractors, analysis of time series, elimination of noise, fractals, orbits, bifurcations, hyperbolic sets, classical and quantum chaos, simplectic methods, KAM theorem, renormalization group, etc., and their applications, as well as some other topics of current interest in Statistical and Condensed Matter Physics such as materials physics, fluids, growth models, kinetics of phase transitions, nonequilibrium ordering phenomena, diffusion, quantum chromodynamics, galaxies, and quasars.

The present book contains the invited lectures notes and an account of communications by participants at the meeting. An effort has been made to write the notes pedagogically for the lowest graduate level, and to give quite a comprehensive description of each topic. Moreover, some practical computer sessions are illustrated. In this way, the book may serve to introduce the beginner to the above-mentioned areas of Computational Physics, and as a work of reference for teachers, researchers and students.

I wish to express my gratitude here to all those who have collaborated in making the 1992 edition of the Seminar a success. I would first like to mention the invited lecturers who established such a good rapport with the participants and whose personal effort helped elaborate the format and enabled us to accomplish the objectives of the Seminar, and Prof. P. L. Garrido, who acted as a scientific coordinator and co-editor. The smoothness of the organization was ensured by the collaboration of Dr. J. J. Alonso, Mr. A. Achahbar and Mr. M. A. Muñoz. We would like to acknowledge financial support given by the *Dirección General de Investigación Científica y Técnica* of the Spanish Ministry of Education and Science and the *Consejería de Educación y Ciencia* of the Andalusian administration *Junta de Andalucía*. Last but not least, the meeting has greatly benefited from the University of Granada, from the help of the previous and the present rectors, both of whom decisively endorsed our initiative, as well as from the excellent framework provided by the University of Granada's International Summer Courses in Almuñécar, directed by Prof. J. F. García Casanova, whose constant support is gratefully acknowledged, and from the local authorities *Excelentísimo Ayuntamiento* of Almuñécar, the capital of the *tropical coast* of Granada. I hope that our combined efforts will be shown to be sufficiently worthwhile to merit the continuity of this Seminar.

Granada
September 1992

Joaquín Marro

CONTENTS

CONTRIBUTIONS

Invited Lectures

APPLICATIONS OF DYNAMICAL SYSTEMS TO THE STATISTICAL ANALYSIS OF TIME SERIES

JAVIER CABRERA
Department of Statistics, Rutgers University,
New Brunswick, NJ 08904, U.S.A.

ABSTRACT

A comprehensive dynamical systems approach is presented for the modeling of time-series. The method of time-delay embedding for strange attractor reconstruction from a time-series is reviewed and several diagnostic tools for time-delay selection are presented in detail. In particular, special attention is given to the possible effects of random measurement noise and dynamic noise. The methodologies of MARS and SMART are applied for the elimination of noise, and efficient algorithms are proposed for the computation of the Grassberger-Procaccia dimension. Finally, a graphics interface which incorporates important parts of the methodology is presented.

1. Dynamical Systems.

1.1 Introduction

For the last forty years the study of time series – When are two time series instances of the same phenomenon ?, Is it possible to extrapolate to guess the future behavior? – has occupied an important place in statistical research.

For many series appearing in real life problems in which no obvious pattern was apparent, the most standard method has been to try and fit autoregressive models, specially linear ones and the more general ARIMA models. In many cases, the main reason to choose autoregressive models is the convenience of use among the models that produce complicated behaviors similar to the ones observed in real phenomena. They are not justified by an underlying physical theory.

In recent years, however it has become increasingly recognized that very simple deterministic models can have also complicated dynamical behavior. (The current buzzword is *"chaos"*.) Then, it becomes very natural to fit complicated time series with non-linear deterministic models. This has the advantage that for many physical systems – e.g. turbulence in fluid mechanics Ruelle e.a.[26] – it is possible to argue physically that the underlying model should be deterministic. This belief is reinforced by the fact that the experiments present a degree of reproducibility higher than what one would expect from stochastic models.

Perhaps the most obvious case of deterministic models that have complicate behavior is that of pseudo-random number generators. In particular, the linear congruential ones are defined by the equation

$$x_{t+1} = A + Bx_t \ MOD \ C$$

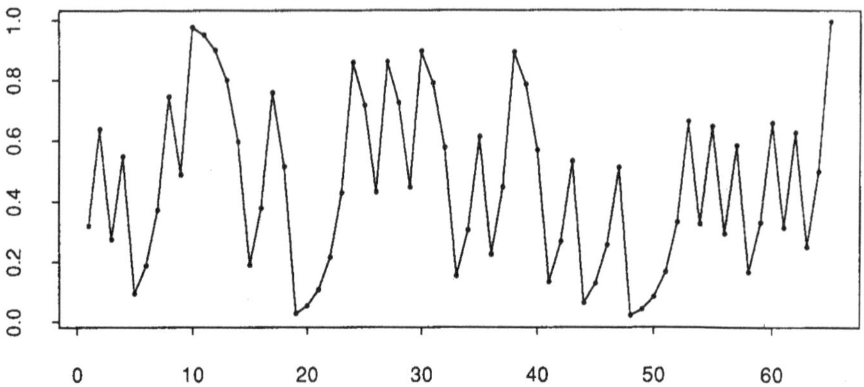

Figure 1. Bernouilly Shift Series.

where A, B, and C are fixed constants. The deterministic sequences generated by RNG's are used for Montecarlo simulations because in most practical situations they behave like truly uniform random sequences.

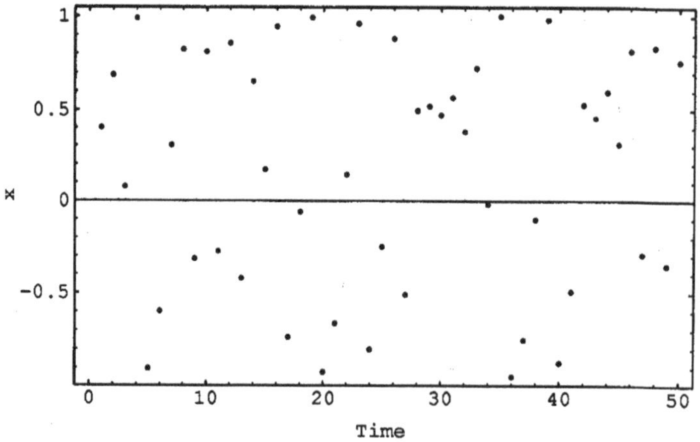

Figure 2. $x_{t+1} = 1 - 1.95\ x_t^2$ Series.

In Fig.1 we show a graph of such a sequence starting at $x_0 = \pi - 3$, for the case $A = 0$, $B = 2$, and $C = 1$, which is also known as the Bernouilly shift.

A second trivial example is shown in Fig.2 where the series

$$x_{t+1} = 1 - 1.95\ x_t^2$$

is plotted against t.

Of course, in both cases, if we plot x_{t+1} versus x_t then the nonlinear function which generated the sequence is easily detected. Figure 3 shows the plot for the first example.

These examples also show that in some cases it is reasonable to model complicated time series with non-linear deterministic models. In some practical applications the justification of the modeling will be based on theoretical grounds, but in other cases it is possible to justify the modeling on the empirical evidence obtained from the data. In such cases we hope to find a dynamical system that will reproduce the same non-linear behavior of the time series.

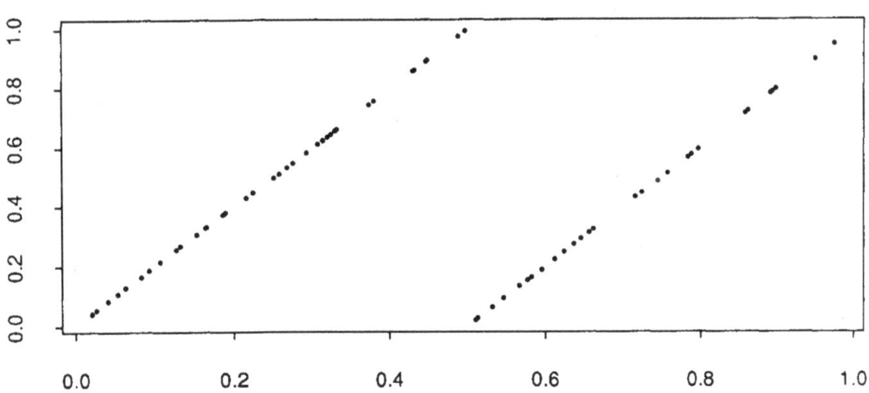

Figure 3. Plot of x_{t+1} versus x_t for the Bernouilly Series.

One question that arises immediately is that, besides the intrinsic chaotic behavior due to the non-linear dynamics, the observed time series are affected by experimental errors that can be considered as truly stochastic.

In this work, we describe the dynamical systems approach to the study of time series, as well as several tools we have developed for this study. In addition, we present methodology for estimating observables such as the dimension which characterize the dynamical system.

We will start with a review of the basic ideas on dynamical systems followed by the method of the delays for the reconstruction of a dynamical system from a time series. In section three we study the problem of selecting the statistical model, and in section four we consider several new methods for model fitting. Section five is dedicated to the problem of estimating the dimension of strange attractors, and finally section six we present a graphics interface and other programs which implement all the above methodology.

We refer to Eckmann e.a.[12] for background on dynamical systems theory and more details on their applications to model experiments.

We also refer to Eckmann e.a.[13], Farmer e.a.[14], and Kostelich e.a.[23] for reviews of the basis and applications of non-interactive versions of algorithms similar to ours.

1.2. Some Basic Ideas on Dynamical Systems and Strange Attractor Reconstruction

A dynamical system is a rule that given one point in a state space, it tells us where it is going to be a unit of time later. For example let x_n be the density of flies in year n then,

$$x_{n+1} = a\ x_n(N - x_n).$$

The basic mathematical model of dynamical systems is a manifold M, a map $f : M \to M$ and a function $\varphi : M \to \mathbf{R}$. f is a diffeomorfism, that is f is differentiable and with differentiable inverse.

Physically, M represents all the possible states of the system — or at least those explored by the evolution of the system we are observing. f represents the time evolution, a rule that determines what is the state one unit of time later. Hence, an orbit of the system starting in x_0 is $\{x_n\}_{n=0}^{\infty}$; $f(x_n) = x_{n+1}$. φ is an observable that, depending on the state of the system takes different values.

The output of the experimental device is a sequence of observations which, in the absence of experimental error, will be $\{\varphi(x_n)\}_{n=0}^{\infty} = \{\varphi(f^n(x_0))\}_{n=0}^{\infty}$.

For many practical cases, after some time the system converges to a subset of M of low dimension, usually called the "attractor".

This attractor is the only relevant piece of M for the long term behavior of the system. Very frequently these attractors have very complex geometrical structures, in particular, they may be fractal.

The motion on the attractor can exhibit different types of behavior from periodic and easily predictable to chaotic.

If we could reconstruct the attractor and the map on it, we would then be able to extrapolate the time series.

It was pointed out by Packard e.a.[24], Eckmann e.a.[12], and Kostelich e.a.[23], that, in many cases, it is possible to find $d_1, \ldots, d_k$ positive integers in such a way that $\varphi, \varphi \circ f^{d_1}, \ldots, \varphi \circ f^{d_k}$ provide an independent set of coordinates on the manifold — (mathematically precise definitions and theorems that substantiate this belief were given by Takens[29]).

The importance of this observation lies in the fact that all these coordinates are available from the time series, since $\varphi \circ f^{d_i}(x_t) = y_{t+d_i}$. Secondly, the point $P_t = \{(\varphi(x_t), \varphi \circ f(x_t)^{d_1}, \ldots, \varphi \circ f(x_t)^{d_k})\}$ is the point x_t in this coordinate system, and

P_{t+1} is the image of P_t by f. Finally, if the manifold M is of sufficiently low dimension, it is possible to plot P_t in $\mathbb{R}^{k+1}$, hence getting an image of the manifold and of f.

In the next two sections we will focus on the selection of k and of the delays $d_1, ..., d_k$, and on noise reduction. But first we will study four examples.

1.3 Examples

In Figure 4 there are four examples of time series that are going to be revied along these notes. Three of them are real data from real problems corresponding to very different areas, Fluid dynamics, electroencephalogram(EEG), and IBM stock prices. The other example is the Lorenz system that was generated from a set of three differential equations and we are going to utilize as testing ground for these the methods.

1.3.1 Physical Example: Couette Flow (H.Swiney, D.Lathrop)

In Fig. 4.2, we present a series generated in a hydrodynamic experiment with a weakly turbulent flow. This experiment consists of a large cylindrical container filled with water whose outer boundary rotates at constant speed. Every unit of time the velocity at one spot is measured using a laser beam.

The state space is the space of all possible velocity fields and there is a dynamical system, the rules of hydrodynamics described by the Navier-Stokes equations.

The observable $\varphi(x)$ is the functional that to every velocity field associates the value of the speed at one point $|v(x)|$.

A detailed observation of Figure 4.2 shows a clear regular pattern that is not periodic and therefore this series is a good candidate for dynamical system modeling.

A second diagnostic tool that is a good indicator of whether a series could be modeled by a dynamical systems is the periodogram(or log-periodogram) of the series. This is defined as

$$I(\omega_j) = \frac{1}{N} \, |\sum_{t=1}^{N} x_t e^{-it\omega_j}|^2 \quad , \qquad \omega_j = \frac{2\pi j}{N}.$$

For a series generated from one of the linear autoregressive models or the more general ARIMA models that were mentioned in the introduction, the periodogram is a rational function–this is a well known fact that can be found in any good time-series text book[4]. Furthermore, if the periodogram of a stationary time-series is a rational function then the series can be modeled by an ARIMA model[4].

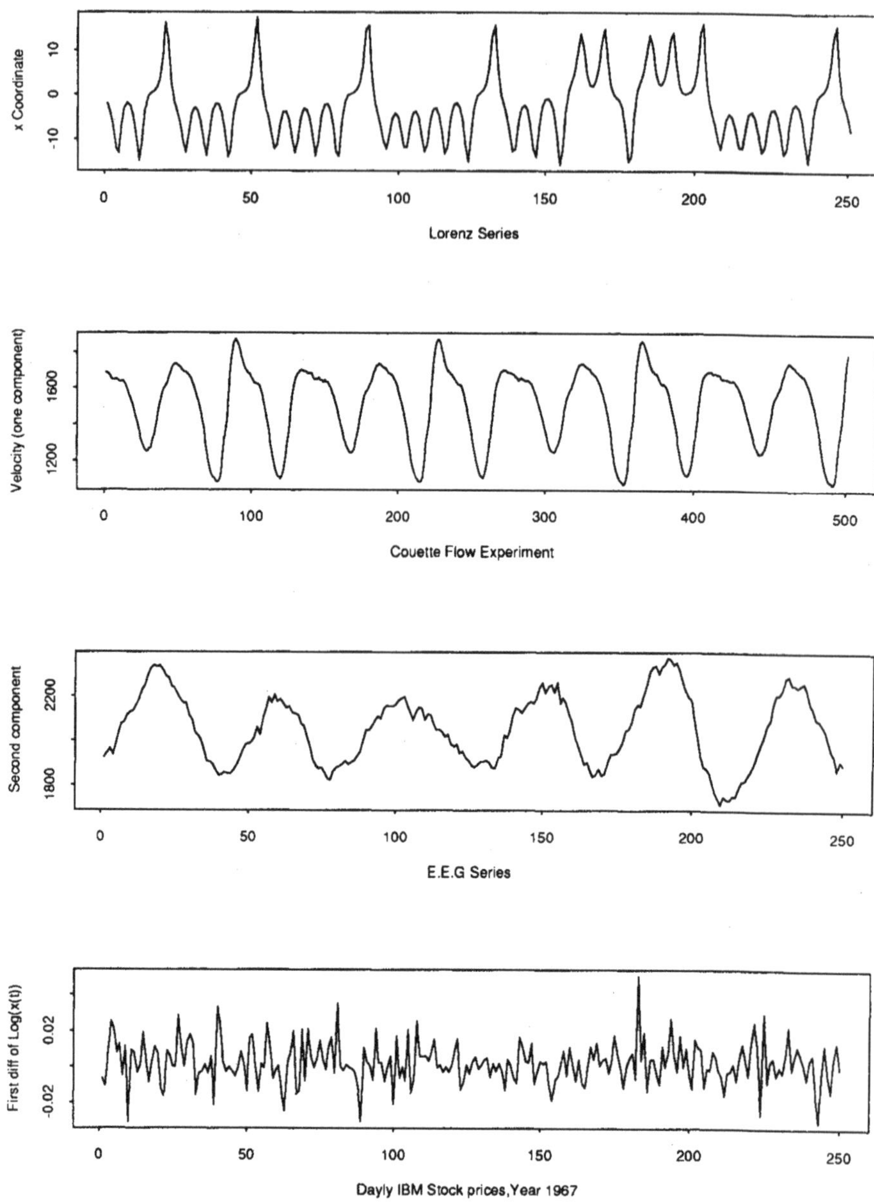

Figure 4. Four examples of time-series. (1) Lorenz Series, (2) Couette Flow Series,(3) E.E.G. Series, (4) Daily IBM Stock, First Diff. of Log Series.

Therefore we are only interested in modeling series for which the periodogram is not a rational function. For example if the periodogram has exponential decay then we can try a dynamical system approach for modeling the time series.

For the example of the Couette Flow Series Fig. 5.2 shows that the Log-Periodogram has linear decay and therefore the Periodogram has exponential decay, and it confirms our prior observation.

1.3.2 The Lorenz Attractor

A well known example of dynamical system is the Lorenz attractor which is given by the following three differential equations:

$$\begin{aligned}\dot{x} &= \sigma(y - x)\\ \dot{y} &= -xy - \gamma - y\\ \dot{z} &= xy - bz \ .\end{aligned}$$

Figure 4.1 shows a series generated from the above equations by observing the x coordinate of a solution at discrete equally spaced intervals of time. The values of the parameters for that series are $\sigma = 10$, $b = \frac{8}{3}$, and $\gamma = 28$.

We will utilize this series as a ground for testing and comparing the various methods that will be presented here, since we already know most of the answers.

Figure 4.1 shows a clear regular non-linear pattern, and the Log-Periodogram in Figure 5.1 shows a clear linear decay, and hence both diagnostics point to the right answer.

1.3.3 The EEG Series

The series in Figure 4.3 was provided by Luciano Mollinary and it is one of seventeen time-series obtained from electrodes placed in different parts of the head. The rate of observation is close to 25000 observations per minute.

Figure 4.3 shows clearly that this data has a far stronger noise component than the prior examples. However it is possible to see a regular non-linear pattern. This is a good example of a time-series with a large noise component but at the same with a strong non-linear component.

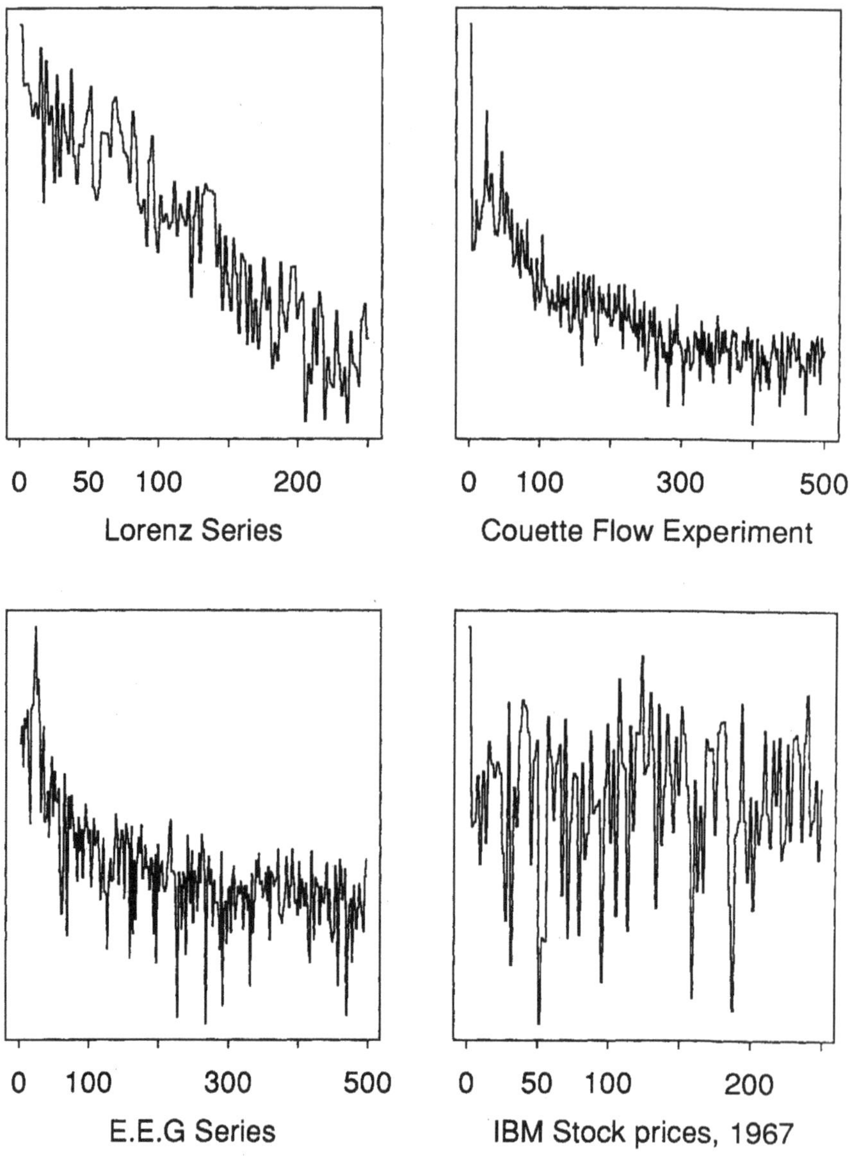

Figure 5. Log-Periodogram of the series. (1) Lorenz Series, (2) Couette Flow Series,(3) E.E.G. Series, (4) Daily IBM Stock, First Diff. of Log Series.

For this series Fig. 5.3 shows that the Log-Periodogram has linear decay near the origin and therefore the Periodogram has exponential decay, and it confirms our prior observation.

The first objective of this analysis is to eliminate the noise and model the data with a dynamical system. The long term goal of the analysis is to be able to find a characterization of these series that are meaningful from the medical point of view, and that provide useful information on the patient.

1.3.4 Daily IBM Stock Price

Figure 4.4 shows a time-series obtained from the daily IBM stock price. In order to eliminate the trend and to stabilize the variance of the series, a transformation was applied. The resulting series plotted in Figure 4.4 is the first differences of the logarithm of the original series, that is

$$y_t = log(x_t) - log(x_{t-1})$$

One reason for including this series is that in the past few year there have been rumors in the financial circles of the NY stock exchange that these series could be modeled by the methodology of dynamical systems. We will argue here that it is highly unlikely that this series can be model by a dynamical system. Another reason for including this time-series is that it is always useful to study an example were the methodology fails, in contrast with other examples where it works.

For instance, Figure 4.4 does not show any structure at all, and it is very different than the other three.

Moreover the Log-Periodogram of this time-series – see Fig. 5.4 – shows no pattern and again it confirms our prior observation.

Finally we end this section with a third diagnostic tool which confirms everything that we said about the examples. In Fig. 3 we showed a picture of a rule from the on the one dimensional real space. In two or more dimensions we can not draw a picture of the rule but we can draw an orbit.

Therefore, the analogous of Fig. 3 for these examples is to draw a curve through the points y_{t+k} versus y_t for consecutives values of t but fixing k to a suitable value. Figure 6 shows the four plots for the 4 examples which confirm everything that was said above.

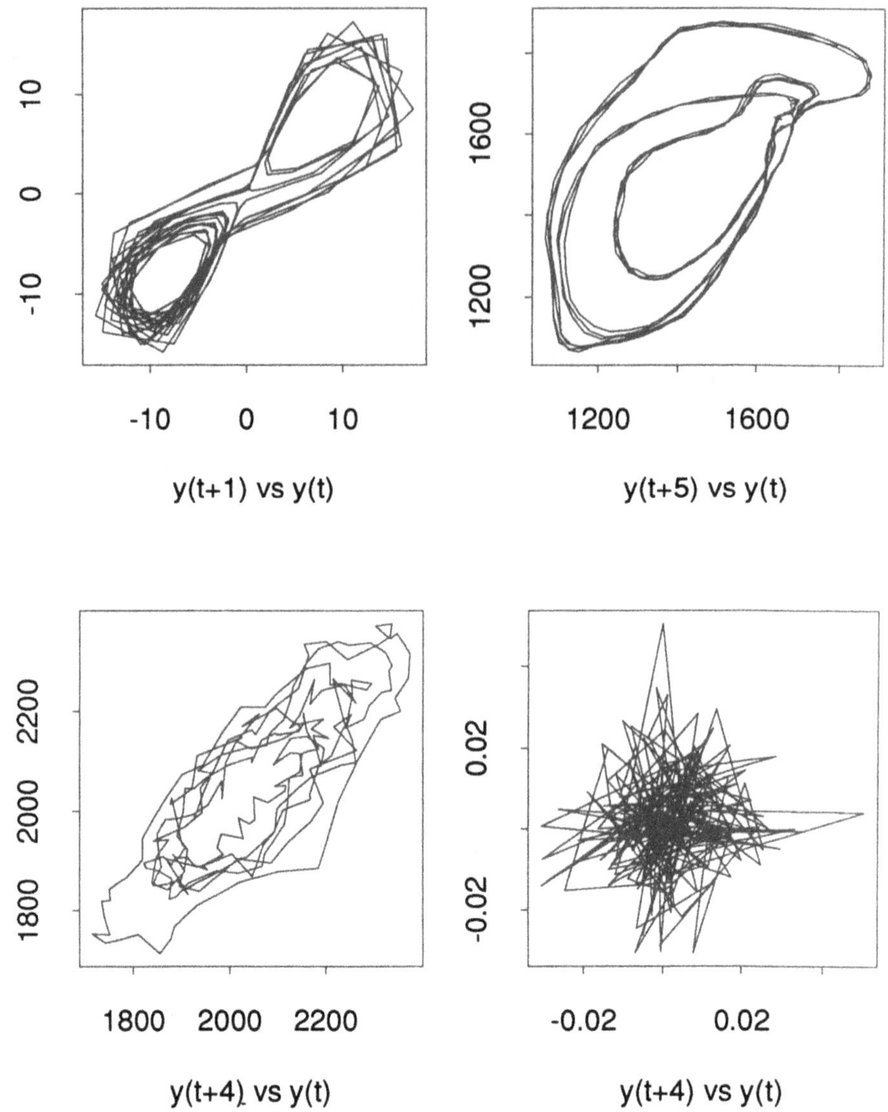

Figure 6. Plot of y_{t+k} versus y_t. (1) Lorenz Series, (2) Couette Flow Series, (3) E.E.G. Series, (4) Daily IBM Stock, First Diff. of Log Series.

The noise in the EEG data is more clearly exposed by the irregularities of the curve, and the orbit obtained from the IBM series shows no regular pattern.

In summary, the first three examples are good candidates for fitting a dynamical system model, but the IBM series is not since it fails all the preliminary diagnostics.

2. Statistical Models.

In practice most experimental data comes with a certain amount of observational error and the goal of the statistical analysis is to assess its level and eliminate it as much as possible.

If the only source of noise is measurement error it is sensible to assume that the observed time series y_t can be modeled by

$$x_t = f(x_{t-1}) \ ,$$
$$y_t = \varphi(x_t) + \varepsilon_t$$

where ε_t's are independent and identically distributed(i.i.d.) random errors with some unknown distribution. For now we do not make any assumptions about ε_t, except that it has zero expected value. We will refer to this model as the Observational Error(OE) Model.

In other cases it could happen that the source of noise is that the system is subject to random forces, such as external random forces, or interactions with another chaotic subsystem, in which case it is not unreasonable to assume that the dynamics are given by the following model.

$$x_t = f(x_{t-1}) + \varepsilon_t \ ,$$
$$y_t = \varphi(x_t)$$

where the ε_t's are i.i.d. random vectors of the same dimension as x. As in the case of the OE model, we assume that the distribution of the random vector ε_t is unknown, but has zero expected value. We will refer to this model as the Dynamical Error(DE) Model. It is also possible to combine both models by adding an observational error term to the DE Model.

Most known statistical model for time series are special cases of these models. In particular the ones mentioned in the introduction are of interest. In order to compare the DE and OE models with the conventional statistical models we will give some of their definitions.

We say that a time-series $\{y_t\}$ follows an autoregressive model of order p (AR(p)) if

$$y_{t+1} = \sum_{j=1}^{p} a_j y_{t-j+1} + \varepsilon_t \ ,$$

where the errors ε_t's are i.i.d. normal random variables with zero mean and variance σ^2 .

The more general autoregressive moving average model (ARMA) of orders (p, q) is defined by

$$y_{t+1} = \sum_{j=1}^{p} a_j y_{t-j+1} + \sum_{l=1}^{p} b_l \varepsilon_{t-l+1} ,$$

where the errors ε_t's are i.i.d. normal random variables with zero mean and variance σ^2 .

We say that a time-series $\{y_t\}$ follows an ARIMA model (autoregressive integrated moving average model) with parameters (p, d, q) if the time-series obtained by taking differences of $\{y_t\}$ d times follows an ARMA(p, q). The operation of taking first differences of a series $\{y_t\}$ consists of constructing a new series $\{z_t\}$ where $z_t = y_t - y_{t-1}$.

The above three models are linear models. A nonlinear version of the AR models is known as nonlinear autoregressive model (NLAR) of order p, and it takes the form

$$y_{t+1} = f(y_t, \ldots, y_{t-p+1}) + \varepsilon_t ,$$

where the errors ε_t's are i.i.d. normal random variables with zero mean and variance σ^2 .

For a complete reference on these models and their applications see any good time-series text book[4,6]. There are other non-linear model that we will not mention here, in particular threshold models[30,31] are also of wide use.

One important point is that the conventional models are special cases of the OE, and DE models. It is left to the reader as an exercise to construct dynamical systems which produce the four standard statistical models.

From the practical point of view one important advantage of the dynamical systems models is that we may choose delays $d_1, \ldots, d_k$ that are not necessarily consecutive, while in the conventional – linear and non-linear – models delays are always consecutive. Another important difference is that in the dynamical systems models we estimate the function f from the p-dimensional real space into itself whereas in the conventional statistical models the image of f is the real line. For the time-series that we are trying to model here these differences are very important as we will see in the analysis of our examples.

The methodology advocated here for fitting the OE and DE models consists of two steps that are analogous to the usual methods for the conventional models. First of all is what we call model selection which consists of selecting the delays $d_1, \ldots, d_k$ and the order $k+1$ of the model, and secondly is model fitting or in other words estimation of the non-linear rule f. These two topics are the subject of the next two sections.

3. Model Selection.

In this section we explore the methodology of selecting the delays $d_1, \ldots, d_k$ and the order $k+1$ of the model. Each set of delays determines a different reconstruction of the strange attractor and of the dynamical system which will be represented by the orbit in the coordinate system of the delays, which we represent by the finite orbit $\{P_1, \ldots, P_N\}$.

Since we are allowed to choose nonconsecutive delays, the number of possible choices is very large. Tu illustrate the point notice that if $k = 3$ and $d_k \leq 20$ the we have 1351 possible ways of choosing the delays and the same number of models. And this is one of the simpler cases!.

This means that any careful algorithm that automatically checks all possible models in order to find the one which minimizes a certain criteria, would require either extensive computations or a more refined theory. At the end of this section we will discuss possible numerical criterias, that maybe used to compare a reduced number of models.

The alternative is to provide a graphics interface which combines methods of exploratory data analysis (E.D.A.) and more precise diagnostics. An important advantage of the E.D.A. methods is that, when the attractors have small dimensions (up to 5, say), it is possible to obtain a geometric intuition of them and of their local properties.

The diagnostic tools give us an idea of the goodness of reconstruction and of the noise level. The delays thus obtained will be the input of a more elaborate algorithm that will produce the fitted model and an assessment of the goodness of the fit. Therefore, the correctness of the final results does not depend entirely on the exploratory phase.

Next we propose a procedure, or even a systematic method in order to reduce the number of models that have to be checked and(or) fitted, which comes from the experience of using them. This is by no means a unique or best procedure, but it is a useful guideline.

(i) Decide on the time scale by checking the Autocorrelation function.

(ii) Choose a relatively large number of delays and check with the diagnostics that the reconstruction is good enough.

(iii) Reduce the number of delays until the diagnostics fail and fix the number of delays $k+1$.

(iv) Improve the reconstruction by changing the delays slowly.

In order to carry out these steps we need a good graphics interface which combines all the necessary tools for the analysis. Next we will describe all the specific tools that are incorporated in this methodology of model selection, and in section 6 we will introduce and discuss the graphics interfaces.

3.1 Autocorrelation Function

Given the time series $\{y_t\}_{t=1}^N$ we define the autocovariance and autocorrelation functions by

$$c(r) = \frac{1}{N-r}\sum_{i=1}^{N-r}(y_i - \bar{y})(y_{i+r} - \bar{y}) \ ,$$

$$\rho(r) = \frac{c(r)}{c(0)} \ .$$

If the time interval between observations is too short then, for small values of r, $\rho(r)$ is nearly one. This means that if two coordinates of the coordinate system correspond to delays less than r then the level curves that they generate on the attractor are nearly parallel, and points which are very far apart in any good coordinate system will be very near in this one. Therefore this is a bad coordinate system and we should either use larger delays, or move to a higher scale by selecting only every k'th observation.

Figure 7 shows a plot of the autocorrelation function for the Lorenz Series, after which it was recommended to select every one hundred'th observations.

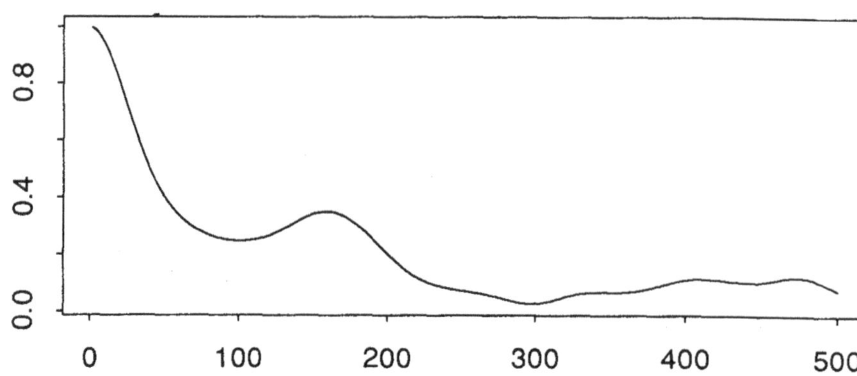

Figure 7. Autocorrelation function for the Lorenz Series with a small time interval.

3.2 Diagnostics for Dynamical Systems.

Even if exploration can identify quickly good candidates for delays, it is quite desirable to have quantitative objective criteria that test whether the series is adequately modeled.

In many applications, it is also necessary to find out whether a theoretical model – for which time series are readily generated numerically – can describe the physical system whose time series are experimentally observed. Direct comparison of the reconstructions is impossible since the delay method only reconstructs the attractor up to a smooth change of variables.

A natural way of solving this problem is finding observables whose value is not altered by making a smooth change of variables in the system and compute them both in the theoretical model and in the experimental time series.

In this section we discuss two methods to verify the basic hypothesis that there is a dynamical system underlying the experimental time series and one observable that can be used to discriminate between two dynamical systems.

The list is not exhaustive and is restricted to the ones we have implemented. Notably, we do not discuss resampling methods to assess the validity of reconstruction nor the computation of Lyapounov exponents.

3.2.1 Recurrence Plots

This diagnostic tool was introduced by Eckmann e.a.[12]. Its goal is to test the basic hypothesis that the state of the system at a certain time determines those at future times. Thus, if two states are very close on the attractor the orbits starting at them should remain close for a certain time.

Suppose our time series is y_t. Fix $d_1, \ldots, d_k$ positive integers and denote the coordinates of x_t by $P_t = (y_t, y_{t+d_1}, \ldots, y_{t+d_k})$. If $\varphi, \varphi \circ f^{d_1}, \ldots, \varphi \circ f^{d_k}$ were a good system of coordinates, P_t would determine the state of the system at time t. It would also determine P_{t+1}. Therefore if P_t and $P_{t'}$ are close, we should expect P_{t+i} and $P_{t'+i}$ would remain close for small values of i.

Recurrence plots are obtained by putting a black dot at the point (i, j) if $d(P_i, P_{j+i}) \leq \varepsilon$.

The parameter ε should be on one hand, small enough that the map f is roughly linear in balls of radius ε and on the other hand, big enough that there are many pairs of points at distance less than ε. (If this condition cannot be met, it is obvious that there is not enough data to attempt a reconstruction.)

If the system were deterministic, and if the set of delays provided a good system of coordinates and if ε were as above, we would observe an abundance of horizontal lines and very few isolated points.

By performing this analysis over many sets of delays it is possible to select promising ones.

The conclusions are not affected by noise levels smaller than ε.

One nice property of this plot is that it let us compare different pieces of the orbit within the same plot. This is useful for detecting slow changes of the dynamical system and it let us pick pieces of the time series that maybe easier to model.

On the other hand it is easy to have resolution problems on the graph, and for moderately large series the user may need to generate many graphs in order to cover all the series. Moreover it may give the wrong information if one uses consecutive delays, for if we choose delays $1, \ldots, k$ the distance

$$|P_{i+1} - P_{j+i+1}| \leq |P_i - P_{j+i}| + |y_{i+k+1} - y_{i+j+k+1}|$$

and if k is moderately large, the distance between the closer points on the attractor may not be sufficiently small compared with the variability of one coordinate. For this diagnostic to work we would have to observe an orbit long enough so the distance between the closer points on the attractor would be sufficiently small.

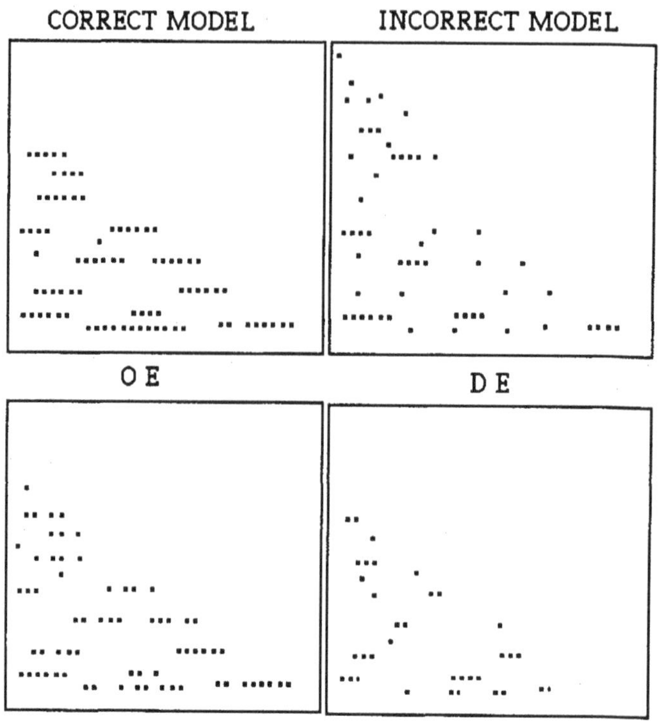

Figure 8. Recurrence Plots Examples.

Figure 8 shows four examples of what is expected of the plot for the correct and incorrect model. We also see the effects of OE and DE in the plot. We see that OE would eliminate points at random from the plot, whereas DE would make for shorter runs.

Figure 9 shows the plot for the reconstruction obtained from the Couette Flow time-series in five dimensions with delays 0, 2, 5, 8, and 12. This is one of the reconstruction that produced better diagnostic plots for this particular series, although there were other reconstructions that where almost as good.

We refer to Eckmann e.a.[12] and our paper in preparation for a fuller discussion of how to diagnose several situations with the help of this plot.

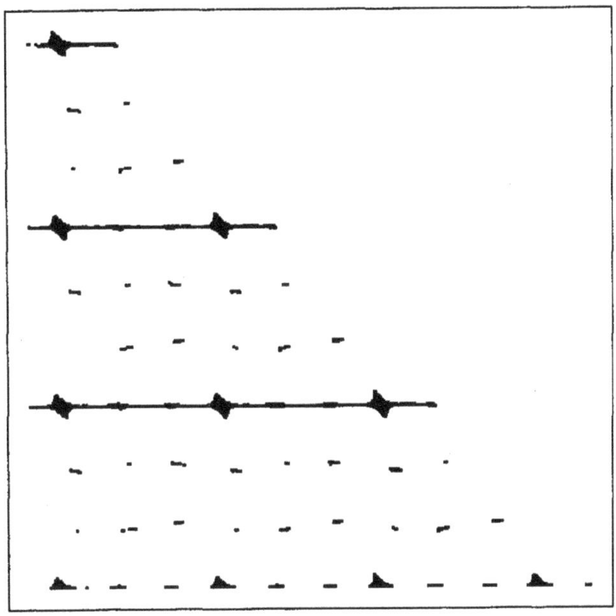

Figure 9. Recurrence Plot of the Couette Flow Series.

3.2.2 Wedge Plots

This diagnostic tool is based on a quantitative statement of the implication of deterministic dynamics.

For a differentiable dynamical system, the mean value theorem implies

$$d(x,y)\Big(\sup_{z\in\gamma_{f(x)f(y)}}|f^{-1\prime}(z)|\Big)^{-1} \leq d(f(x),f(y)) \leq \sup_{z\in\gamma_{xy}}|f'(z)|\,d(x,y)$$

where $\gamma_{x,y}$ is the geodesic joining x, y.

Hence, if we plot $d(f(x),f(y))$ versus $d(x,y)$, the resulting points will fall between two straight lines passing through the origin.

If there is noise present — either on the measurements or on the dynamics themselves, no such bounds are possible since points arbitrarily close can be moved to an amount proportional to the noise level.

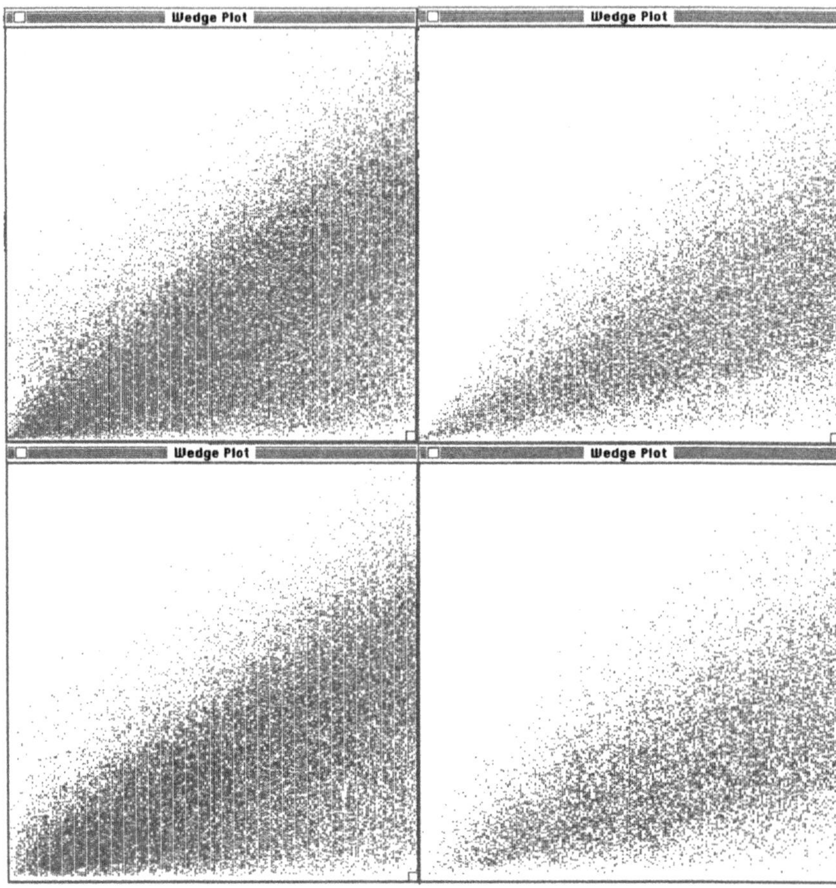

DIM = 2 **DIM = 3**

Figure 10. Wedge Plot of Lorenz Series Without Error(Top), and with DE added(Bottom), for dimension 2(Left) and 3(Right).

We expect inequalities of the form

$$\begin{aligned} d(f(x), f(y)) &\leq Md(x,y) + \alpha \\ d(f(x), f(y)) &\geq md(x,y) - \alpha \end{aligned} \tag{1}$$

These plots are more informative if the only pairs of points (x, y) used are points that are within the domain of applicability of the linear approximation. In that case, we expect that the bounds will be saturated and a characteristic wedge will appear in the plot. (Hence, we will refer to them as "*wedge plots*".)

In the presence of noise whose amplitude is smaller than the range of validity of the linear approximation we also expect that the bounds (1) are saturated and, hence we will observe all the points lying on a region bounded by straight lines with non-trivial intercepts. Unfortunately, in the case of noise, the boundaries tend to appear blurred.

Figure 10 shows four example of wedge plot for the Lorenz series with and with error added. The top two plots are without error added, and the bottom ones correspond to the series with DE from a uniform $(-0.5, 0.5)$ distribution. The plots on the left correspond to delay 1, and the ones on the right are using the correct delays 1, and 2.

In both cases, the plots on the right hand side of the graph show a clear wedge, while the ones on the left show many points with small x coordinate and large y coordinate. Of course the plots at the bottom of the graph are not as sharp as the top ones due to the presence of DE but still shows a clear preference for the model with delays 1 and 2.

Observe that it is also possible to include in the plot only those points that lie on a certain region. Hence, it is possible to diagnose the quality of the linear approximation in different parts of the manifold where the motion takes place. This tool combined with brushing is a very powerful tool to examine the changes in motion in different regions of the manifold.

Finally Figure 11 shows three wedge plots obtained from brushing in three regions of a reconstruction of the attractor from the Couette Series with delays 0, 2, 3, 4, 5, 7. It is possible to observe from the bottom figures that there is some amount of error, and one could possibly get a rough estimate of the level of the error. Following the procedure that was specified at the beginning of Section 3 it was found that we could reduce the dimension of the reconstruction to 5 with delays

0, 2, 5, 8, and 12.

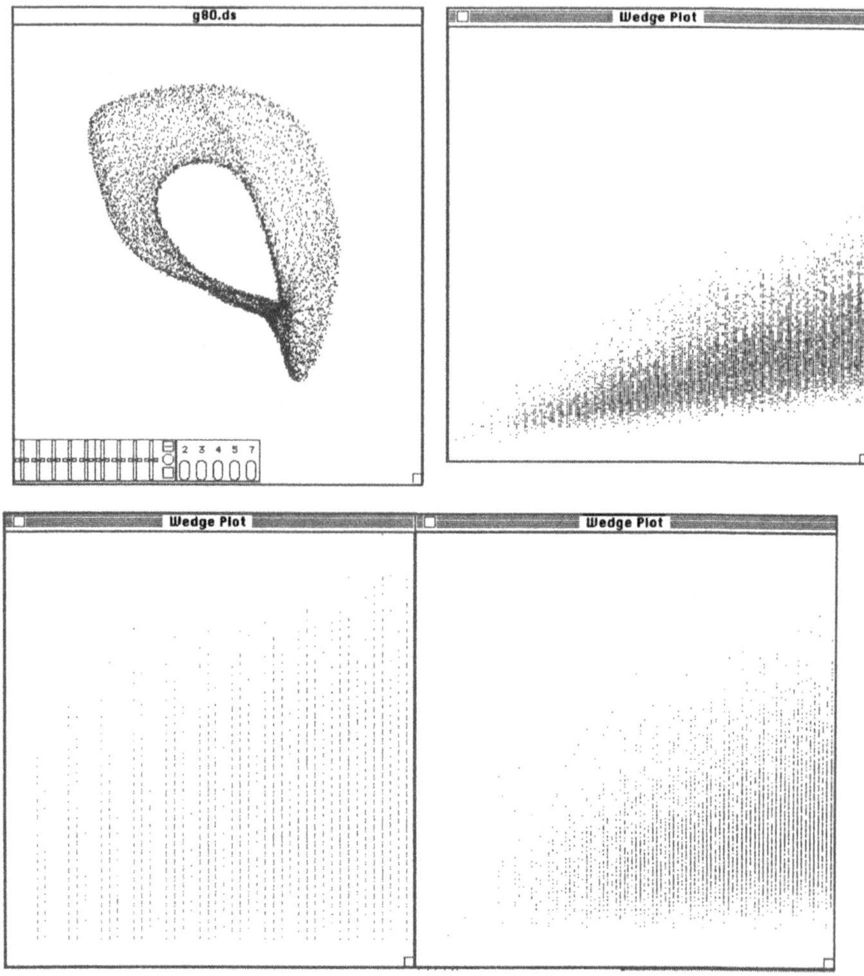

Figure 11. Wedge Plots of three regions from a reconstruction from the Couette Series for dimension 2(Left) and 3(Right).

3.3 Criterias for Model Selection

In this section we will discuss the basic criterias for model selection for the conventional models, which could be also applied for the OE and DE models. The computation of these criterias involves the estimation of the map which is computationally very slow, and hence it is not feasible to do it for too many models.

The basic criteria for model selection was proposed by Akaike in 1970[1] – see reference for details–, it is known as AIC (Akaike Information Criteria). An improved version of AIC– AICC where the last C stands for Corrected– was proposed by Hurvich e.a.[20].

We will use the basic model

$$y_{t+d_k+1} = f_\theta(y_t, \ldots, y_{t+d_k}) + \varepsilon_t$$

where θ is the vector of parameters, and the ε_t's are i.i.d. gaussians with zero mean and variance σ^2.

The Kullback-Liebler information criteria measures the distance from any model to the true model F under the true model and it is given by

$$\Delta(\theta, \sigma^2) = E_F\{-2log l_{\theta,\sigma^2}(y)\}$$

where

$$l_{\theta,\sigma^2}(y) = (2\pi\sigma^2)^{-\frac{N}{2}} Exp\{-\frac{1}{2\sigma^2}(y - h(\theta))'(y - h(\theta))\}$$

is the likelihood function of the data. The values of θ and σ^2 that maximize the likelihood function are called maximum likelihood estimators (MLE) and ar represented by $\hat{\theta}$ and $\hat{\sigma}^2$.

Then, the following approximation holds.

$$\Delta(\hat{\theta}, \hat{\sigma}^2) = N log(\hat{\sigma}^2) + N\frac{\sigma_0^2}{\hat{\sigma}^2} + \{\frac{1}{2\sigma^2}(\mu - h(\hat{\theta}))'(\mu - h(\hat{\theta}))\}$$

where where θ_0 and σ_0^2 are the values of the parameters under the true model, and $\mu = h(\theta_0)$.

A second approximation yields the two criterias.

$$AIC = N(log(\hat{\sigma}^2) + 1) + 2(m + 1),$$

and

$$AICC = AIC + \frac{2(m+1)(m+2)}{N - m - 2},$$

where m is the number of constrains of the model and N is the number of observations.

These criterias require the calculation of $\hat{\sigma}^2$ which itself requires the estimation of the rule.

4. Estimation of the Rule.

Let us assume that we have already made our choice of delays and we have a reconstruction $\{P_1, \ldots, P_N\}$ of the attractor. In order to estimate the rule f we make use of the fact that f maps the point P_t into P_{t+1} plus maybe an error term. This is

$$P_{t+1} = f(P_t) + \varepsilon_t \ , \tag{2}$$

where the vector of errors ε_t has zero expected value. This is a minimal assumption that makes sense since if ε_t had nonzero expected value, equal to $g(P_t)$ then we would replace f by $f + g$.

The naive method of estimating f consists of dividing the attractor into small pieces in which the rule is linear and estimating the rule separately on each piece by a linear function. This produces a fit which is not continuous and it depends very much on an arbitrary partition of the attractor.

There are many other more sophisticated ways of generating a fit, there are methods based on Neural Nets, or radial functions, but we are going to concentrate here into two recent nonparametric nonlinear regression methods which have been been compared very favorably with Neural Nets in several examples of different nature[10].

4.1 Smooth Multiple Additive Regression Technique (SMART).

The original idea which gave rise to this method first proposed by Friedman e.a.[15], under the name of Projection Pursuit Regression, and it was based in the following expansion. For any mapping $f : \mathbf{R}^m \to \mathbf{R}^n$ were $m \geq n$, for M sufficiently large we can approximate $f(x)$ by the sum

$$f(x) \approx \sum_{i=1}^{M} \beta_i f_i(\alpha_i' x),$$

where the β_i's are vectors on $\mathbf{R}^n$, the α_i's are unit vectors in $\mathbf{R}^m$ and the f_i's are functions from $\mathbf{R}$ to $\mathbf{R}$.

The properties of this approximation were studied by Diaconis e.a.[11], where it was found that for many functions –e.g. polynomials– the equality is also true, but for others –e.g. $h(x_1, x_2) = e^{x_1 x_2}$ – the sum needs an infinite number of terms in order to be exact.

In our case m and n are both equal to $k+1$ and the SMART estimator of the rule f takes the form

$$\hat{f}(P_t) = \bar{P} + \sum_{i=1}^{M} \beta_i \hat{f}_i(\alpha_i' P_t), \tag{3}$$

where P_t, β_i, and α_i are all in $\mathbf{R}^{k+1}$, and $f_i : \mathbf{R} \to \mathbf{R}$ are smooth functions.

The estimator $\hat{f}$ is obtained by the method of least squares, which in our case corresponds to minimizing the criteria

$$L_2 = \sum_{t=1}^{N-1} |P_{t+1} - \hat{f}(P_t)|^2$$

over the three groups of parameters $\{\beta_i\}_1^M$, $\{\alpha_i\}_1^M$, and $\{\hat{f}_i\}_1^M$.

A more general form of this criteria is the following.

$$L_2^* = E|P_* - \hat{f}(P)|^2$$

where P_* is the observed image of P, and the operator E– expected value – is the weighted average

$$Ez = \frac{\sum_i \omega_i z_i}{\sum \omega_i}$$

The weights ω_i are arbitrary and should reflect the variability of the error term, which sometimes is associated with the size of the jump. A reasonable value for ω_i is $\hat{\sigma}(P)^{-1}$, were $\hat{\sigma}(P)$ is an estimate of the standard deviation of ε at P or proportional to it. In many cases $\hat{\sigma}(P)$ is proportional to the change of $f(P) - P$, such that a rough estimator estimator could be obtained.

The algorithm proposed by Friedman[16] for the estimation of f consist of a global loop, and at each pass the following steps are performed:

STEP 1: For each j obtain β_j that minimizes L_2 given that all the other parameters – $\{\beta_l\}_{j\neq l}$, $\{\alpha_l\}$, and $\{\hat{f}_l\}$– are known. This is the linear regression estimator.

STEP 2: The same as step 1 but for α_j . For each j find α_j that minimizes L_2 given that all the other parameters – $\{\alpha_l\}_{j\neq l}$, $\{\beta_l\}$, and $\{\hat{f}_l\}$– are known. In order to minimize L_2 we may write

$$\begin{aligned} P_{t+1} - \hat{f}(P_t) &= P_{t+1} - \bar{P} - \sum_{l=1}^{M} \beta_l \hat{f}_l(\alpha_l' P_t) \\ &= R_{tj} - \beta_j \hat{f}_j(\alpha_j' P_t), \end{aligned}$$

where

$$R_{tj} = P_{t+1} - \bar{P} - \sum_{l\neq j} \beta_l \hat{f}_l(\alpha_l' P_t)$$

The resulting optimization criteria is

$$L_2(\alpha_j) = \sum_{t=1}^{N-1} |R_{tj} - \beta_j \hat{f}_j(\alpha_j' P_t)|^2 \ , \tag{4}$$

and the α_j's are obtained by a Gauss-Newton algorithm.

STEP 3: For each j find $\hat{f}_j$ that minimizes L_2 given that all the other parameters – $\{\alpha_l\}$, $\{\beta_l\}$, and $\{\hat{f}_l\}_{j\neq l}$ – are known. In order to minimize L_2 we take from Eqn.(4)

$$\hat{f}_j(\alpha_j' P_t) = \frac{E(\beta_j' R_{tj})}{\beta_j' \beta_j}$$

were the expected valued is conditioned on the value of $\alpha_j' P_t$. The expected value can be calculated using a local average or a spline.

These three steps are repeated until the change is less than ϵ in two consecutive iterations.

The only task left is to choose M, the number of terms. In the examples that are shown here we found that fixing M to 30 to 50 terms gives satisfactory results, however certain amount of trial is necessary. Fewer than 30 terms did not produce satisfactory results because the bias of $\hat{f}$ made hard to produce orbits that show a pattern similar to the one from the original time-series.

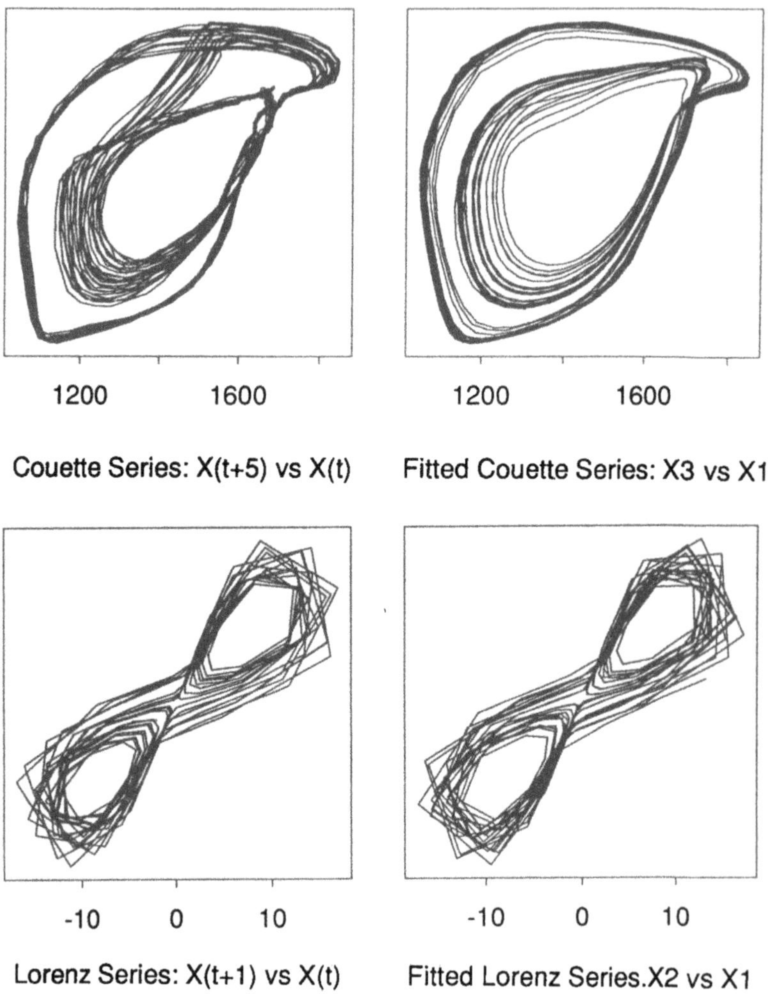

Figure 12. Comparison of original orbits and orbits fitted with SMART. (1) and (2) Couette Flow Series. (3) and (4)Lorenz Series.

4.2 Multivariate Additive Regression Splines (MARS)

Bear in mind that the standards of goodness of fit that we are hoping to achieve here are very high, because we would like $\hat{f}$ to produce orbits that resemble the originals produced by f, and preserve their properties. Therefore if fewer terms are fitted, small inaccuracies in the estimator may appear in regions where the function is not smooth enough, or where there are sharper changes, resulting in a rule that does not produce a full orbit, but a rather simpler orbit or perhaps converging to a point.

Therefore moderate amounts of bias are not acceptable and it is better to have a small amount of overfitting in some regions in order to obtain a good fit everywhere.

There are fancier automatic criterias in the statistical methodology–cross validation and others– that can be applied to this problem for smaller numbers of terms and of observations than what we deal with here, but unfortunately more computational power is needed. Just to give an example of the computational difficulties, it took 12 hours to a Sparcstation I to fit a model with 50 terms to the dynamical system reconstructed from the Couette Series for delays 0, 2, 5, 8, 12 and were the series had 7000 observations.

Figs. 12.1 and 12.2 show the original orbit from the Couette series and the orbit generated by the fitted rule $\hat{f}$, and Figures 12.3 and 12.4 show the same plots but from the Lorenz series. The plots in Figs. 12.2 and 12.4 are quite spectacular because they show that the fitted rule reproduces the dynamics of the original rule. In particular, the one fitted to the Lorenz series reproduces the original dynamic behavior very well. The rule fitted to the Couette flow series shows two loops instead of the three shown by the reconstruction with the original series, but it is still very remarkable.

We have seen that the main criticism of the SMART algorithm for fitting dynamical systems models is that is expensive computationally. Another criticism is that has appeared in the literature is that it does not work very well id the function we are estimating is not very smooth.

In this section we will study a new statistical methodology for nonparametric function estimation knows as Multivariate Additive Regression Splines (MARS) which was introduced by Friedman[17]. As we will see here MARS does overcome the problems of the SMART algorithm.

The ideas of MARS come from the one dimensional case, where the method of splines is the most widely used method for function estimation. Consider the one dimensional model

$$y = f(x) + \varepsilon . \tag{5}$$

The method of splines consists of dividing the domain of x in $K+1$ regions by the ordered points $t_1, \ldots, t_K$, and consider the basis of functions $\{ \{x^j\}_1^q, \{x - t_j\}_1^K \}$ which will be denote by $\{B_j(x)\}_1^{K+q}$. Then the spline estimator of f is

$$\hat{f}(x) = \sum_{i=1}^{K+q} a_i B_i(x) ,$$

where the a_i's are fitted by least squares.

The generalization of this method to dimensions greater than one is conceptually trivial, and it works in two dimensions, but for dimensions greater than two is computationally impractical in most situations.

In the remaining of this section we will describe the algorithm for recursive partitioning which is the basis for MARS, we will describe the MARS algorithm, and we will show examples.

Recursive partitioning is a method of estimating f, which consists of fitting a step function by sequentially partitioning the domain of f into regions and assigning a value to each region. It was introduced as part of the methodology of classification and regression trees (CART[5]). The estimator of f is

$$\hat{f}(P) = \sum_{i=1}^{m} a_i B_i(P) ,$$

where the B_i are indicator functions over sets that make a partition of a the domain of f. The criteria that will be optimized is

$$L_2(\hat{f}) = \sum_{t=1}^{N-1} (y_{t+k+1} - \hat{f}(P_t))^2 .$$

Where y_{t+k+1} is one component of P_{t+1}. The parameters $\{a_i\}$ are estimated by least squares. The algorithm goes as follows

STEP 1: Set $m = 2$ and find the component of P_t, namely y^*, a fixed value z_1 and the basis functions

$$B_1(P_t) = H(-(y^* - z_1))$$
$$B_2(P_t) = H(+(y^* - z_1))$$

that minimize $L_2(\hat{f})$. $H(x)$ is the heavy side function which takes value one for non-negative argument and zero otherwise.

STEP i: Set $m = i + 1$ and find the component of P_t, namely y^*, z_i, and replace the basis function B_j by

$$H(+(y^* - z_i))\ B_j(P_t)$$

and add the basis function

$$B_{i+1}(P_t) = H(-(y^* - z_i))$$

that minimize $L_2(\hat{f})$.

Figure 13 shows two graphs of a simple recursive partitioning. The left one side graph shows the tree representation of the recursive partitioning and the right hand side graph shows the actual partitioning of the domain of dimension 2.

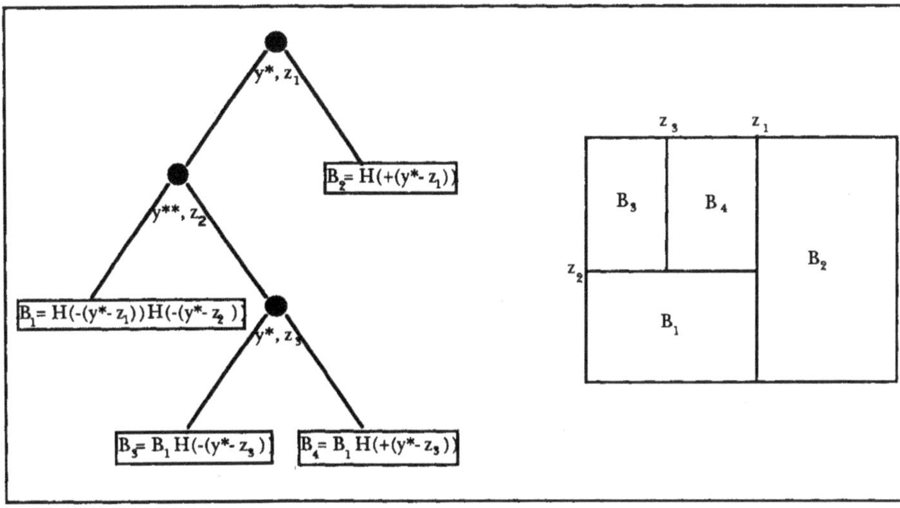

Figure 13. Recursive Partitioning Graphs. (1) Tree Representation. (2) Two dimensional example .

The MARS algorithm is essentially the same as recursive partitioning but there are a few modifications.

(i) Replace the twos-sided basis functions $H(\pm(y*-z*))$ by two-sided truncated power basis functions

$$[\pm(y*-z*)]_+^q$$

where the $[x]$ is gives x for non-negative values of x and zero otherwise.

(ii) Intermediate functions are not replaced.

(iii) Products are restricted to factors with different variables.

Since in our case we are fitting a rule in $(k+1)$ we apply the algorithm to each of the coordinates. This is not an ideal method if ones wants to produce an orbit generated with the fitted rule, because we need to keep in memory all the fits at the same time. This is problematic because MARS uses a large amounts of memory for each fit. Work in progress will introduce a modification to the algorithm that would allowed us to fit all the responses at the same time.

The most important advantages of MARS over SMART are first of all that will give better result when the true map is not very smooth, and in addition the computing time used by MARS is much less than the one taken by SMART.

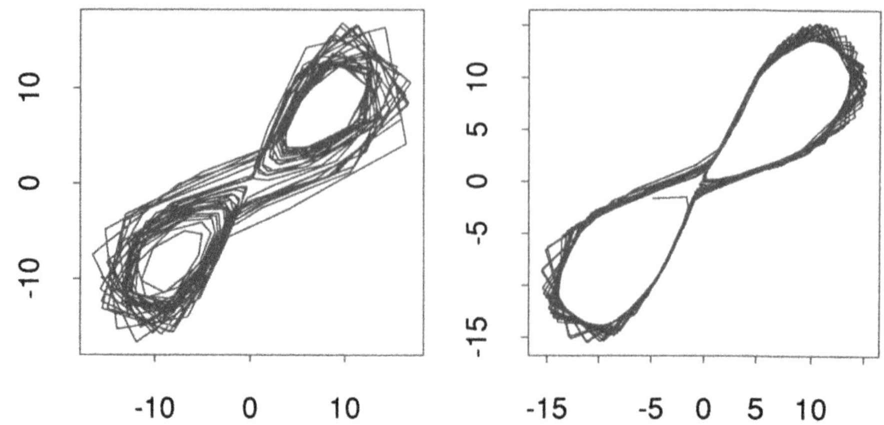

Figure 14. Comparison of orbits fitted with MARS. (1) Lorenz Series. (2) Lorenz Series with Uniform(-0.5,05) error added.

The algorithm stops after m steps or after certain stopping rule is satisfied and the final coefficients are obtained by least squares. The stopping rule proposed by Friedman[18] is based on cross-validation. We refer to the work of Friedman[18] for a more detail discussion of the algorithm and examples in other areas. De Vaux e.a[10] did a comparison of MARS with Neural Nets in several examples for very different applications, and for each of them MARS always seemed to give better results.

Figure 14 shows the orbits generated from the rule fitted –with MARS– to the Lorenz time-series. The one on the left was obtained from the original time-series, while the one on the right was obtained from a new time-series created by adding observational error from a uniform distribution $(-\frac{1}{2}, \frac{1}{2})$ to the Lorenz series. In both examples, the three dimensional embedding with delays 0, 1, and 2 was selected for the reconstruction of the attractor.

The graph on the left of Figure 14 shows an orbit very similar to the original data that was shown in Figure 6.1. This is an indication of an excellent reconstruction. However the graph on the right of Figure 14 shows the effect produced by moderately large noise to the MARS estimator. It is possible that using more data, and tuning up the estimation better we would be able to improve the estimator but it is also possible that MARS needs to be modified in order to account for errors in both sides of the model equation, namely errors and variables[22].

Figure 15 shows the orbits generated with MARS, for the Couette Flow series. The delays that were selected for the reconstruction were 2, 5, 8, 12, 18, 24, and 29. The orbit produced by MARS is a very simplified version of the original orbit, although it preserves the main feature of the orbit.

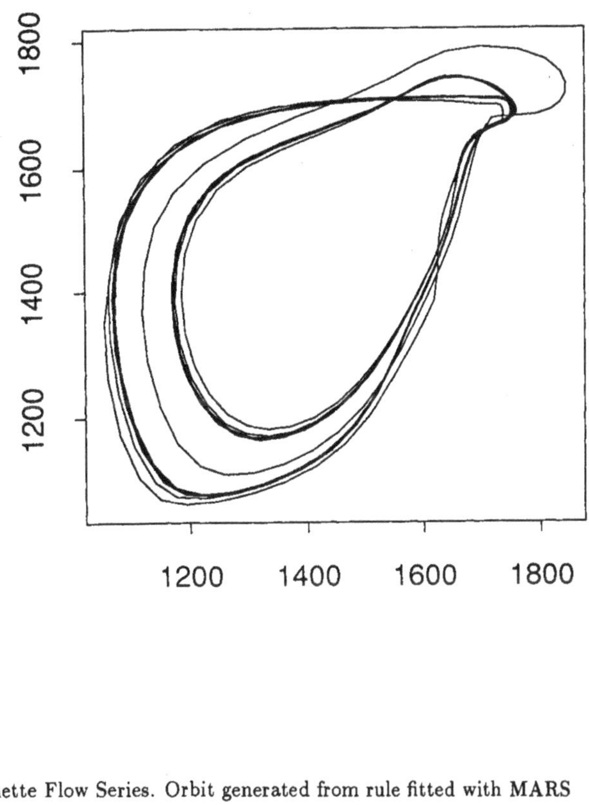

Figure 15. Couette Flow Series. Orbit generated from rule fitted with MARS

Figure 16 shows the corresponding orbit for the EEG series. Delays 6, 12, 18, 24, 30, 33, and 35 were used for the reconstruction of the attractor.

The orbit produced by MARS is a very simplified version of the original orbit because of the large noise component of the time-series. If we compare Figure 16 with the the graph on the right of Figure 14 we see some similarities, and perhaps this is also the effect of not accounting for errors and variables[22].

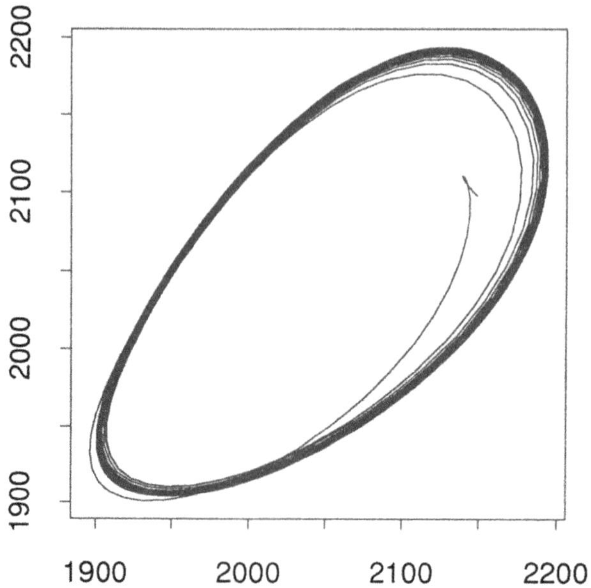

Figure 16. EEG Series. Orbit generated from rule fitted with MARS

5. Dimension.

Once we have eliminated the noise from our reconstruction we proceed to the next objective of the analysis. In many cases this consists of estimating the dimension of the strange attractor.

There are many observables that go under the name "dimension." The one that will discusses here, the Grassberger-Procaccia dimension, was introduced by Grassberger e.a.[19], and some related references are by Takens[29], Cabrera e.a.[8] and Smith e.a.[27].

We will start with the definition of dimension by Hausdorff-Besicovitch for a set in $\mathbf{R}^p$, which is the limit

$$D = \lim_{\epsilon \to \infty} \frac{ln N(\epsilon)}{ln \frac{1}{\epsilon}}$$

where $N(\epsilon)$ is the least number of cubes(or hypercubes) that are necessary in order to cover the set. This also means that

$$N(\epsilon) \approx \epsilon^{-D} \ .$$

We will now explore a few simple examples.

(i) For a single point $N(\epsilon)$ is 1, therefore the dimension is 0.
(ii) For a line of longitude L, $N(\epsilon)$ is $\frac{L}{\epsilon}$, and it follows that the dimension is 1.
(iii) For a surface of area S, $N(\epsilon)$ is $\frac{L}{\epsilon^2}$, and the dimension is 2.
(iv) For the Cantor set, $N(\frac{1}{3})$ is equal to 2, $N(\frac{1}{9})$ is 4, and in general $N((\frac{1}{3})^m)$ is 2^m, and therefore the dimension is

$$D = \frac{ln(2^m)}{ln(3^m)} = 0.63 \ .$$

However, for dimension grater than 3 it is difficult to compute and the definition becomes impractical.

This problem motivated the Grassberger-Procaccia definition of dimension. In principle, it is a local quantity that could also depend on the orbit. If $\{P_i\}_0^\infty$ is an orbit, we define:

$$d(P) = \lim_{r\to 0} \frac{1}{\log r} \lim_{n\to\infty} \frac{1}{n} \#\{P_i | i \le n, \ d(P_i, P) \le r\}. \tag{6}$$

Notice, that even if manifestly invariant under smooth changes of coordinates, this dimension is not a geometric property of the attractor but rather of the measure defined by the orbit according to

$$N(A) = \lim_{n\to\infty} \frac{1}{n} \#\{P_i | i \le n, \ P_i \in A\}.$$

It is possible to find easy dynamical systems for which their orbits do not define measures. It could also happen that d depends on the point but these examples do not seem to appear in practical applications often.

However, it was quickly found that for many systems, $d(x)$ existed and was independent of the point and the orbit (see Eckmann e.a.[12]). This can be now proved under rather general hypothesis Pesin e.a.[25].

Eqn.(6) implies that for large n we may use the approximation:

$$N_n(P, r) \approx C \cdot r^{d(P)}, \ 0 \le r \le \epsilon, \tag{7}$$

where ϵ is a sufficiently small value and C is a constant which does not depend on r.

This means that, in practice, the calculation of the dimension is done by selecting a neighborhood (using brushing) and measuring the slope of a log/log plot of the number of pairs of points at a distance less than r versus r. Clearly this calculation should be done only for a range of r's which are sufficiently small to be close to the limit but which are large enough to contain a statistically significant number of points, therefore it is convenient to try it for several values of ϵ and decide on the appropriate value.

One can make sure that the dimension is a well defined number for the system by computing it in different neighborhoods and for different lengths of the time series. Furthermore, if the delay reconstruction is successful the dimension should not change by much if we add some extra delays to the reconstruction.

The approximation in Eqn.(7) suggests that if we restrict the measure defined by the sequence $\{P_i\}_1^\infty$ to the ball with center at P and radius ϵ, namely $B_\epsilon(P)$ we obtain the probability measure:

$$Prob[B_r(P)] = \lim_{n\to\infty} \frac{N_n(P,r)}{N_n(P,R)} = \left\{\frac{r}{\epsilon}\right\}^{d(P)}$$

Then the function:

$$F(r) = \left\{\frac{r}{\epsilon}\right\}^{d(P)} I_{[0,\epsilon)}(r) + I_{[\epsilon,\infty]}(r)$$

is a distribution function for the random variable r. We can obtain a simpler form for this function by applying the transformation $u = -log(r/\epsilon)$. Then the variable u has an exponential distribution with parameter $d(P)$.

In practice we only observe a finite piece of the sequence, namely $\{P_i\}_1^n$, and only m of them $\{P_{(i)}\}_1^m$ are within a distance ϵ of P. Their corresponding values of r are $\{r_i\}_1^m$, where $r_i = d(P_{(i)}, P) < \epsilon$, and let $u_i = -log(r_i/\epsilon)$. Under certain assumptions about the original sequence we can treat the u_i's as an independent sample and therefore we can estimate $d(P)$ by maximum likelihood[22]:

$$\hat{d}_2(P) = \{\bar{u}\}^{-1} = \left\{\sum_{i=1}^{m} \frac{log(r_i)}{m}\right\}^{-1}$$

If the dimension is constant across the attractor it can be estimated by taking an average over the set of points. We define the estimators $\hat{d}_1$ and $\hat{d}_2$ as the average over the set of points of $\hat{d}_1(P)$, and $\hat{d}_2(P)$ respectively.

In practice it turns out that both estimators are highly variable and the calculation should be done very carefully. One suggestion is to replace the average over the data points by a trimmed mean were 5% of the observations are trimmed from each tail of the sample, thus avoiding the instability introduced by unusually large values. A general methodology for obtaining estimators that are not affected by bad data – coined by the term "robust" – was proposed by Huber[21].

In order to study the performance of estimators $\hat{d}_1(P)$, and $\hat{d}_2(P)$, an orbit of 50000 three-dimensional observations was generated from the Lorenz equations, where the true value of the dimension is believed to be 2.06. The orbit was divided in 25 pieces of size 2000, and the two estimators were computed on each piece. The mean, standard deviation (SD), and mean square error (MSE) where calculated for the sample of 25 values. The same calculations where repeated for sample sizes 5000, 10000, 15000 and 50000, and for different values of ϵ, 1%, 2%, 5%, 10%, 10%, 25%, and 50%. The results are shown in Table 1.

Sample Size	N of Samp.	$\hat{d}_1$	SD($\hat{d}_1$)	MSE($\hat{d}_1$)	$\hat{d}_2$	SD($\hat{d}_2$)	MSE($\hat{d}_2$)
			$\epsilon = 1\%$				
2000	25	2.219	0.364	0.158	4.604	2.093	10.859
5000	10	2.177	0.211	0.058	2.626	0.228	0.373
10000	5	2.194	0.100	0.028	2.330	0.084	0.080
15000	3	2.183	0.100	0.025	2.230	0.079	0.035
50000	1	2.309	.		2.094	.	
			$\epsilon = 2\%$				
2000	25	2.167	0.270	0.084	2.433	0.272	0.213
5000	10	2.156	0.174	0.039	2.169	0.040	0.013
10000	5	2.167	0.066	0.015	2.101	0.004	0.001
15000	3	2.161	0.088	0.018	2.085	0.005	0.0006
50000	1	2.212	.	.	2.067	.	.
			$\epsilon = 5\%$				
2000	25	2.137	0.201	0.046	2.045	0.019	0.0005
5000	10	2.131	0.127	0.021	2.029	0.009	0.001
10000	5	2.144	0.064	0.011	2.023	0.005	0.001
15000	3	2.133	0.063	0.009	2.021	0.006	0.002
50000	1	2.198	.	.	2.018	.	.
			$\epsilon = 10\%$				
2000	25	2.095	0.159	0.026	1.870	0.035	0.037
5000	10	2.105	0.105	0.013	1.865	0.028	0.038
10000	5	2.114	0.041	0.004	1.864	0.022	0.038
15000	3	2.114	0.058	0.006	1.856	0.015	0.041
50000	1	2.097	.	.	1.859	.	.
			$\epsilon = 25\%$				
2000	25	2.031	0.120	0.015	1.535	0.022	0.275
5000	10	2.054	0.076	0.005	1.519	0.017	0.292
10000	5	2.072	0.041	0.001	1.511	0.018	0.300
15000	3	2.059	0.045	0.002	1.506	0.009	0.306
50000	1	1.992	.	.	1.492	.	.
			$\epsilon = 50\%$				
2000	25	1.929	0.088	0.024	0.845	0.016	1.474
5000	10	1.970	0.063	0.012	0.831	0.011	1.510
10000	5	1.989	0.026	0.005	0.825	0.013	1.524
15000	4	1.094	0.585	1.274	1.090	0.577	1.272
50000	1	1.850	.	.	0.811	.	

Table 1. Dimension Estimation. Study of the dimension of the Lorenz Attractor.

Sample Size	N of Samp.	$\hat{d}_1$	SD($\hat{d}_1$)	$\hat{d}_2$	SD($\hat{d}_2$)
		$\epsilon = 1\%$			
1000	32	3.202	1.326	2.105	0.007
2000	16	3.395	1.801	2.305	0.063
5000	6	3.564	2.269	2.420	0.129
10000	3	3.760	2.900	2.480	0.176
32648	1	4.187	.	2.377	.
		$\epsilon = 2\%$			
1000	32	2.728	0.455	1.556	0.254
2000	16	2.978	0.850	1.851	0.044
5000	6	3.144	1.182	2.198	0.019
10000	3	3.459	1.960	2.210	0.022
32648	1	3.731	.	2.176	.
		$\epsilon = 5\%$			
1000	32	2.307	0.070	1.613	0.200
2000	16	2.565	0.264	1.762	0.088
5000	6	2.855	0.633	2.175	0.013
10000	3	2.977	0.845	2.175	0.013
32648	1	3.275	.	2.182	.
		$\epsilon = 10\%$			
1000	32	2.165	0.015	1.873	0.035
2000	16	2.398	0.118	1.886	0.030
5000	6	2.656	0.358	1.928	0.017
10000	3	2.857	0.636	1.924	0.018
32648	1	3.099	.	1.919	.
		$\epsilon = 25\%$			
1000	32	2.023	0.005	1.489	0.325
2000	16	2.215	0.027	1.482	0.333
5000	6	2.449	0.152	1.468	0.350
10000	3	2.533	0.224	1.464	0.354
32648	1	2.712	.	1.457	.

Table 2. Dimension Estimation. Study of the dimension of the Couette Flow Attractor.

The Mean Square Error(MSE) of an estimator is defined as the variance of the estimator plus the square of the bias. This criteria was chosen in order to compare the various cases because the estimators maybe biased, in which case it is the appropriate criteria. Notice that in order to compute the MSE we need to have an estimator of the bias. In this case this is not a problem since we assume that the true value of the dimension for the Lorenz attractor is 2.06. The value of ϵ is measured as a percent of the maximum distance of two points on the attractor.

Table 1 shows some interesting properties of both estimators. First of all, as it was mentioned before, both estimators are highly variable and they are very sensitive to the value of ϵ. For $\hat{d}_1$ it is better to use small ϵ near 5%, whereas for $\hat{d}_2$ the better values of ϵ are around 25%. But the main point of the table is that these estimators are far more variable than was expected.

An analogous study was performed for the Couette flow series, where the orbit was generated by using an embedding in 5 dimensions, with time-delays 0, 30, 65, 105, and 135, on the entire series. This reconstruction is different than the one shown on Fig. 9 but the dimension calculations give similar results. The orbit consists of 32648 points in the phase space, which is the complete series.

Table 2 shows the same high variability that was observed in Table 1, and high sensitivity of the estimators to the value of ϵ. If we use the same criteria as before – for $\hat{d}_1$ ϵ near 25%, and for $\hat{d}_2$ ϵ near 5% –, the dimension of this system should be around the 2.2, 2.5 range. Of course, many more experiments are needed in order to obtain more accurate values.

Finally the third example that was studied is the E.E.G. series, where the the orbit was generated by using an embedding in 8 dimensions, with time-delays 0, 6, 12, 18, 24, 30, 33, and 35 , on the entire series. The orbit consists of 22000 points in the phase space, which is the complete series.

One property of this series that was observed during the exploration phase is that the dynamical systems which produces it, may be changing slowly over time, since reasonable pictures of the orbits where produce for sub-series of 2000 or so, whereas if we increase the number of points to 4000 it is harder to see the orbits and the shape of the attractor is not as clear. Therefore it is better to pay more attention to the rows of Table 3 which correspond to a sample size of 2000.

From Table 3, the dimension of this system should be around 3.5 plus or minus one. Of course, this is not a very accurate estimate since the time-series has a large noise component.

One of the factors that induces the high variability is the boundary effect, because the reasoning that lead to the estimators assumes that there are no boundaries. However if the ball of radius ϵ where the estimator was computed intersects with the boundary of the attractor then the estimator for that point would be biased.

A naive solution to this problem is to minimize this effect by making ϵ very small, but this will increase the variability of the estimator because the estimation at each point would involve few points. A better solution is to combine the estimation with brushing within the graphics interface, by which the user would select an interior region of the attractor where the computation of the dimension would be performed.

Figure 17 shows an example for the attractor reconstructed from the Lorenz Series with delays 0, 1 and 2. The clearer interior points where selected with the

Sample Size	N of Samp.	$\hat{d}_1$	SD($\hat{d}_1$)	$\hat{d}_2$	SD($\hat{d}_2$)
		$\epsilon = 1\%$			
1000	22	6.116	18.150	48.081	3643.490
2000	11	6.171	17.617	25.373	661.555
5000	4	6.413	19.438	18.257	338.732
10000	1	6.385	.	22.466	.
		$\epsilon = 2\%$			
1000	22	5.311	10.932	6.193	19.152
2000	11	5.637	13.127	5.542	12.949
5000	4	5.975	15.453	5.091	10.031
10000	1	6.372	.	5.627	.
		$\epsilon = 5\%$			
1000	22	4.494	6.021	3.401	1.881
2000	11	4.780	7.488	3.343	1.796
5000	4	5.083	9.347	3.024	1.339
10000	1	5.682	.	3.222	.
		$\epsilon = 10\%$			
1000	22	3.967	3.690	2.114	0.045
2000	11	4.231	4.807	2.010	0.090
5000	4	4.445	5.884	1.751	0.324
10000	1	4.988	.	1.973	.
		$\epsilon = 25\%$			
1000	22	3.158	1.248	0.906	1.337
2000	11	3.353	1.725	0.844	1.489
5000	4	3.494	2.270	0.744	1.753
10000	1	4.102	.	0.806	.

Table 3. Dimension Estimation. Study of the dimension of E.E.G. Attractor.

brush and the dimension calculation was performed on those points. Since the value of ϵ was 5.2% one expects the estimator $\hat{d}_2$ to be better than $\hat{d}_1$. In Figure 17, the values of the estimators appear under the labels "GP dim/Est1" and "GP dim/Est2".

6. Computational Methods and Graphics Interface.

The computational aspects of the methodology presented here are divided in several categories. First of all is the graphics interface with the implementation of graphical diagnostics, secondly are the algorithms for doing function estimation,

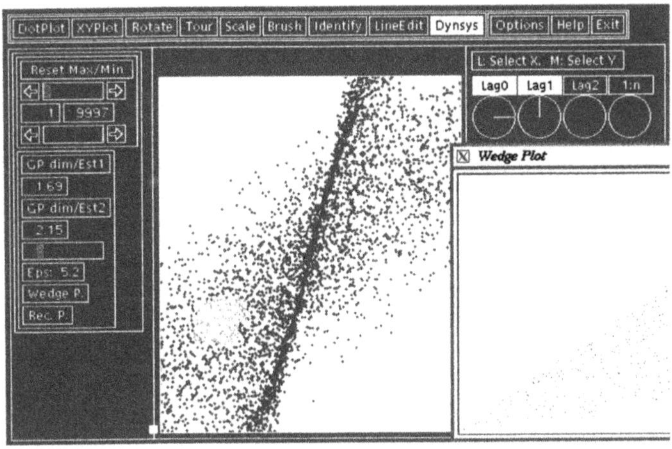

Figure 17. Estimating the dimension of a brushed subset of the Lorenz attractor.

based on MARS and SMART, and finally the algorithms for the computation of the dimension.

The source code of these three groups of programs can be obtained by anonymous *ftp* over the internet network from *stat.rutgers.edu*. If the reader does not have access to *ftp* the code can also be obtained by electronic mail request to *cabrera@stat.rutgers.edu*, but the first option is strongly encouraged.

6.1 Graphics Interface

For the type of problems considered here, the main goals of exploration are to obtain a detailed geometric intuition of a geometric object and of its local properties. Finding outliers and clusters seems to be of secondary interest.

With those goals in mind, it seems useful to have the capability to perform rotations that can be intertwined with diagnostics. It is also important that the diagnostics can be localized to parts of the data selected interactively.

xgobi is a public domain statistical software for the analysis of multivariate data using graphical methods [28], which runs under the X-windows system on unix workstations. In order to make *xgobi* a useful graphics interface for the reconstruction of dynamical systems, we incorporated new features such us diagnostic plots, dimension calculations, and subsetting that were interfaced with the existing brushing capabilities of *xgobi*. Figure 17 shows an example of *xgobi* with dynamical systems with the Lorenz Series. The main window shows the display of the phase space, another window shows a wedge plot of the brushed area. The dimension computations are displayed on the left hand side panel.

There are two modes for calling *xgobi*, the first one is for a k-dimensional orbit dynamical system in which case the data must be placed on a text file as a matrix with n nrows and k columns, were each row contains the coordinates of a point separated by spaces. To call *xgobi* type

$$\% \ xgobi \ filename$$

The second mode for calling *xgobi* is for a time-series, in which case the option $-d$ is used. The time series must be place in a text file. The option $-d$ must be followed by the number of observation that are to be reed from the time series, the number of delays including the delay zero, the delays and the file name. For example

$$\% \ xgobi \ -d \ 2000 \ 5 \ 0 \ 2 \ 5 \ 8 \ 12 \ filename$$

Other features of interest that are part of *xgobi* let us apply a wide variety of methods. For example, the *scaling* panel allows to zoom into and out of any particular region of the attractor, and to focus on the structure at different scales. Also, the time-series can compressed and extended, thus exposing its different features. The *tour* panel incorporates the gran tour visualization tool for high dimensional data sets, and the data can be transformed into principal components.

We have also written a program for the Macintosh which implements in an interactive way the graphic diagnostics described above, namely *dynsys*.

A main window displays a two dimensional projection of the $k+1$ dimensional reconstruction. This projection can be rotated continuously. A set of $2k+1$ buttons controls the velocity of rotation. Another set of buttons controls the values of the delays. The delays and other parameters of the algorithms can be selected from an options dialogue box.

A menu activates windows for the recurrence, wedge plot, Fourier transform, dimension calculation, and the real time display of the series.

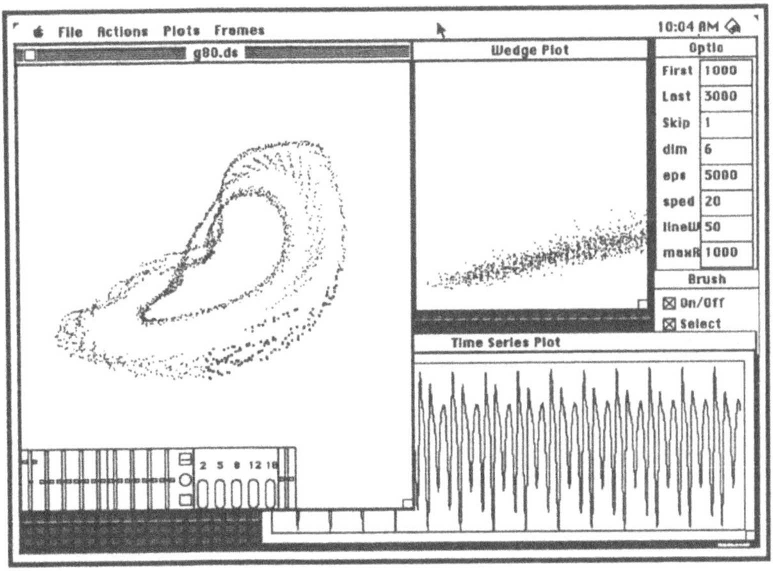

Figure 18. Dynsys, the macintosh graphics interface, with the Couette series. Display showing Time-series, Phase Space, and Wedge Plot.

The same menu allows to select brushing and focusing operations explained in Section 6.2. and both can be combined by zooming on a region selected with the brush.

One of the interesting aspects of this interface is that there is an implementation for playing the orbit in real time.

Figure 18 shows an example of the macintosh graphics interface with the Couette time-series where different windows show the time-series, phase space for delays 2, 5, 8, 12, and 18, and wedge plot of the brushed area on the phase space.

In the following sub-section, we discuss the implementation of the two main tools of the graphics interface, rotations, and brushing.

6.2 Implementation of the Graphics Interface

For multivariate data, two dimensional projections suitable for displaying on a computer screen can be obtained by computing the first two coordinates of the points with respect to an orthogonal basis.

Rotations can be implemented by multiplying the basis by an orthogonal matrix between consecutive redisplays. If the rotations are close enough to the identity the display may be made to look continuous.

The standard way in statistics to do explorations with multivariate data has been the "grand tour" or random jump walk tour[2,3,9], which is the way that rotations are implemented in *xgobi*. The grand tour consists of starting at a two dimensional projection, selecting another two dimensional projection at random and moving by small rotations towards the new projection geodesic interpolation between the two. When the new projection is attained a different one is selected and the path is continued. The sequence of all intermediate projections is displayed giving the appearance of continuous motion. Perhaps a more precise definition of this is a random walk on a Grassmann manifold in the k dimensions.

An alternative method that relays in a more local control of rotations was implemented in *dynsys*.

Since only the two first coordinates are displayed, it is only necessary to consider rotations that affect the first two coordinate vectors.

An efficient way of parametrizing the rotations close to the identity that affect the plane $1,2$ is as $G(\theta_{12}\ldots\theta_{1k+1},\theta_{23}\ldots\theta_{2k+1}) = G_{12}(\theta_{12})\ldots G_{2k+1}(\theta_{2k+1})$, where G_{ij} means the rotation of angle θ in the plane spanned by the ith and jth coordinate axis.

It is a well known fact that by using a rotation of the form above, we can transform any pair of orthonormal vectors into the two first coordinate vectors. That is, we can reach any two dimensional projection.

Another important part of the implementation of the graphics interface is "brushing." In the middle of exploring visually the attractor one frequently finds visually patterns and regions for which a more detailed analysis would be particularly desirable. Most of the diagnostic tools considered here are local in nature and require essentially the selection of a neighborhood.

Since the interesting places are identified visually, it is natural to use a pointing device such as a mouse to select points that can be subject to further analysis. This is usually referred to as "brushing."

If "brushing" and "unbrushing" operations can be intertwined with rotations, one can select rather precisely neighborhoods in the space.

Since many of the details of interest may be on a scale much smaller than that of the whole data set, it is useful to be able to focus only on a smaller set selected by brushing. While we focus on this set, it becomes the whole data set.

One instance where this feature becomes very useful for dynamical systems is when we try to study in detail local singularities of the reconstruction that manifest themselves as trajectories coming close to crossing in one projection.

In addition to the graphics interfaces, several algorithms were implemented separatedly in order to perform computations that are not feasible to perform in real time. All of them are available as well by anonymous *ftp*.

6.3 Computation of the Autocorrelations (`correl.c`).

The routine used for the computation of the autocorrelations is based on a Fast Fourier Transform (FFT) function. It computes the autocorrelations by applying the inverse FFT to the Periodogram of the time series.

The resulting vector is the vector of sums of pairs of autocorrelations of symmetric lags $(c_1 + c_{n-1}, c_2 + c_{n-2}, \ldots)$.

The routine will add zeros to the time series until its length is a power of 2. If k zeros are added, the autocorrelations $(c_{n-1}, c_{n-2}, \ldots, c_{n-k})$ are zero, so the first k values returned by the routine are the exact autocorrelations for those lags. The remaining values are only approximations to the autocorrelations of those lags. If the difference between the length of the series and the next power of two is too small (k is too small), it will be safer to add zeros to the series until its length is 1 plus the power of two. This will result in exact autocorrelations for all lags, but it will increase computation time by a factor slightly bigger than 2. This is usually of no concern, since the computation of autocorrelations takes much less time than the other routine.

It should be pointed out that c_k is the biased estimator of the autocorrelation of lag k. The unbiased estimator of the autocorrelation of lag k is $\frac{n}{(n-k)}c_k$, which is recommended when $\frac{k}{n}$ is not small.

To use this routine type:

```
correl < filename
```

6.4 Computation of Recurrence and Wedge Plots (`wedge.c, rec.c`).

Although both plots were implemented within the graphics interfaces, stand alone versions of the routine are also provided, which return a list of points to be plotted. This is specially useful when the number of points is very large and it may not be convenient to run the diagnostics in real time.

The naive implementation of these plots would be to keep all the data in an array and then, consider all possible pairs. For the wedge plot, if both the distance of the two points and the distance of the their images are within the bounds that we have requested, the point is plotted. It is also possible to produce a list of the points rather than a plot which can be used to produce contour or scatter plots with another graphics interface or package. However the naive algorithm has a running time of the order of N^2 where N is the number of points in the time series.

The optimal algorithm in terms of number of computations uses a nearest neighbor scheme by generating a data structure of nearest neighbors. However this method is memory intensive, specially for our case since the number of points can reach the tens of thousands.

A good compromise that reduces the number of computations but requires litle additional memory is obtained by sorting the points by their first coordinate.

It becomes quite easy to decide that points that are far in the array, do not verify the requirement that their distances are small. So that, for every point, we only have to consider an small fraction of the points as possible candidates.

STEP 1 : Generated $OR =$ vector of indices of ordered first coordinate. SET $I = 0$.

STEP 2 : SET $I = I + 1$. SET $J = I$.

STEP 3 : SET $J = J + 1$

STEP 4 : IF $|Y_{OR(I)} - Y_{OR(J)}| \leq \varepsilon$ DO CALCULATION

ELSE GO BACK TO STEP 2 .

STEP 5 : IF $J \leq N - 2$ GO BACK TO STEP 3 .

STEP 6 : IF $I \leq N - 1$ GO BACK TO STEP 2 .

Table 4. Sorting by one coordinate algorithm.

This should result in notable savings in the case that the attractor is quite extended and we only are using an small fraction of the pairs of points. Of course, the sorting, the use of more complicated data structures, etc. can cause an overhead that could make the implementation counterproductive, but fortunately this happens only in very rear cases. For example if the bounds for the distances are big - greater than the maximum distance within that one coordinate- the algorithm produces no savings, but the number of comparisons is so large anyway that the overhead caused by sorting is negligible compared with the total. The extra memory requirements consists only of a vector of integers of length N containing the indices of the sorted coordinate.

An example of usage of the routine *wedge* is

wedge 0.5 1.0 lorx.out 1 1 10000 lorx 3 0 1 2

where the arguments are first bound, second bound, output file, number of input files, first time, last time, first input file, number of delays, delay 1, ..., delay n, second input file, ... Notice that the routines allow reconstructions with more than one time series.

Upon starting execution the program opens a file named LOG and leaves on it - respecting the text already on it - some records of the time of execution, the parameters used and the error messages given by the program.

This could be useful since a typical use of this program will consist in running a batch job that tries to eliminate many possibilities.

Another important example where the sorting by one coordinate was applied is in the implementation of real time brushing where it makes a big difference in term of response of the brush. This was implemented in the macintosh version *dynsys* only, where the brushing is much faster than it is in *xgobi*.

6.5 Computations of the Rule with MARS *and* SMART *(*`mars_ds.f`, `smart_ds.c`*)*

The implementation of the estimation of the rule f is done by two shell routines which call MARS and SMART respectively.

The routine that calls SMART is a C routine which interfaces with the original FORTRAN program `smart.f`. The user should be aware that interfaces between C and FORTRAN are very much system dependent. The implementation that is provided was developed for a SUN workstation and it may not work under other versions of unix or non unix operating systems.

The output generated by `smart_ds` consists of the fitted values to the reconstruction and, if requested, of an orbit starting at some initial point on the attractor generated by the estimated rule.

An example of usage of the routine `smart_ds` is

smart_ds 40 50 lorx.out 1 1 10000 lorx 3 0 1 2 1 1000

where the arguments are minimum and maximum number of terms in expansion, output file, number of input files, first time, last time, first input file, number of delays, delay 1, ..., delay n, second input file, ..., and finally, index of starting point, and length for orbit generated by the fitted rule.

In case the user does not want to generate an orbit from the fitted rule, one of the last two arguments must be set to zero. If the user wants to start the orbit from a specific point not in data, a small modification o

The routine that calls MARS is a FORTRAN program that call the original program `mars.f`. Because of the large memory allocation that is necessary in order to call `mars.f`, the implementation of `mars_ds.f` is dependent on the particular reconstruction and the user may need to modify the source code in `mars_ds.f` for each run.

6.6 Computation of the Dimension (`gp.c`, `gp_ts.c.`*)*

The calculation of the dimension is done by a call to `gp` which returns the two estimators of the Grassberger-Procaccia dimension.

The algorithm that was implemented is again based on is obtained by sorting the points by their first coordinate, which reduces the number of computations but requires little additional memory.

For the estimator $\hat{d}_2$ the user may provide a parameter giving the amount of trimming for the trimmed mean.

The reader should notice that since we are using a finite number of points it is possible that for some points the calculation of the dimension will give very odd results, because there may not be many points nearby etc. Therefore we choose to trim a small percent at each tail so the calculation is less affected by does odd values. This is a common practice in many statistical calculations.

The two routines provided by the author are **gp.c**, for obtaining the dimension estimates of the attractor of a k-dimensional dynamical system, and **gp_ts.c** for obtaining the dimension of the reconstruction from a time-series.

An example of usage of the routine **gp_ts** is

gp_ts 5 2 0 lorx.out 1 1 10000 lorx 3 0 1 2

where the arguments are the radius of the ball used for obtaining the first an second estimators–as a percent of the maximum distance–, the percent of trimming for the second estimator, output file, number of input files, first time, last time, first input file, number of delays, delay 1, ..., delay n, second input file, ..., etc.

Since the sorting algorithm reduces dramatically the number of computations, in theory one could apply this routine to series of length in the millions even in a typical unix workstation.

Acknowledgments.

This work was supported in part by various grants from the National Science Foundation. The author wishes to thank the organizers of the workshop for publishing this volume. Much of the work presented in here and some of the algorithms and software were developed jointly with Rafael de la Llave. The author wishes to thank Deborah F. Swayne, Dianne Cook for making the source code of *xgobi* available, and allowing that the dynamical systems version of *xgobi* be distributed.

References.

1. H. Akaike: *Statistical Predictor Identification. Ann.Inst.Satist.Math.* **22**, (1970) p. 203-220.
2. D. Asimov: *The Grand Tour: A tool for viewing multidimensional data,* SIAM J. Sci. and Stat. Comp. **6**, 128-143 (1985).
3. A. Buja, D. Asimov and C. Hurley: "Methods for Subspace Interpolation in Dynamic Graphics," Bellcore Technical Memorandum TM-ARH-015639
4. P.J.Brockwell, R.J. Davis. *Time series : theory and methods.* (Springer-Verlag, New York, 1991).
5. L.Breiman, J.H. Friedman. Estimating Optimal Transformations for Multiple Regression and Correlation. *J.Amer.Statist.Assoc.* **80** (1985) p. 580–619.
6. G.E.P. Box and G.M. Jenkins: "Time Series Analysis: Forecasting and Control," (Holden Day, San Francisco 1976).
7. D.S. Broomhead and G.P. King: *Extracting Qualitative Dynamics From Experimental Data,* Physica **20D**, 217–236 (1986).
8. J. Cabrera, D. Cook Projection Pursuit Indices based on Fractal Dimension. *Proc. of the 24th Symp. on the Interface between Comput. Sci. and Statist.*, (Springer-Verlag, New York, 1991) .
9. D. Cook, A. Buja, J. Cabrera. Direction and Motion Control in the Grand Tour. *Proc. of the 23rd Symp. on the Interface between Comput. Sci. and Statist.*, (Springer-Verlag, New York, 1991).

10. R.D. De Vaux, D.C. Psichogios, L.H. Ungar. A Comparison of Two Non-Parametric Estimation Schemes: MARS and Neural Networks. Technical Report SOR-92-01, Dept. Stat. and O.R. Princeton University(1992).
11. P. Diaconis, M. Shahshahani. On nonlinear functions of linear combinations. *SIAM J. Sci. Statist. Comput.* **5** (1984) p. 175–191.
12. J.P. Eckmann, S.Oliffson-Kamphorst and D. Ruelle : *Recurrence Plots of Dynamical Systems*, Europhys. Lett. **4**, 973–977 (1987).
13. J.P. Eckmann, S.Oliffson-Kamphorst, D. Ruelle and S. Ciliberto : *Lyapounov Exponents from Time Series*, Phys. Rev. **34A**, 4971–4975 (1987).
14. J.D. Farmer and J.J. Sidorovich: *Exploiting chaos to predict the future and reduct noise*, Los Alamos Preprint (1989).
15. J.H. Friedman, W.Stuetzle. Projection Pursuit Regression. *J. Amer. Statist. Assoc.* **76** (1981) p. 817–823.
16. J.H. Friedman. SMART: User's guide. Technical Report LCS01, Dept. Stat., Stanford University(1984).
17. J.H. Friedman. Classification and multiple regression through projection pursuit. Technical Report LCS012, Dept. Stat., Stanford University(1985).
18. J.H. Friedman. Multivariate Adaptive Regression Splines, *The Annals of Statistics*, **19:1** (1991) p. 1–141.
19. P. Grassberger and I. Procaccia, *Dimensions and entropies of strange attractors from a fluctuating dynamics approach*, Physica, **13D**, 34-35 (1984).
20. C.M. Hurvich and C.L. Tsai, *Regression and time series model selection in small samples*, Biometrika **76**, p.297-307 (1989).
21. P.J. Huber: *Robust Statistics*, (Wiley, New York 1979).
22. M. Kendall and A. Stuart: *The Advanced Theory of Statistics* (McMillan, New York 1979).
23. E. Kostelich and J.A. Yorke: *The analysis of experimental data using time-delay embedding methods*, Preprint (1990).
24. N.H. Packard, J.P. Crutchfield, J.D. Farmer and R.S. Shaw: *Geometry from a time series*, Phys. Rev. Let. **45**, 712–716 (1980).
25. Y. Pesin and C. Yu: *Dimension of Attractors. Proof of the Eckmann Ruelle Conjecture*, Penn State Preprint.
26. D. Ruelle and F. Takens: *On the nature of turbulence*, Comm. Math. Phys. **20**, 167–192 (1971); **23**, 343–344 (1971).
27. R.L. Smith. Estimating Dimension in Noisy Chaotic Time Series. *J.R. Statist. Soc. B* **54:2** (1992) p. 329-351.
28. D.F. Swayne, D. Cook XGobi: A Dynamic Graphics Program Implemented in X with a Link to S. *Proc. of the 22th Symp. on the Interface between Comput. Sci. and Statist.*, (Springer-Verlag, New York, 1990) .
29. F. Takens: *Detecting strange attractors in fluid turbulence*, in "Lecture Notes in Mathematics" **898** (Springer-Verlag, New York 1981).
30. H. Tong: *Threshold Models in Non-linear Time Series Analysis*, in "Lecture Notes in Statistics" **21** (Springer-Verlag, New York, 1983).
31. H. Tong: *Non-linear time series : A dynamical Systems Approach*, (Oxford University Press, New York, 1990).

INTRODUCTION TO THE THEORY OF COMPUTATIONAL DYNAMICS

RAFAEL DE LA LLAVE
Mathematics Department, The University of Texas at Austin
Austin, Texas 78712-1082, USA

1. Introduction

It is a basic philosophical principle that for many physical systems, the state at a certain time determines the state one unit of time later. If the laws governing the system are independent of time we can write a function f, so that, if a system is in an state x at time zero, $f^n(x) = \underbrace{f(f \cdots f}_{n\text{-times}}(x))$ will be the state occupied at n times the basic unit of time. The sequence of $f^n(x)$ is usually called the orbit of x and represents the states that the system will occupy at successive times. The map f is called a discrete dynamical system in the Mathematical literature.

If such a system were given to us in a practical problem (e.g., the solar system, whose laws of evolution are well known) there are some natural questions to ask. For example, we may inquire if there are certain types of orbits (e.g., starting near the Earth and going close to all the planets, remaining close to the Earth but not drifting away due to the periodic pull of other planets and moon) or, in a more philosophical mood we may inquire whether all the orbits are going to keep their shape forever except for small oscillations. For the solar system, the later is a non-trivial question, the planets do eject and capture small bodies such as meterorites all the time. Could it happen that the Earth captures the moon due to the influences of the other planets? Which is the time scale of such phenomena? This type of questions, which even now are not completely answered, were among the first posed to Mathematical Physics.

To the previous questions, we could also add the worry of what would happen if the laws of evolution were slightly modified. Of course, in a physical system there is always the possibility that we have neglected some effect. More importantly, when we introduce a system in a computer or treat it theoretically, we make modifications to it to accommodate the limitations of the computer or the methods. Does this invalidate the conclusions?. Notice for example, that even if a harmonic oscillator has periodic solutions, the introduction of even the smallest friction drives it to equilibrium.

Let us start by discussing away two naive approaches. The first is to try to find a solution in "closed" form and the second is to observe that since f can be written in a computer we can easily iterate any initial condition we like.

To the first we just answer that finding such closed form solutions turns out to be impossible for many dynamical systems (see e.g., [1] for a discussion of impossibility of finding closed form solution of $\ddot{y} + y - y^2 + f(t) = 0$). Even if we found those solutions, given that many of the questions we are interested in are qualitative in nature, it is not clear that we would profit much from such solutions as long formulas.

For the second approach we observe that when the number of parameters is even as small as 6 — 3 positions, 3 velocities — a very crude examination of 1% precision in each coordinate (nowhere near enough the precision required for an orbit of a satellite) would require 10^{12} orbits which is well beyond the reach of today's fastest computers. Even if by some remarkable progress in computer technology — it is a very parallelizable problem — we got to the level where enough orbits could be generated, we would still have the problem of selecting those that perform the motions we want or more difficult one of realizing that some of motions which we are computing may be useful for some goal.

Neither of these two naive methods offers any clue on how to start considering the effects caused by studying modifications of the system.

The fact that simple methods do not work is not a coincidence. Many people heuristically identify deterministic with predictable and predictable with simple. This is merely an abuse of language. It is becoming increasingly recognized that even very simple systems can have very complicated behaviour. Naive iteration of many initial conditions can make one realize very quickly how complicated things can be.

The best answer to the type of questions that we have asked seems to be a program that could be attributed to Poincaré. It roughly, proceeds along the following steps

* Find landmarks that organize the long term behaviour. For example:
    - periodic orbits
    - invariant tori
    - attractors
    - bifurcations
* Perform a local analysis around those landmarks so that we control a non-negligible part of phase space.
    - normal forms
    - local stable manifolds and foliations
* Join all the local information together into global structures
    - index theory
    - global bifurcation
    - basins of attraction

Part of the beauty of this program is that in it, one can make progress using at the same time theoretical analysis and numerical exploration.

If the observer has the right theoretical training, even the naive direct iteration of random points may suggest the presence of features or landmarks which can then be computed precisely with more detailed numerical methods. After some more work, one can use some global topological argument to discover that the features already found imply the existence of another one, which once it is found and its details elucidated, implies some others, etc.. After a few iterations of the process, one may hope to arrive to an understanding sufficient for the applications.

In this lecture, we would like to give some of the flavor of these techniques, emphasizing as corresponds to the title of the school the computational techniques that can be used to find the objects we describe theoretically. Given the limitations of time and space, we can only hope to discuss the most elementary phenomena and be just a guide to the literature. In particular, we will not discuss much about Global Behaviour. To a large extent, the global phenomena is not understood systematically yet, hence it is one of the most interesting areas of research and it is certainly one of the areas in which computer explorations is having a larger effect. Good introductions to some global problems from the point of view of applications exist in the literature [2 3]. It is also instructive to see how this methods can be brought to bear on a concrete example [4].

With the availability of powerful computers at affordable prices it has become possible to develop programs that display the graphical information in real time and allow the user to perform some of the standard diagnostics.

Two such programs that are particularly appropriate for the level of these lectures are DSTOOL and DYNAMICS which are available with the source code. Reading the code of these programs can be a very instructive process that we recommend.

Exercise 1. Take one of the interactive packages mentioned in the text and run at least three examples observing the behaviour of different orbits.

Remark. As a historical note we may add that topology — under the name of "Analysis situs" — was developed by Poincaré as part of this program. This included simplicial homology and many other index theories. Even now topology is defined sometimes as the science of passing local information to global one.

Remark. It goes without saying that this program we have outlined is not the only one that is useful to study dynamics. (Not even if one wants to study the systems computationally.) We just mention the "ergodic program" or the Smale's generic program. In ergodic theory, we try to study the statistical properties of orbits. In the topological classification program one tries to identify properties that are "typical" and obtain a topological classification that applies to all systems. There are already good books and review papers [5 6 7 8] about these subjects. In particular [7 8] contain a very modern point of view that uses ergodic theory and the geoetric program at the same time.

Remark. Very often the laws of dynamics appears given as a differential equation in an space of finite dimensions. Standard theory of ordinary differential equations[9] shows that under very mild regularity assumptions, the two formulations are equivalent. There is a well developed branch of numerical analysis devoted to the computation of evolution maps out of the differential equations. In this lecture we will emphasize maps. The issue of how to approximate differential equations by maps – the only thing that is easy to implement on the computer – is a large field of numerical analysis. Up to date reviews are the books [10] [11].

2. Reliability of Computer Exploration.

Unfortunatley, if one is working with computers, all computer operations involve approximations. Computers work with a finite numbner of digits and, hence, all the operations are approximate. The most naive answer to this problem is just to increase the precision till we can obtain sufficiently many significant digits.

This is not a satisfactory answer on two counts. First of all, it is very difficult to analyze theoretically what is the precision lost at every iteration. Second, it is true that in many cases the number of figures lost is proportional to the number of iterations. When one has to perform operations with numbers which are not supported directly by the hardware the efficiency drops very quickly.

There are indeed methods to work with computers in such a way that one can get definitive answers to numerical computations. More important for dynamical systems is the fact that many systems satisfy the so-called shadowing property. That is:

Definition. We say that a system f has the shadowing property whenever given a sequence of points $\{x_n\}$ such that

$$|f(x_n) - x_{n+1}| \leq \varepsilon \quad (\varepsilon \leq \varepsilon_0)$$

then, there exist a point y such that

$$|f^n(y) - x_n| \leq \delta(\varepsilon) \quad (\delta(\varepsilon) \to 0 \ \text{ when } \ \varepsilon \to 0)$$

A sequence x_n such that $|f(x_n) - x_{n+1}| \leq \varepsilon$ is usually called an ε pseudo-orbit. Notice that a computer is perfectly capable of producing ε pseudo-orbits with ε about the round-off error.

This result is quite satisfactory from the point of view of an analysis based — as Poincaré program — in the qualitative properties of some orbits. Nevertheless, it is important not to read too much from it. For example, it is not clear at all if properties considered "typical" on a system will remain typical in the system of pseudo-orbits produced by introducing approximations. Hence, this result is somewhat useless from the point of view of the ergodic program.

For the so-called Anosov systems — which we will define later — in two dimensions, it is possible to show that if two of them assign positive probabilities to the same observables both of them differ just by a change of variables as smooth as they are.

Example. $x \to 2x \bmod 1$.

This system satisfies the shadowing property. It is possible to show that a typical orbit of this system is equally distributed over $[0,1)$. Nevertheless, on a binary machine, all computed orbits end at zero.

Exercise 1. Compute reliable 10 figures of $f^{100}(0.3)$ where $f(x) = 1 - 2x^2$.

Exercise 2. The magnification of the error in one data after N iterations can be estimated by $|\frac{d}{dx} f^N(x)|$. For one dimensional maps this can be computed by

$$\ln\left|\frac{d}{dx} f^n(x)\right| = \ln \prod_{i=0}^{N-1} |f'(f^i(x))| = \sum_{i=0}^{N-1} \ln |f'(f^i(x))| \ .$$

Compute this estimate of the error for several functions of the form

$$f_\lambda = 1 - \lambda x^2 \qquad \lambda \in [0,2] \qquad (\text{Take } N = 500,\ x = 0.3)$$

Exercise 3. How does the previous result change if you take different x? What happens if you take a larger N?

Exercise 4. Consider $f(x) = 2x \bmod 1$. Show that if $f(x_n) - x_{n+1} = \varepsilon_n$, then $y = x_0 + \sum(\varepsilon_n/2^n)$ is a shadowing orbit.

If rather than starting with initial conditions x, we start with initial conditions $x + \varepsilon e$, after N iterations $f^N(x + \varepsilon e) \approx f^N(x) + \varepsilon Df^N(x)v$ so that

$$\lim_{N\to\infty} \frac{1}{N} \ln(\|Df^N(x)v\|/\|v\|) = \Lambda(x, v)$$

is an adequate measure of the effect of an error in the initial conditions.

Even if it is not true that these limits always exist, a deep theorem of Osledec's — see [8] for a more modern proof — tells us that for a "typical" initial condition x, these limits exist for all vectors. Moreover, if we define $E_\lambda = \{v \mid \Lambda(x,v) \le \lambda\}$ we can see it is a linear space $E_\lambda \subseteq E_{\lambda+\delta}$, $\delta > 0$, so that we can except to have numbers $\lambda_1, \ldots, \lambda_k$ ($k \le$ dimension) that appear. Such numbers are called the Lyapunov exponents.

Notice also that $\Lambda(f(x), Df(x)v) = \Lambda(x, v)$ so that the set — and multiplicity — of Lyapunov exponents is invariant under the map. In many cases this implies that the exponents are constant over large region of phase space. (This is related to the "ergodic" hypothesis for this region.)

Intuitively, positive Lyapunov exponents mean that small causes will have large effect – growing exponentially in time – and, hence, we expect very widely varying types of long term behaviour intermingled. This intuition can be made precise and quantitative and there are theorems that relate the Lyapunov exponents to the "entropy" — a measure of the complication of the set of orbits — and to the dimension of the attractors — a measure of the complication of the geometry of the set described by the orbits[13]. Making all this more precise is a very active area of research now.

For a typical orbit, we can find a basis $v_1, \dots, v_n$ in such a way that $\|Df^N(x)v_i\| \approx e^{\lambda_i N}C$ both for $N \to \pm\infty$ (several λ_i's here could be the same). This suggest that we can compute the largest Lyapunov exponent by a method very similar to the "power method" to compute the modulus of the largest eigenvalue of a matrix.

We pick a vector u_0, $|u_0| = 1$ and recursively set $u_{n+1} = Df(f^n(x))u_n$. Unless v_0 does not have any component on the direction of the leading eigenvalue, we will have $\lambda_{\max} \approx \frac{1}{N} \ln \|u_N\|$. In practice, even the tiniest round-off errors will take us to this situation.

Computation of other Lyapunov exponents is much harder. A good algorithm — used in [14] to give an alternate proof of Osledec's theorem but discussed in detail in discussed in [7] — is: set $Df(x) = Q_0R_0$ where a_0 is orthogonal R_0 is upper triangular (this is the well known QR factorization of linear algebra, implemented reliably in many good numerical packages such as EISPACK). Then recursively set $Df(f^i)Q_{i-1} = Q_iR_i$. It is clear that then we have $Df(f^{N-1}x)\cdots Df^n(x) = Q_NR_{N-1}\cdots R_0$. For low dimensions, there are shortcuts. For example, in dimension 2, just notice that $\det Df^N(x) = \det f(f^{N-1}x)\cdots \det f(x)$ and that the rate of growth of the determinant should be the sum of the two Lyapunov exponents.

We also point out that the concept of Lyapunov exponent can be generalized for approximate orbits. One can show that for approximate orbits with non-zero Lyapunov exponents can be shadowed (we will discuss this in more detail for periodic orbits). If a system has abundance of true orbits with non-zero Lyapunov exponents, these conditions will be satisfied. This seems paradoxical because we cannot compute well individual trajectories, we should believe qualitatively what we compute! Nevertheless (see exercise 4) what we are saying is that for these chaotic systems, anything can happen, in particular what we computed!

3. Periodic Orbits

Periodic orbits are the simplest landmarks to compute. On the other hand, many features of dynamical systems can be approximated by periodic orbits. This was already predicted by Poincaré [15] Vol I p. 82.

Il y a même plus: voici un fait que je n'ai pu demontrer rigorousment, mais qui me parait pourtant très vrisemblable. Ètant données des èquations de la forme definie dans le n° *13 [Hamilton eq.] et une solution particulière quelconque de ces equations, on peut toujours trouver une solution périodique (dont la période peut, il est vrai, être très longue), tel que la différence entre les deux solutions soit aussi petite qu'on le veut, pendant un temps, assi long qu'on le veut. D'allieurs, ce qui nous rende ces solutions périodiques si precieuses, c'est qu'elles sont, pour ansi dire, la seule brèche par où nous puissions essayer de pénétrer dans une place jusqu'ici réputée inabordable.*

The simplest method to compute periodic orbits is to search for solutions of $f^N(x) = x$ using, e.g., a Newton method or any other non-linear equation solver. As mentioned before, the numerical calculation of the derivative becomes delicate since numerical errors tend to line up all the vectors into the dominant Lyapunov exponents.

A more reliable method is based on the study of the operator acting on sequences $\mathbf{x} = \{x_i\}_{i=1}^N$

$$\mathcal{C}(\mathbf{x}) = (f(x_N), f(x_1), \ldots, f(x_{N-1}))$$

We just point out that $|\mathcal{C}(\mathbf{x}) - \mathbf{x}| \leq \varepsilon$ (with the supremum norm) is the same as saying that $\mathbf{x}$ is an ε pseudo-orbit.

If one could show that $D\mathcal{C}(x) - \mathrm{Id}$ is invertible and ε is sufficiently small, then, a Newton method would converge to a true solution. This is a version of the shadowing theorem mentioned before.

Notice that $D\mathcal{C}(x) - \mathrm{Id}$ is a very sparse matrix. If one picks an adequate solver for sparse matrices, it could be that the method is better conditioned than the naive one.

As it turns out $D\mathcal{C}(x) - \mathrm{Id}$ is invertible when and only when the pseudo-orbit has no zero Lyapunov exponents. Hence, if there are recurrent orbits and non-zero Lyapunov exponents, there should be periodic orbits close by. That is, periodic orbits appear as tracers of chaotic behaviour. (A precise version of this are the "ergodic" closing lemmas[16,17,18].)

For many systems that have special structures, one can use more effective methods. For example, the symplectic integrators considered in Professor Jauslin's lectures when applied to Hamiltonians $H = \frac{p^2}{2} + V(q)$ lead to maps of the form

$$\begin{aligned} p_{n+1} &= p_n - h\nabla V(q_n) \\ q_{n+1} &= q_n + p_{n+1} \end{aligned} \tag{1}$$

One can see that these equations are variational equations for:

$$S(\mathbf{q}) = \sum_{i=1}^{N} (q_{n+1} - q_n)^2 \tfrac{1}{2} + hV(q_n)$$

(where $q_{N+1} = q_1$).

This variational principle is rather well behaved and using a minimization algorithm it is possible to obtain good solutions even for fairly long periods.

Of course, the existence of this variational principle is a discrete version of Hamilton's stationary action principle.

For maps f that have the property that $f^{-1} = I \circ f \circ I$ with $I^2 = \mathrm{Id}$, usually called reversible — very often in Physics this arises when I is just reversing the velocity and keeping the positions— one can find a very effective method to compute periodic orbits

Notice that $f = I(If)$ and that $(If)(If) = \mathrm{Id}$ so that we can write $f = IJ$ where $I^2 = J^2 = \mathrm{Id}$.

Exercise. Compute explicitly the maps I, J for (1).

The following algorithm can be found in [19].

Algorithm. Let Σ be the set of points x such that $J(x) = x$. If $x \in \Sigma$ and $f^N(x) \in \Sigma$ then $f^{2N}(x) = x$.

Proof.

$$\begin{aligned} f^N(x) &= (IJ)^{N-1} IJ(x) = \\ &= J(IJ)^{N-1} I(x) = \\ &= f^{-N}(x) \end{aligned}$$

where in the first equality we have used $J(x) = x$, $Jf^N(x) = f^N(x)$ and in the second we have used $J^{-1} = J$, $I^{-1} = I$.

Notice that Σ is a manifold of half the dimension of the space and the equation $f^N(x) \in \Sigma$ amounts to a system of equations of half the dimension. In particular, for 2-dimensional reversible systems, this leads to solving 1-equation with one unknown which is quite easy since in one dimension finding zeros of a function can be done very systematically exploiting the fact that a change of sign implies the existence of a zero in between. For a complete discussion of algorithms for periodic orbits for symplectic and reversible mappings we refer to [20].

The computation of periodic orbits for differential equations is essentially more difficult than for discrete systems since the period also has to be computed. Moreover, all the points in the periodic orbit are periodic points with the same period so that the solutions are never isolated, i.e., a Newton method will not work.

As an added complication, computing the derivatives of the time T map — needed for a Newton method — requires to integrate the so-called variation equation (see [9]).

A method which is quite analogous to the fixed point method we discussed for maps is the so-called "multiple shooting method." One tries to compute a vector of points in which we impose that each one is in the image of the next. A convenient way of fixing the period is to impose that some point belongs to a transversal section of the flow and the last step is not a fixed time, but rather a projection on the transversal surface. (See [21] for a discussion of the multiple shooting method for boundary value problems.) A good numerical integration that includes study of the variational equations is ODESSA.

For Hamiltonian systems one still Lagrange's variational principle for periodic orbits:

$$\delta \int_0^T L(\dot{q}, q)\, dt = 0$$

among all functions which satisfy $q(t+T) = q(t)$.

For Hamiltonian systems, one also has that the energy is a conserved quantity and often the orbits appear in families for which the period depends on the value of the energy. That is, we can prescribe the period of any preassigned value.

If we discretize the space of periodic functions, e.g. using Fourier series,

$$q(t) = \sum_k \hat{q}_k e^{2\pi i k t/T}$$

we can transform

$$\int_0^T L(\dot{q}, q)\, dt = S(\hat{q}_k)$$

which when truncated to a finite number of modes becomes a function of finitely many variables whose critical points can be computed by conventional methods.

See [22,23] for a more complete discussion of these variational methods for Hamiltonian systems.

Notice also that, once we know that we can prescribe the period, studying a discrete version of the system (1) may produce a solution which is approximate up to the truncation error of the integration (related to the step h) — one needs precautions against the problem of multiple solutions.

If we require a solution with a prescribed energy, e.g., if we are computing several families of orbits to study their interactions, we may want to refer them to the same energy we still need to study the function $E(T)$ and invert it.

Exercise. For equations of the form $y'' = f(y)$, denote by $y_n = y(nh)$ and analogously for other functions notice that

$$y_{n+1} + y_{n-1} - 2y_n = h^2 y_n'' + \frac{h^4}{12} y_n^{(iv} + o(h^6)$$
$$y_{n+1}'' + y_{n-1}'' - 2y_n'' = h^2 y_n^{(iv} + o(h^4)$$

Show that

$$\left(y_{n+1} - \frac{h^2}{12} f(y_{n+1})\right) + \left(y_{n-1} - \frac{h^2}{12} f(y_n)\right) = 2\left(y_n - \frac{h^2}{12} f(y_n)\right) = f(y_n)$$

is equivalent to the original equation up to errors of order h^6 and that if we consider the y_n's as unknowns, this gives a system with sparse coefficients that can be solved efficiently. (This is a non-linear version of the so-called Numerov method for Schrödinger equations.)

Before computers were available, one of the few methods to compute periodic orbits were perturbative methods. The most natural one for is the so-called

Lindstedt-Poincaré method. We will discuss it in detail for the case of differential equations, where one has to guess not only the solution but also the period.

The basic idea is that we search for a solution of the differential equation $y_\varepsilon(y) = u_\varepsilon(T_\varepsilon t)$ where $u_\varepsilon = u_0+\varepsilon u_1+\varepsilon^2 u_2+\cdots$, $u_\varepsilon(t+2\pi) = u_\varepsilon(t)$, $T_\varepsilon = T_0+\varepsilon T_1+\varepsilon^2 T_2+\cdots$. We substitute in the original equations and require that the equations have solutions at all orders, the coefficients in the expansion of T_ε will appear as compatibility conditions in the expansion of u.

For example, if we consider

$$y'' = y + \varepsilon N(y)$$

(a non-linear oscillator with friction). The Lindstedt-Poincaré equations become:

$$(T_\varepsilon)^2 u_\varepsilon''(t) = -u_\varepsilon(t) + \varepsilon N(u(t))$$

At order zero we have

$$(T_0)^2 u_0'' = -u_0$$

This equation says $u_0 = A\sin(T_0 t+\varphi)$. We will set $\varphi = 0$ since it physically corresponds to choosing the origin of time. We also obtain $T_0 = 1$ from the periodicity conditions. At order 1, we have

$$2T_1 u_0 + u_1'' = -u_1 + N(u_0)$$

If we consider this as a (linear!) equation for u_1 we see that it is forced with the term $N(u_0)-2Tu_0$. If this forcing term contained a term of the form $\sin(t)$ or $\cos(t)$ the theory of linear systems with constant coefficients tells us that the solution should contain a term of the form $Ct\sin(t)$ or $Ct\cos(t)$. This would violate the condition that the solutions are periodic. If we impose that these resonant forcings do not appear, this amounts to conditions on A, T_1. If these equations can be solved, then, we proceed to next order.

It is by no means obvious that these equations can be solved. Indeed for

$$y'' = -y - \varepsilon(y')^3$$

they cannot. Physically, this is a system with friction and, indeed, as soon as friction appears and is not balanced by an external force, periodic motion becomes impossible.

Nevertheless, it is quite frequently the case that, after a few order, the recursive equations develop some structure that makes it clear that they can be solved for all higher orders. This sometimes can even be proved using more elaborate theory.

We just mention that there are symbolic programs that can perform these calculations [24]. Usual numerical programs do not work very well since it is necessary to perform Taylor expansions, gather terms, etc.

Exercise. Compute by hand the Lindstedt-Poincaré series of to order 2 for the Van-der Pol oscillator

$$y'' = y + \varepsilon(1 - y^2)\dot{y}$$

Exercise. Obtain one of the symbolic manipulation programs mentioned above and compute the same series.

A complication on periodic orbits is quasi-periodic orbits. Recall that a quasi-periodic orbit is one that has an expansion:

$$y(t) = \sum_{k_1,\cdots,k_n} e^{2\pi i(k_1\omega_1+\cdots+k_n\omega_n)t}\hat{y}_{k_1\ldots k_n}$$

where $\omega_1,\ldots,\omega_n$ are independent frequencies. Unfortunately, the sense in which this expansion converges is quite delicate. We will just mention that the numerical computation of such orbits seems to be best done by transforming the equation for these orbits into an equation for the coefficients $\hat{y}_{k_1\ldots y_n}$. When $n > 2$, they are practically intractable. Nevertheless, given the importance of quasi-periodic orbits, there is considerable effort done in their calculation.

4. Local Theory Near a Fixed Point

If we have a fixed point — we will assume for simplicity it is the origin — our goal is to obtain a detailed description of the dynamics of the map in a neighborhood of the origin.

Since $f(x) = Df(0)x + o(|x|^2)$, it is natural to start by comparing the dynamics of the full map to that of the linear part.

The dynamics of the linear part is very easy to study. Since $Df(0) = V^{-1}JV$ where J is a (real) Jordan form we see $Df(0)^n = V^{-1}J^nV$ so that the long term behaviours in the dynamics of the linear approximations can be studied by studying those of Jordan blocks. The blocks corresponding to eigenvalues of modulus bigger than 1 will blow up exponentially fast, those corresponding to eigenvalues of modulus smaller than 1 will shrink exponentially fast and those corresponding to modulus 1 will remain bounded or blow up (polynomially) both in the future and in the past.

Sample examples such as $f(x) = x + x^{2n+1}$, $f(x) = x - x^{2n+1}$ show that no matter how high order the perturbation, the dynamics of maps with eigenvalue 1 can be radically changed. We define a map to be hyperbolic if no eigenvalue has modulus 1.

A very effective way to compare the dynamics of two maps is to find a change of variables that sends one into the other. The situation for hyperbolic maps is very clean thanks to the following theorem of Hartman (differential equation) and Grobman (maps). A modern proof can be found in [25].

Theorem. Let f be a C^1 map on a hyperbolic fixed point. Then, there exists a continuous change of variables such that

$$h^{-1}fh = Df(0)$$

in a neighborhood of the origin.

The map produced the the Hartman-Grobman theorem is, unfortunately, not very computable and it is, in general not very smooth. To uderstand why it is instructive to try to compute it formally. This computation will also give hints of some other phenomena that we will need to study in the chapter on bifurcations.

For simplicity, we will assume that the matrix $Df(0)$ is diagonal and will denote the components by subindices. Here we will have

$$f^i(x) = \lambda^i x^i + f^i_{jk} x^i x^k + f^i_{jk\ell} x^j x^k x^\ell + \cdots$$

(No convention summation for $\lambda^i x^i$, but we use the convention for all the other indices.)

If we set $h^i(x) = x^i + h^i_j x^j$ and try to solve $hf = Df(0)h$ up to third order we obtain

$$(h^i_{jk} \lambda^i \lambda^k + f^i_{jk} - \lambda^i h_{jk}) = 0$$

so that we can solve the equation provided $\lambda^i - \lambda^j \lambda^i \neq 0$. When equalities such as this happen, they are usually called "resonances" and it is clear that, if we choose examples for which $f^i_{jk} \neq 0$, it is impossible to obtain h which are C^2 and which reduce the map to its linear part. (In [26] one finds obstructions for C^1 maps!)

The map $\tilde{f} = hfh^{-1}$ satisfies

$$D\tilde{f}(0) = Df(0) \quad , \quad \tilde{f} = Df(0) + o(|x|^3)$$

We will work by induction and show that if

$$f^i(x) = \lambda^i x^i + f^i_{j_1 \ldots j_n} x^{j_1} \cdots x^{j_n} + o(|x|^{n+1})$$

and if $\lambda^{j_1} \cdots \lambda^{j_n} \neq \lambda^i$ we can find $h^i(x) = x^i + h^i_{j_1 \ldots j_n} x^{j_1} \cdots x^{j_n}$ in such a way that

$$hf = Df(0)h + o(|x|^{n+1})$$

Hence $hfh^{-1} = \lambda^i x^i + \tilde{f}^i_{j_1 \ldots j_{n+1}} x^{j_1} \cdots x^{j_{n+1}} + o(|x|^{n+2})$. The composition of all the changes of variables obtained by the induction process will achieve the desired reduction.

If we expand $hf - Df(0)h$ we obtain that the coefficient of order n is

$$0 = h^i_{j_1 \ldots j_n} \lambda^{j_1} \cdots \lambda^{j_n} + f^i_{j_1 \ldots j_n} - \lambda^i h^i_{j_1 \ldots j_n}$$

so that, we can solve provided that $\lambda^{j_1 \cdots j_n} \neq \lambda^i$ and, if not, it is possible to find examples.

A more complete treatment can be found in [27], [28] among other places, but we recommend the reader that he/she tries the following exercises.

Exercise. Study the case where the matrix has a non-trivial Jordan block. Show that the equation always has solution provided that the product of eigenvalues is not an eigenvalue (i.e., the same as in the case discussed).

Exercise. Discuss the case of complex eigenvalues and show that even in that case, the conclusions are not altered.

Exercise. Show that in the case that we have one resonant term, we can still eliminate all others provided that they are not resonant.

All this leads to the Poincaré-Dulac theorem.

Theorem. Let f be a formal power expansion. Then, there exists a power expansion h such that

$$h \circ f \circ h^{-1} = Df(0) + R$$

where R is a poser series expansion that only contains resonant terms (i.e., terms $R^i_{j_1 \dots j_n}$ where $\lambda^i = \lambda^{j_1} \cdots \lambda^{j_n}$).

The fact that these formal power series have a real — not purely formal meaning is given by the celebrated Sternberg theorem.

Theorem. Assume that f as before is non-resonant, and hyperbolic. Then, one can find a C^∞ change of variables reducing it to the linear part.

There are many variants of the Sternberg theorem which cover cases in which there are resonances or finite number of derivatives. We will not discuss them here. Standard references are [27,28]

We note that there are many physically natural reasons that force the existence of resonances. For example, preservation of volume requires that the product of eigenvalues is 1 and the Hamiltonian structure of equations imposes much more severe restrictions.

Exercise. A linear map A is called Hamiltonian (or symplectic) if $A^{-1} = JAJ$ where J is the antisymmetric matrix $J = \left(\begin{pmatrix} 0 & 1 \\ -1 & 0 \end{pmatrix}\right)$. Show that if λ is an eigenvalue of a real Hamiltonian matrix, so are $\bar{\lambda}$ and $1/\lambda$.

We also point out that even if the existence of resonances is atypical for a map, it may be unavoidable for families of maps. That is, if a map is resonant, we can make an small perturbation that makes it non-resonant. If we have a family one of whose eigenvalues starts in -0.8 and ends up in -1.2, all nearby families are going to contain a point that has an eigenvalue -1.

Notice also that these normal forms are also discontinuous. An small modification in the eigenvalues may make a particular term become non-resonant. (This is quite similar to the Jordan blocks.) Since the structure of the terms depends on exact comparisons calculations in which roundoff is introduced cannot be used to compute normal forms. One has to use exact arithmetic.

If one is using inexact arithmetic, the only possible course is not to eliminate the terms for which the product of eigenvalues is close to another eigenvalue.

We refer to [24] for programs for the exact calculation of normal forms using symbolic manipulators. We warn that the inductive method we have sketched before is not very efficient and that one has to use a more efficent method.

The theory of normal forms for Hamiltonian maps is quite interesting. The last exercise shows that there are neccesarily lots of resonances. Also, it shows that for Hamiltonian maps, there can be no linearly stable equilibrium points. The best that one can hope is to have linearly neutral equilibrium points. For these points, one has to study the stability using higher order terms so that the problem has a considerable interest. It is conjectured that most fixed points for maps are eventualy unstable when the number of variables is four or greater – this is related to the famous Arnol'd diffusion – but neverthelss, one can study the normal forms in great detail and obtain estimates for practical times of stability. Since the fixed points are the places where one tends to leave satelites, this is of considerable practical interest [29].

5. Invariant Manifolds

By Hartman-Grobman theorem, in a neighborhood of a hyperbolic point, we expect that there is a continuous image of a subspace that converges to the origin, under iteration and another one of points Δ that converge to the origin under inverse iteration.

It is surprising, but true that these sets are manifolds as smooth as the map.

The precise statement is:

Theorem. Let $f : \mathbf{R}^n \to \mathbf{R}^n$. Assume $f(0) = 0$, $f \in C^r$, $r \in \mathbf{R} \cup \{\omega, \infty\}$,

$$\mathbf{R}^n = E^s \oplus E^u \ , \quad Df(0)E^s = E^s \ Df(0)E^u = E^u$$
$$||Df(0)^N|_{E^s}|| \leq \lambda < 1 \ , ||Df(0)^{-N}|_{E^u}|| \leq \lambda < 1 \ ,$$

Then, in a neighborhood of zero we can find a C^r function $W : B_\delta \subset E^s \to E^u$ whose graph is invariant under f $f(0) = 0$, $Df(0) = 0$ (analogously for E^u). Moreover, if x is such that $|f^n(x)| \leq \delta$ then $x \in \text{graph}(W)$).

In [30] one can find a very readable proof reproduced in [31].

We will denote these points as W^δ, the "local stable manifold". Using this local stable manifold, one can characterize the set of all points that converge to the origin as the union of the inverse images of the local stable manifold. These stable and unstable manifolds tend to wander around large regions of the phase space of the system so that they organize a lot of the behavior.

Many calculations of the unstable manifold — the stable manifold is the unstable manifold for the inverse — are based on the following result, usually called the λ-lemma. (See [6] for a proof of an slightly weaker result.)

Theorem. Let D be a C^r disk of the same dimension of E^u, sufficiently small, and transversal to E^s. Then, $f^n(D) \cap B_\delta$ converges to $W^{u,\delta}$ in the C^r sense.

The reason for this convergence is that the direction along E^u stretching and the direction along E^s is contracting. The non-linear effects are negligible in sufficiently small neighborhoods.

A practical algorithm in the case that the manifolds are one-dimensional is to pick an small segment, fill it with points, iterate the map, discard the points that go out of the neighborhood and interpolate with the remaining ones — e.g., using splines — to make up for those lost.

The algorithm to compute the local manifold stops when there is no change in these points. Once this local manifold has been computed, we can interpolate more densely — refine if desired by iterating again — and iterate and plot without discarding the points that move out of the neighborhood.

This seems to be an extraordinarily effective and reliable algorithm. It, of course fails in situations in which the stable direction is very close to the unstable one (this happens often in Hamiltonian systems near to integrable where the angles are often exponentially small in the perturbation parameter). The only drawback is that one has to select the length of the segment, the number of points. It seems difficult to devise rules that select these parameters without human intervention.

Even if the mathematical justification of the λ-lemma works in any number of dimensions, this algorithm is hard to adapt to two or bigger dimensions. The main source of trouble is that if the two unstable directions have very different eigenvalues, a large number of iterations tends to cluster the points along the one with largest modulus. If one knew how many iterations were necessary, one could counteract this problem by iterating an ellipse which is much larger in the weak unstable direction than in the strong unstable one. Another complication is that interpolation of two dimensional functions is notoriously more difficult than for one-dimensional functions. Of course, the problem becomes much harder as the dimension increases.

Even if it is quite possible to compute two dimensional invariant manifolds for some concrete systems, it seems that there is no generally available implementation that is robust enough to be left in the hands of general users. It would most welcome.

Another completely different algorithm for 1-dimensional manifolds is trying to find out the power series expansion for the function $\varphi : \mathbf{R} \to \mathbf{R}^n$ that satisfies

$$\varphi(\lambda t) = f\varphi(t)$$
$$\varphi(t) = \sum \varphi_n t^n$$

We see that $\varphi_0 = 0$, $Df(0)\varphi_1 = \lambda\varphi_1$ so that φ_1 is an eigenvalue and that to order n, the expansion becomes $\lambda^n\varphi_n = Df(0)\varphi_n + R_n$ where R_n is an expression involving the previously computed φ. Since λ^n is not an eigenvalue this equation can be solved for all φ_n.

It turns out that one can estimate the recursion in the series under the assumption that f is analytic. This was done in [32] which is, presumably the first proof of the stable manifold theorem.

The algorithm is easy to implement in the case the f is a polynomials. A very nice modern discussion of the implementation and discussion of computational errors can be found in [33].

I have found it more computationally efficient and more stable to use an accelerated convergence method that I will describe now.

If we have $\varphi(\lambda t) = f(\varphi(t)) + R$ where $R = O(t^N)$ we try to find $\Delta(t) = \sum_N^{2N} \Delta_N t^n$ in such a way that $(\varphi + \Delta)(\lambda t) = f((\varphi + \Delta)t) + \tilde{R}$ with $\tilde{R} = O(t^{2N})$. The equations for Δ are

$$\Delta(\lambda t) = Df(\varphi(t))\Delta(t) - R^{[\leq N]}$$

where $R^{[\leq 2N]}$ is $\sum_N^{2N} R_n t^n$.

If $Df(\varphi(t)) = \sum_{n=0}^N S_n t^n + O(t^{N+1})$ the equation for Δ amounts to

$$\Delta_n \lambda^n = Df(0)\Delta_n + \sum_{k=1}^{n} S_n \Delta_{n-k} + R_n$$

We see that this can be solved much faster than the non-linear recursions that involve recomputation of the remainders. More importantly, it is much more stable numerically. Similar tricks can be used to accelerate the calculation of normal forms.

Notice that the only thing that is used in this algorithm is that λ is an eigenvalue but that λ^n is not. In those cases, one can construct invariant manifolds which are tangent to the eigenspaces just the same. The algorithm can also be extended easily to the case that λ is a diagonal matrix. With some more complication it can also be extended to other cases.

6. Bifurcations

Many physical systems involve external parameters and it is a natural question to try to study the behavior for all possible values of the parameter.

The implicit function theorem guarantees that if the derivative of the return map at a periodic orbit does not have an eigenvalue 1, the periodic orbit will also exist for sufficiently close values of the parameter and, moreover can be written as an smooth function of these parameters. Hartman-Grobman theorem says that, unless some eigenvalues cross the unit circle the topological class of the periodic point will not change.

From the numerical point of view, this following of non-degenerate periodic orbits is relatively easy to implement. If we know periodic orbits for a value of the parameter, we can take it as initial guess for a Newton method that tries to find the periodic orbit for an slightly bigger value (how to choose it?). Since the solution $x(s)$ of $f(x,s) = x$ satisfies

$$\frac{d}{ds}x(x) = -\left(\frac{\partial f}{\partial x}(x,s) - \mathrm{Id}\right)^{-1} \frac{\partial f}{\partial s}(x,s)$$

we can try to use a high quality O.D.E. solver (with adaptive step!) to obtain very good approximate solutions that can then be refined by an equation solver.

These methods — usually called "homotopy" or "continuation" are very effective and frequently one finds solutions by connecting the physically relevant problem to another problem that can be solved exactly. Of course, finding suitable systems that approximate the required one is an art.

Notice also that the considerations above apply to fixed point equations that involve parameters. Since we reduced many equations for invariant objects to fixed point equations, these methods are widely applicable. We will, nevertheless, discuss only periodic orbits.

Severe complications arise when some eigenvalue of the periodic orbit cross the unit circle and, hence the local picture changes quite drastically — even the topological picture changes.

There are three main simple ways in which this can happen:

A) one eigenvalue becomes 1

B) one eigenvalue becomes −1

C) a pair of complex eigenvalues crosses the unit circle.

Even if there are other possibilities such as several eigenvalues coming onto the unit circle at the same value of the parameter, we will ignore them for the moment for the sake of simplicity.

Notice also that, for a "typical" family, one does not expect that these things happen since arbitrarily close families can break the degeneracy. Of course this is a rather poor justification for restricting our attention to these cases. In Physics, one often has systems with symmetries or geometric properties that force these situations or we may be particularly interesting in studying the system in the situation where this happens — analogous to putting a thermodynamic system at the tricritical point! The true argument for restricting ourselves to these cases is just lack of space.

In these lectures we can only underline the main results. We refer to standard treatises such as [34] for full proofs and precise formulations — including the explicit form of the "non-degeneracy" conditions. A good treatment of the elementary bifurcations suitable for practitiones is [35].

The behavior in A) can be understood by looking at what happens for the family $f_\lambda(x) = X^2 + \lambda$ at $\lambda = 1/4$. We see that the fixed points are $X = \frac{1}{2} \pm \sqrt{(1/4-\lambda)(1/4+\lambda)}$. We see that at $\lambda = 1/4$ we obtain a double root and that, for $\lambda > 1/4$ there is no root an for λ smaller there are two roots at a distance commensurate with $(\lambda - 1/4)^{1/2}$.

Notice that the fixed point start to change very quickly with the parameter $(\frac{d}{d\lambda} x_\lambda \approx \frac{1}{\sqrt{\lambda - 1/4}})$ but with a well defined law.

It is possible — but non-trivial — to show that qualitatively similar things happen for "typical systems" in higher dimensions. The proof roughly goes as follows: First, we show that, by a suitable change of variables — that depends on the parameter λ — we can reduce our system to:

$$f\begin{pmatrix} s \\ u \\ x \end{pmatrix} = \begin{pmatrix} A_\lambda(s) \\ B_\lambda(u) \\ C_\lambda + (1+\lambda)x + D_\lambda x^2 \end{pmatrix} + o(|x| + |u| + |s|)^3$$

where s, u correspond to the stable and unstable directions of $Df(0)$ and $A_\lambda(s)$, $B_\lambda(u)$ are contractive and expansive mappings respectively.

This change of variables can be computed exactly as in the normal form case, but we just leave the terms that we cannot eliminate. Second, we study the dynamics of the main part of the normal form above. Third, we show that the features of existence of periodic orbits etc. are not modified by the high order terms.

If we believe that the perturbation argument can be carried out, we see that the conclusions we wanted to argue are true provided that $C_0 \neq 0$, $D_\lambda \neq 0$. This is what we meant by a "typical system".

The fact that the stable or unstable components do not play any role in the bifurcation is quite a general phenomenon. It can be understood at the level of normal forms as we have sketched here (this is the so-called Lyapunov-Schmidt reduction) or one can invoke a high powered invariant manifold theorem to show that all the interesting behavior happens on a two dimensional set. (See [31] for a detailed discussion of both methods.)

Exercise. In population dynamics it is very natural to assume that zero population in one generation implies zero population for future generations so that one is lead to consider models of the form $f_\lambda(x) = x g_\lambda(x)$. Show that these models violate the genericity assumptions and study the bifurcations that one gets. (An often used example is $f_\lambda(x) = \lambda x(1-x)$. For this example the critical value is $\lambda = 1$.)

It is very tempting to say that B) reduces to A) just by considering f^2. Unfortunately this does not work as can be seen by noticing, by the implicit function theorem we know that there is a family of fixed points for f — hence for f^2 — on both sides of the critical point. Nevertheless A) predicts that there are no fixed points on one side.

The reason why this does not work is that when $f^1_{\lambda_0}(0) = -1$ $f^2_{\lambda_0}$ violates the non-degeneracy conditions.

Exercise. Show that if $f_{\lambda_0}(0) = 0$, $f'_{\lambda_0}(0) = -1$ then $f^{2}_{\lambda_0}{}''(0) = 0$.

Nevertheless, it is possible to show that, given some non-degeneracy conditions in f_λ we can obtain normal forms up to one order more and the results produced by the main part are not affected by the high order remainder terms.

Exercise. Compute explicitly the non-degeneracy conditions.

Suffice it to say that the main result is that the fixed point changes from stable to unstable but that also a periodic orbit of period 2 appears on one side of the bifurcation. This periodic orbit has an amplitude proportional to the square root of the difference of the parameter to bifurcation and it has opposite stability to that of the fixed point on the branch. Except for that, all the possibilities can occur. This can be seen very simply by studying the family

$$f_\lambda(x) = (-1 \pm \lambda)x + Ax^2 + Bx^3$$

for different values of A, B.

Exercise. Study — algebraically — or with the help of an small computer the periodic orbits of the example above and get a classification of all the pictures.

The most spectacular bifurcation happens in case C). This is the so-called Hopf bifurcation. It is important to note that the non-degeneracy conditions include that the eigenvalues are not roots of unity up to order 4.

In that case, one can show that taking polar coordinates, one can reduce the normal form

$$f_\lambda(r,\theta) = \big(\varphi_\lambda(r), \theta + \omega(r)\big) + o(r^5)$$

where $\varphi_\lambda(r) = (1+\lambda)r + A_\lambda r^2 + B_\lambda r^3$ and ω has a Taylor expansion that starts with the argument of the eigenvalues that cross the unit circle.

Again, under genericity assumptions we can see that the equation for r has a fixed point that grows with the parameter as $\sqrt{r}$. Notice that this corresponds to an invariant circle. Again, it is possible to show that this is not modified by the high order terms in a substantial way.

Notice that this is quite a remarkable result since it produces an invariant circle for a map, quite a non-trivial statement.

The theory of bifurcation of differential equations can be reduced to the theory we have just developed by taking time-one maps. (Notice that then, we do not get an analogue of case B).) The proofs also can be done directly. In particular, the treatment of case C) is much simpler since a periodic orbit of a differential equation is a much less surprising statement than an invariant circle for a map.

Even if only elementary bifurcations are involved, the complete bifurcation diagrams may be quite complicated, periodic orbits may appear and disappear through any combination of those bifurcations. Actually, some bifurcations tend to appear together. For example, after a differential equation has experienced a Hopf bifurcation, the return map of the resulting periodic orbit may experience a Hopf bifurcation giving rise to a solution with two frequencies (this is part of the so-called Ruelle-Takens proposal for the origin of turbulence). It is also known that period doubling bifurcations tend to appear together and to lead to a cascade with remarkable properties. (See the collection [35,7] for reviews on these properties.)

The idea that there are several cascades of bifurcations that can lead from ordered states to complicated motions has been discussed in [7,36] where these phenomena have been given the name of "scenarios for the transition to chaos".

One can understand the existence of scenarios with the help of the following picture. We may think of the set of bifurcating maps as a submanifold — with some corners corresponding to degenerate cases — of the set of maps. In some regions in the set of maps, these manifolds accumulate towards the boundary of a region — somewhat ill defined — of chaotic systems.

Of course a good numerical package for studying solutions as functions of the parameters should include some provisions to try to defeat the complications introduced by the bifurcations and the encounter with critical points. This is somewat delicate because in the neighborhood of the bifurcations, the new periodic orbits move very fast with respect to the parameter. So, one has to take special precautions to extrapolate for new values in a way which is consistent with the laws of the bifurcation and not using a linear extratpolation method.

A popular method is the arc-length parameterization method [37]. In this method we introduce a fictitious parameter (call it s) and we try to write both the real parameter and the solution as a function of s while at the same time we impose some normalization condition.

If we were interested in solving $f(x,\lambda) = x$ we would introduce the system

$$\begin{aligned} &f(x,\lambda) = x \\ &\frac{\partial f}{\partial x}(x,\lambda)\dot{x} + \frac{\partial f}{\partial \lambda}\dot{\lambda} = 1 \end{aligned}$$

(where $\dot{}$ denotes the derivative with respect to s). If we had a solution $x^*\lambda^*$ we would declare it a good solution for the parameter $s = 0$ and we could try to find solutions x, λ for small s by solving:

$$\begin{aligned} &f(x,\lambda) - x = 0 \\ &\frac{\partial f}{\partial x}(x^*,\lambda^*)(x - x^*) + \frac{\partial f}{\partial \lambda}(x^*,\lambda^*)(\lambda - \lambda^*) = s \end{aligned}$$

If we choose s sufficiently small, this will produce a new pair of solutions.

There are several publicaly available implementations of these continuation methods. One of them is PITCON whose theory is discussed in detail in [38]. A really useful program is AUTO which not only follows the solutions but also tries to follow the different branches when one that appear as bifurcations.

These two programs are relatively easy to use. One just needs to supply the function on is trying to solve and its derivatives as well as some starting guesses and some global parameters. Even if they fail sometimes, they usually provide enough information that one can try to design better ways.

The study of bifurcations typically requires to do quite extensive manipulations. Many of them can be done using symbolic manipulators such as Reduce, Macsyma (we have often used the version of DOE MAXIMA, which is quite affordable and comes the the source code), Maple or Mathematica.

These manipulators are often invaluable in substituting expansions into each other. Unfortunately, they are quite bade at trying to discuss the existence of real solutions as functions of the parameter. Even in the case that the equations can be solved just using quadratic equations, the existence or not of solutions involves that certain expressions are positive or not. Of course, since the expressions involve common symbols, the assumptions are not independent. It does not seem that there is any symbolic manipulator capable of dealing with such things. MAPLE, MAXIMA and MACSYMA have ways of assigning attributes to the objects — such as positiveness — that can be used to control some simplifications. For example, in many calculations it is quite important to disable simplifications such as sqrt $[x^2] = x$ — which are wrong if x is negative! — or $(x-1)A = (x^2-1) \Rightarrow A = (x-1)$ (it misses $x = 1$). In MAXIMA this can be done by controlling the application of ratsimp.

A very important algorithm to know about when one has to consider systems of algebraic equations with several unknowns is the Grobner-basis algorithm. This algorithm is, roughly, an analogue of gaussian elimination for systems of non-linear polynomial equations. One can operator systematically on the equations in such a way that they appear in a upper-triangular way. That is, we obtain a system of equations of the form

$$F_i(x_1, \ldots, x_i) = 0$$

The discussion is very similar to gaussian elimination. It could well happen that $F_1 \equiv 7$, hence, there is no solution or that $F_1 \equiv 0$, and in that case we have more than one solution. In the ideal case, we can get to solve $F_1(x_1) = 0$ and obtain a finite set of values, which when substituted in $F_2(x_1 x_2)$ will produce a finite set of values for x_2, etc.

The symbolic manipulators mentioned above contain implementations of the Grobner basis algorithm. Nevertheless, their speeds are quite different. For a problem that MAXIMA did in 10 minutes, Mathematica took more than 20 hours and it did not finish! In any case, this algorithm seems quite impractical for more than say 6 equations and 6 variables. For a fuller discussion of the Grobner basis algorithm we refer to [39]. One should note that in this area, even small modifications of the algorithm may make enourmous differences in speed for concrete problems.

If one knows which normal forms one wants to compute there are several FORTRAN or C programs available. A nice one is discussed in [40]. Unfortunately, the inexact arithmetic makes it impossible to detect degenerate normal forms since this identification depends deciding whether exact equalities take place.

References

1. A. Dou, *Ecuaciones Diferenciales Ordinarias* (Dossat, Madrid, 1968).
2. S. Wiggins, *Global Bifurcations and chaos* (Springer, New York, 1988).
3. F. C. Moon, *Chaotic and Fractal dynamics* (Adison Wesley, New York, 1992)
4. C. Sparrow, *The Lorenz equations* (Springer, New York, 1982).
5. V. I. Arnol'd, A. Avez, *Ergodic Problem of Classical Mechanics* (Benjamin, New York, 1968).
6. J. Palis, W. de Melo, *Geometric Theory of Dynamical Systems* (Springer, New York, 1982).
7. J.-P. Eckmann, D. Ruelle, Ergodic theory of chaos and Strange attractors. *Rev. Mod. Phys.*, **57**, 617–656 (1985).
8. D. Ruelle, *Chaotic evolution and strange attractors* (Cambridge Univ. Press, Cambridge, 1989).
9. P. Hartman, *Ordinary Differential Equations* (Birkhauser, Boston, 1983).
10. E. Hairer, G. Wanner, E. NØrset, *Solving Ordinary Differential Equations I* (Springer Verlag, New York, 1987).

11. E. Hairer, G. Wanner, *Solving Ordinary Differential Equations II* (Springer Verlag, New York, 1991).
12. Ergodic theory of differentiable dynamical systems, *Pub. Mat. I.H.E.S.* , **50**, 27–58, (1979).
13. L.S. Young, Dimension, entropy and Lyapunov exponents in differentaible dynamical systems, *Physica*, **124A**, 639-645 (1983).
14. R. A. Johnson, K. Palmer, G. Sell, Ergodic properties of linear dynamical systems. *SIAM Jour. Math. Anal.* **18** 1–33, (1987).
15. H. Poincaré, *Les méthodes nouvelles de la mécanique céleste*, (Gauthier Villars, Paris 1891-1897).
16. R. Mañé, An ergodic closing lemma, *Ann. of Math.*, **116**, 503-541, (1982).
17. A. Katok, Lyapounov exponents, entropy and periodic orbits for diffeomorphisms, *Pub. Mat. I.H.E.S.*,**51**, 137–173, (1980).
18. C. Falcolini, R. de la Llave, *A rigorous partial justification of Greene's criterion Jour. Stat. Phys.*, **67**, 609–643, (1992).
19. R. DeVogelere, On the structure of symmetric periodic solutions of conservative systems, with applications, *Contributions to the theory of Nonlinear oscillations IV* S. Lefschetz ed. (Princeton Univ. Press, Princeton, 1958).
20. H-T. Kook, J.D. Meiss, Periodic orbits for reversible symplectic mappings, *Physica* **35D**, 65–86 (1988).
21. J. Stoer, R. Burlisch, *Introduction to numerical analysis* (Springer Verlag, New York, 1980).
22. R. G. Helleman, Variational Solutions of Non-integrable Problems, *Topics in Noonlinear Dynamics* (S. Jorna, ed) (A. I. P. New York, 1978).
23. J. D. Meiss, Symplectic maps, variational principles and transport. *Rev. Mod. Phys.*, **64**, 795–847, (1992).
24. R. H. Rand, D. Armbruster, *Perturbation methods, bifurcation theory and computer algebra*, (Springer–Verlag, New York, 1987).
25. Z. Nitecki, *Differentiable Dynamics*, (M.I.T. Press, Cambridge, 1971).
26. P. Hartman, On local homeomorphisms of Euclidean spaces, *Bol. Soc. Mat. Mex.*, **5**, 220–241, (1960).
27. E. Nelson, *Topics in Dynamics: I Flows*, (Princeton Univ. Press, Princeton, 1968).
28. S. Sternberg, *Celestial Mechanics*, (Benjamin, New York, 1969).
29. A. Giorgilli, A. Delshams, E. Fontich, L. Galgani, C. Simó, Effective stability for a hamiltonian system near an elliptic equilibrium point with an application to the restricted three body problem *Jour. Diff. Eq.*, **77**, 167–198, (1989).
30. O. E. Lanford III, Bifurcation of periodic solutions into invariant tori: the work of Ruelle and Takens, *Lect. Notes in Math.*, **322**, (1973)
31. J. Marsden, M. McCracken, *The Hopf bifurcation and certain of its applications.* (Springer, New York, 1976).
32. H. Poincaré, Sur une classe nouvelle de transcendantes uniformes, *Jour. de Math.*, **6**, 313–365, (1890).
33. W. Fraceschini, L. Russo, Stable and unstable manifolds of the Henon mapping, *Jour. Stat. Phys.* **25**, (1981).
34. S.-N. Chow, J. Hale, *Methods of bifurcation theory*, (Springer Verlag, New York, 1982).
35. P. Cvitanovic, *Universality in Chaos*, (Adam Hilger, Bristol, 1984).

36. J.-P. Eckmann, Roads to turbulence in dissipative dynamical systems, *Rev. Mod. Phys.*, **53**, 643–654, (1981).
37. H. Keller, *Numerical methods in bifurcation problems*, (Springer Verlag, New York, 1987).
38. W. C. Rheinboldt, Numerical analysis of parametrized nonlinear equations, (Wiley, New York, 1986).
39. A. Akritas Elements of computer algebra with applications, (Wiley, New York, 1989).
40. B.D. Hassard, N.D. Kazarinoff, Y.-H. Wan, *Theory and applications of Hopf bifurcation* (Cambridge University Press, Cambridge, 1981).

INTRODUCTION TO K.A.M. THEORY*

RAFAEL DE LA LLAVE
Mathematics Department, The University of Texas at Austin
Austin, Texas 78712-1082, USA

1. Introduction

In this lecture, we will discuss the K.A.M. theorem about persistence of quasi-periodic motions under small perturbations of the Hamiltonian.

Rather than presenting a proof of the full theorem, we will give a proof of a caricature that goes half-way in the difficulty and give a sketch of the modifications necessary to obtain the complete result. Very pedagogical proofs of K.A.M. theorem exist [10,14,15,35]. A proof that contains very precise information about differentiability properties is [29].

In our model problem we will study diffeomorphisms of the torus $T^{2d} = \{(q_i, p_i), i = 1, \ldots, d\}$ which preserve the symplectic form $\omega = \sum dp_i \wedge dq_i$. Preserving a symplectic form is one of the key features of Hamiltonian systems, and is what gives the characteristic flavor to most of the problems and techniques of Hamiltonian mechanics.

Take a translation in the torus $E_\alpha : T^{2d} \to T^{2d} \quad E_\alpha(\mathbf{x}) = \mathbf{x} + \alpha (\text{mod}\, 1)$. Clearly, this preserves the symplectic form ω. Now consider a perturbation of it, that is, a family of symplectic transformations $f_\varepsilon : T^{2d} \to T^{2d}$, such that it is analytic in ε and $f_0 \equiv E_\alpha$.

The question is: When can this perturbation be eliminated by a change of coordinates; in more analytic terms, when can we find another analytic family of invertible symplectic transformations such that

$$f_\varepsilon \circ g_\varepsilon = g_\varepsilon \circ E_\alpha \tag{1}$$

The most effective way (even computationally) of making perturbation calculations in mechanics is probably to try to absorb as much of the perturbation as possible in a change of coordinates (see [8] for very practical computations not necessarily in Hamiltonian mechanics).

The advantage of such a method with respect to other more straightforward methods — like trying to make perturbation around each orbit — is that we can obtain global information about the whole phase portrait; we also obtain control — uniform in all initial conditions — of the range of validity of the expansions. This takes care of secular terms which are difficult to interpret, and a computational hazard (e.g., if we try to match powers in the problem $\ddot{x} = -(\omega+\varepsilon)^2 x$; $x(0) = 0$; $\dot{x}(0) = \omega + \varepsilon$ we are led to $x = \sin(\omega t) + \omega\cos(\omega t)\varepsilon t + \omega^2 \sin(\omega t)\frac{(\varepsilon t)^2}{2!} - \omega^3\cos(\omega t)\frac{(\varepsilon t)^3}{3!}$ which is, of course, $\sin(\omega+\varepsilon)t$. However, if we try to sum the series for large values of εt, we are led to unstable computations. Notice also that the periodicity of the perturbed solution is not apparent in any order of the perturbation theory. Those terms first appeared in the computation of rotation of perihelia and they become appreciable — for Mercury — when t is of the order of centuries, hence the name secular. The classical ways of dealing with secular terms were to resume most of them in recognizable forms (Lindstet, etc.) [28].

Remark. (Relation to commutator groups) If we rewrite $f_\varepsilon = \hat{f}_\varepsilon \circ E_\alpha$, then Eq. (1) becomes

$$\begin{aligned} \hat{f}_\varepsilon E_\alpha g_\varepsilon &= g_\varepsilon E_\alpha \\ \hat{f}_\varepsilon &= g_\varepsilon E_\alpha g_\varepsilon^{-1} E_\alpha^{-1} \ . \end{aligned} \tag{2}$$

This means that we rewrite $\hat{f}_\varepsilon$ (close to the identity) as a commutator. The problem of characterizing the group of diffeomorphisms generated by commutators is very important, since it is related to geometric properties of the manifold [13]. So, the study of which diffeomorphisms are commutators has considerable interest.

Even if Eq. (1) and Eq. (2) are equivalent, from the point of view of analysis, Eq. (1) is nicer because it has "group structure," so that it can be studied by hard implicit function theorems. The idea of using this strategy for this problem of commutators appeared in [20].

If our interest were to study the problem of the commutator subgroup, it is worth remarking that the K.A.M. strategy we are going to discuss fails in C^r when $r \neq \omega, \infty$ because we will only be able to conclude that $g \in C^{r-r_0}$ $(r_0 > 0)$, that is, we cannot show that g is in the group of C^r diffeomorphisms. However, there is another strategy of proof which works even in this case [6] (see also [23,24,25]). This suggests that K.A.M. strategy is not always the finest tool — as should be expected from a technique that works for so many problems. It would be quite interesting if the analogy between the two problems could be pushed far enough to give a proof of the K.A.M. theorem not based in K.A.M. strategy.

Remark. Looking for analogues of K.A.M. theorem, it would be more natural, perhaps, to study flows rather than diffeomorphisms. The problem for flows is simpler than for diffeomorphisms and handling the latter is useful in mechanics anyway.

2. Formal Considerations

Before stating the theorem that solves the problem, we will pause to develop some formalism; this will help to find some hypothesis that this theorem should have.

We will develop several formalisms in the hope of enabling the reader to understand the relationships between several results in the literature and to serve as a motivation for the last formulation — the one I personally favor. So, it will be quite possible — indeed, even sensible in a first reading — to skip quite a lot of what comes next.

The first difficulty that stands in the way of solving the problem is how to write diffeomorphisms in a way which allows effective computation. In the case of the torus, we have an obvious way: to use coordinates. However, this is a bad solution because there are many manifolds that do not admit such global coordinates; besides in coordinates the symplectic character appears as a set of non-linear non-local constraints. A technique that solves the problem of keeping the symplectic character (but does not help with the problem of requiring a global system of coordinates — it makes it worse) is the use of generating functions; this has the side advantage that rather than dealing with $2d$ functions of $2d$ variables linked by (horrid) constraints, we will only have to deal with one function of $2d$ variables. We will consider the variables as ranging in $\mathbf{R}^{2d}$ and keep track of periodicity conditions.

We observe that E_α admits a generating function of the type $(\mathbf{q},\mathbf{p}')$. (It does not admit one of the form $(\mathbf{q},\mathbf{q}')$ or $(\mathbf{p},\mathbf{p}')$ which are the most natural.

$$\mathbf{q}' = \mathbf{q} + \mathbf{a} = \frac{\partial F_0}{\partial \mathbf{p}'} \quad ; \quad \mathbf{p} = \mathbf{p}' - \mathbf{b} = \frac{\partial F_0}{\partial \mathbf{q}} \ .$$

Hence, $F_0 = \mathbf{p}'\mathbf{q} + \mathbf{a}\mathbf{p}' - \mathbf{b}\mathbf{q}$ up to a constant which we will fix as zero without loss of generality. Since the construction of a generating function out of a diffeomorphism is quite explicit, we can conclude that f_ε will have a family of generating functions $F_\varepsilon(\mathbf{q},\mathbf{p}')$ which will be as smooth both in the coordinates and in ε as f_ε (provided that we choose the constants adequately).

Likewise, g_ε, if it exists at all, will have a family of generating functions $G_\varepsilon(\mathbf{q},\mathbf{p}')$; $G_0(\mathbf{q},\mathbf{p}') = \mathbf{p}'.\mathbf{q}$. We systematically will write

$$G_\varepsilon = G_0 + \varepsilon G_1 + \varepsilon^2 G_2 + \cdots$$

$$G_\varepsilon^{[\leq N]} = \sum_{k=0}^{N} \varepsilon^k G_k \ .$$

We can write Eq. (1) as an equation for G_ε in terms of F_ε and, then, match powers to obtain a hierarchy of equations of the G_n's in terms of the F_n's. We will only observe that the first equation of the hierarchy is

$$G_1(\mathbf{x} + \alpha) - G_1(\mathbf{x}) = F_1(\mathbf{x}) \qquad (3)_1$$

and that the other equations of the hierarchy are

$$G_n(\mathbf{x} + \alpha) - G_n(\mathbf{x}) = F_n(\mathbf{x}) + R_n \qquad (3)_n$$

where R_n is an (awful) expression involving only $G_1, \ldots, G_{n-1}$, $F_1, \ldots, F_{n-1}$ and their derivatives.

From the structure of Eq. $(3)_n$, we can proceed inductively, examining recursively the equations, provided that we have a theory for equations of the form

$$G(\mathbf{x} + \alpha) - G(\mathbf{x}) = F(\mathbf{x}) \ .$$

We should notice that since G_ϵ, F_ϵ are generating functions of a transformation in the torus, they are obtained integrating a periodic function and, hence, they are the sum of a periodic function and a linear one. We should also remember that they are defined up to constants. These restrictions on the class of functions that can appear are a translation of the geometric content of the problem.

If the linear part of F were non-null, then G would have a quadratic term and, hence, be out of the class we are interested in.

The periodic part can be studied by using Fourier components; we are led to

$$(e^{2\pi i \mathbf{k}\alpha} - 1)G_{\mathbf{k}} = F_{\mathbf{k}}$$

so that there is a solution provided $F_{\mathbf{k}} = 0$ whenever $\mathbf{k}\alpha \in \mathbf{Z}$. We will call "resonant" those $\mathbf{k}$'s such that $\mathbf{k}\alpha \in \mathbf{Z}$, $\mathbf{k} \neq 0$.

The solution is not unique for the $F_{\mathbf{k}}$'s with $\mathbf{k} \cdot \alpha = 0$. When $\mathbf{k} = 0$ this does not matter, because it is precisely the natural ambiguity of the G's — as generating functions defined up to constants. Moreover it requires some computation to become convinced, but is nevertheless true, that the ambiguities making different choices for constant terms at some stage do not affect the solvability of the equations of higher order in the hierarchy and only affect the solution up to constants.

Some α, enjoy the remarkable property that there are no resonant terms. This can be expressed as the components being "linearly independent over the rationals." (For example $\alpha = (1, \sqrt{2}, \sqrt{5})\alpha\mathbf{k} \in \mathbf{Z}$, $\mathbf{k} \in \mathbf{Z}^3 \Rightarrow \mathbf{k} = 0$.)

Summing up, the necessary conditions for the existence of a formal solution are that $F_n + R_n$ do not have either a linear part or resonant terms for any n. The absence of resonant terms is automatic if α is such that there are no resonances. The absence of linear part depends on $\hat{f}_\epsilon$ (and, even if mysterious in this formalism, it will turn out to be quite simple when interpreted geometrically).

Discussing convergence, even in the case of rationally independent α seems, at first sight, to be quite hopeless because $G_{\mathbf{k}} = (1/(e^{2\pi i \mathbf{k}\alpha} - 1))F_{\mathbf{k}}$ and the denominators become arbitrarily small, since $\mathbf{k} \cdot \alpha$ will become arbitrarily close to integers as $\mathbf{k}$ ranges over $\mathbf{Z}^{2d}$. Moreover, this is only the linear approximation!.

For the moment, we will postpone all discussions about convergence and proceed to develop other formalisms in perturbation theories. The next method is based on the idea of writing all symplectic diffeomorphisms close to the identity as time one maps of vector fields.

The idea is very analogous to the very successful use of Lie algebras in the study of problems of conjugacy and commutators for finite dimensional Lie groups.

We think of symplectic diffeomorphisms as an infinite-dimensional Lie group (the group operation is composition) then the elements of the Lie algebras (in old fashioned language the "infinitesimal canonical transformations") are the locally Hamiltonian vector fields.

Unfortunately, the relation of infinite dimensional Lie groups and their Lie algebras is much more complicated than for the finite dimensional ones.

For example, if the manifold is not compact, there could be vector fields some of whose trajectories get out of the manifold in a finite time so that those vector fields do not define a flow (i.e., there is no exponential for all the elements of the Lie algebra). This is not a very serious difficulty because for compact manifolds, all vector fields have time one map (i.e., we can define the exponential of the Lie algebra).

Much worse is the fact that, even when the exponential is well defined, it does not need to cover a whole neighborhood of the origin. That is, there are diffeomorphisms as close to the identity as we please which cannot be expressed as time one maps. This shows there is no hope of getting a logarithm (the inverse of the exponential) which works without problems.

Example. We will show that if $f : S^1 \to S^1$ without fixed points is the time one map of a vector field, then the derivatives of f^n are bounded independently of n.

Since arbitrarily close to the identity there are diffeomorphisms with expansive periodic points $f^N(x_0) = x_0$ $|f^{N'}(x_0)| > 1$ we have $f^{nN}(x_0) = (f^{N'}(x_0))^n$ which is arbitrarily large.

In effect, if f without fixed points in the time-one map of the vector field X, X cannot have any zeros. $t_0 = \int_S \frac{d\theta}{|X(\theta)|}$ is finite and, the time-t_0 map is the identity.

So that

$$f^n = T_n = T_{t_0[n/t_0]+r(n)} = T_{r(n)}$$

but $0 \leq r(n) \leq t_0$ so that $T_{r(n)}$ is uniformly bounded in C^1.

Therefore, for the circle, the exponential of the Lie algebra of the group of diffeomorphisms does not cover any C^∞ neighborhood of the identity.

Remark. Something slightly weaker is however true: all diffeomorphisms of class C^r, $r \notin \mathbf{Z}$, $r \neq n+1$ (n dimension of the manifold) can be written as the product of a finite number of time one maps (this number, however depends on the diffeomorphism) of C^r vector fields. This is because of the results of Mather [23,24,25] and Epstein [13], who show that the group of diffeomorphisms homotopic to the identity with this regularity is simple. We observe that the set of all diffeomorphisms that can be written as a finite product of time one maps of C^r vector fields is a group.

Moreover, it is normal (if h is time-one map of the vector field X, $g^{-1}hg$ is the time-one map of the vector field obtain changing coordinates in X) so by the results of Epstein and Mather, it should be the whole component of the identity.

But, even if the machinery of finite dimensional Lie groups does not carry over, it is true that many of the calculations are done in the spirit suggested by it and it can be a very strong guide to know which is the right framework in which to perform analysis (to identity the obstructions and decide which properties should the conjugation elements have) [34,22].

But, if we do not want bona-fide convergence the problems are much easier. Formal power expansions work pretty much like finite dimensional Lie groups.

In particular, we have

Proposition. Given a family f_ε of smooth diffeomorphisms $f_0 = \mathrm{Id}$, we can find a sequence of vector fields $\mathcal{L}_1, \ldots, \mathcal{L}_n, \ldots$ such that

$$d\left(f_\varepsilon, \exp\left[\sum_{i=1}^{n} \varepsilon^i \mathcal{L}_i\right]\right) = O(\varepsilon^{n+1}) \ \forall\, n \ .$$

Proof. Suffices to show that we can match enough derivatives of f_ε at $\varepsilon = 0$. A simple computation shows that

$$\left(\frac{d}{d\varepsilon}\right)^k \exp\left[\sum_{i=1}^{n} \varepsilon^i L_i\right]\bigg|_{\varepsilon=0} = i!\mathcal{L}_k + \mathcal{R}_k$$

where $\mathcal{R}_k$ depends complicatedly on $\mathcal{L}_1, \ldots, \mathcal{L}_{k-1}$ but *not* on any $\mathcal{L}$ of higher index.

So that recursively we can adjust the $\mathcal{L}_k$'s so that those derivatives agree with those of f_ε. The same result holds true when f_ε preserves the symplectic form and we restrict the $\mathcal{L}_i$'s to be locally Hamiltonian.

Once we have a logarithm, (with some reasonable properties with respect to derivatives, etc. which are easy in this case) there are many pieces of machinery that can be imported. Notably Campbell-Hausdorff formula. (It suffices to walk through the proof of [12] reproduced e.g., in [30].)

$$\exp \mathcal{A} \exp \mathcal{B} = \exp \mathcal{C}$$

where

$$\mathcal{C} = \mathcal{A} + \mathcal{B} + \sum_{n>0} \frac{(-1)^{n+1}}{n(n+1)} \sum \frac{\overbrace{[\mathcal{A},[,\ldots,\mathcal{A},}^{r_1}\overbrace{[\mathcal{B},[\ldots\mathcal{B}_1}^{s_1}\overbrace{[\mathcal{A}_1[\ldots\mathcal{A}\ldots\ldots}^{r_2}\overbrace{[\mathcal{A}\ldots\mathcal{A}}^{r_n},\mathcal{B}]\ldots]}{(s_1+\cdots+s_{n-1}+1)r_1!\ldots r_n! s_1!\ldots s_n!}$$

which will hold in the sense of asymptotic expansions when the manifolds are compact.

Taking into account that E_α is a canonical transformation, Eq. (2) can be written as

$$\exp\left[\varepsilon\mathcal{F}_1 + \varepsilon^2\mathcal{F}_2 + \cdots\right] = \exp\left[\varepsilon\mathcal{G}_1 + \varepsilon^2\mathcal{G}_2 + \cdots\right] \exp\left[-\varepsilon E_\alpha^*\mathcal{G}_1 - \varepsilon^2 E_\alpha^*\mathcal{G}_2 \ldots\right] \ .$$

("$*$" is the usual push forward)

Using the Campbell-Hausdorff formula to express the second term, we obtain matching powers

$$\mathcal{F}_1 = \mathcal{G}_1 - E^*_\alpha \mathcal{G}_1$$
$$\mathcal{F}_2 = \mathcal{G}_2 - E^*_\alpha \mathcal{G}_2 - \tfrac{1}{2}[\mathcal{G}_1, E^*_\alpha \mathcal{G}_2] \ .$$

Notice that all the nonlinear terms in the hierarchy are going to be commutators.

Again, we obtain a perturbation theory that has all the features desirable of a perturbation expansion; that is, it can be solved step by set (and renormalizability).

Now, we can investigate a little bit more the obstructions for solution.

Lemma 1. If $\mathcal{G}$ is a locally Hamiltonian vector field in a torus, then $\mathcal{G} - E^*_\alpha \mathcal{G}$ is globally Hamiltonian.

Lemma 2. If $\mathcal{G}, \mathcal{H}$ are locally Hamiltonian vector fields $[\mathcal{G}, \mathcal{H}]$ is globally Hamiltonian with Hamiltonian $\omega(\mathcal{G}, \mathcal{H})$.

Proof. Lemma 1 can be established quickly (we will see another proof later) noticing that $\mathcal{G}$ is globally Hamiltonian in $\mathbf{R}^{2d}$ and the Hamiltonian is periodic and linear. $\mathcal{G} - E^*_\alpha \mathcal{G}$ will also have a Hamiltonian in $\mathbf{R}^{2d}$ and, when we compute it, the linear parts drop. Hence, the Hamiltonian in $\mathbf{R}^{2d}$ is periodic and that is equivalent to being globally Hamiltonian.

Lemma 2 is much more important than Lemma 1; it is true for all manifolds and [36] showed the converse, that is, that all globally Hamiltonian vector fields are commutators of locally Hamiltonian ones. This means that we have found the commutator subalgebra of the Lie algebra tangent to the Lie group.

Here we present a coordinate independent proof similar to the one that can be found in [2].

Given a vector field X we call $i_\omega X$ the one form defined by $(i_\omega X)(Y) = \omega(X, Y)Y$.

X is a globally Hamiltonian vector field of Hamiltonian H and can be written as $i_\omega H = dH$ and X being a locally Hamiltonian vector field can be written as $d(i_\omega X) = 0$.

We can express $\omega(Z[X, Y]) = \alpha_z([X, Y]) = X(\alpha_Z(Y)) - Y(\alpha_Z(X))$. Since $d\omega = 0$, we also have, for all vector fields X_1, X_2, Y

$$\begin{aligned} 0 &= (d\omega(X_1, X_2, Y) \\ &= Y\omega(X_1 X_2) + X_2 \omega(Y, X_1) + X_1 \omega(Y, X_2) \\ &\qquad + \omega([Y, X_1], X_2) + \omega([X_2, Y], X_1) + \omega([X_1 X_2], Y) \ . \end{aligned}$$

If $X_1\, X_2$ are locally Hamiltonian, we can substitute the previous identity and derive

$$\omega(X_1, X_2) = \omega([X_1 X_2], Y)$$
$$d(\omega(X_1, X_2)) = ([X_1 X_2], Y)$$

so that indeed, $[X_1 X_2]$ is the Hamiltonian vector field associated to the function $\omega(X_1, X_2)$.

A coordinate dependent proof is also quite feasible. The theorem can be reduced to show that in any set of coordinate patches covering the manifold, the commutator and the Hamiltonian flow agree. We can take coordinate patches in such a way that $\omega = \sum_i dp_1 \wedge dq_i$ (Darboux theorem) and we can compute both objects and show that they are the same, using the identities implied by local Hamiltonian property for the derivatives of the components of $\mathcal{G}$ and $\mathcal{H}$.

A s a corollary of the Lemmas we clearly obtain

Proposition. In order that a solution of Eq. (2) exists, it is necessary that all $\mathcal{F}_i$ be globally Hamiltonian.

In that case, it suffices to look for $\mathcal{G}_i$ only among globally Hamiltonian vector fields and the equations of the hierarchy are

$$F_1(\mathbf{x}) = G_1(\mathbf{x}) - G_1(\mathbf{x} + \alpha)$$
$$F_2(\mathbf{x}) = G_2(\mathbf{x}) - G_2(x + \alpha) + R_2(\mathbf{x})$$

where F_i and G_i are the Hamiltonians of $\mathcal{F}_i$ and $\mathcal{G}_i$ and the equalities are up to constants and R_n, as before, can be computed from the previous terms.

So far, we have only been working in the formal power series context and the obstructions for solvability can be quite obscure if left in this form. So, we will have to work a bit more of formalism. This formalism will also become quite useful to establish convergence, because a direct proof estimating the coefficients of ε in the expansion is quite impractical and, in the second case, even false, due to the poor properties of log in infinite dimensional Lie algebras.

The idea is to interpolate all the f_ϵ by a vector field

$$\frac{d}{d\varepsilon} f_\epsilon = \mathcal{F}_\epsilon \circ f_\epsilon$$

and similarly for g_ϵ and use the vector fields to express the conjugacy equation (1). As is well known, the $\mathcal{F}_\epsilon$ will be locally Hamiltonian vector fields.

If we take derivatives with respect to ε in Eq. (1), we obtain

$$\mathcal{F}_\epsilon(f_\epsilon \circ g_\epsilon) + f_\epsilon^* \,_\epsilon(f_\epsilon g_\epsilon) = \mathcal{G}_\epsilon(g_\epsilon E_\epsilon) \ . \tag{4}$$

And using Eq. (1) again

$$\mathcal{F}_\epsilon = \mathcal{G}_\epsilon - f_\epsilon^* \mathcal{G}_\epsilon \ . \tag{5}$$

This implies that F_ϵ should not only be locally Hamiltonian, but also globally Hamiltonian. In some particular cases, we had already established that in Lemma 1. A more geometric proof is as follows. Globally Hamiltonian means $[i_\omega(\mathcal{F}_\epsilon)] = 0$ (that is, the result of identifying the vector field with a one-form produces an exact differential -cohomology zero-). On the other hand, we just observe that f_ϵ preserves the symplectic form, so that we can write

$$[i_\omega f_\epsilon^* \mathcal{G}_\epsilon] = [f_\epsilon^* i_\omega \mathcal{G}_\epsilon] = f_{\epsilon\#}[i_\omega \mathcal{G}_\epsilon] \ .$$

Moreover, since f_ε is connected to the identity, its action on cohomologies is the identity

Hence,

$$[i_\omega \mathcal{F}_\varepsilon] = [i_\omega \mathcal{G}_\varepsilon] - f_{\varepsilon\#}[i_\omega \mathcal{G}_\varepsilon] = 0 \ .$$

The previous calculation make it reasonable to introduce the following definition.

Definition. We can say that f_ε is a globally Hamiltonian isotopy (G.H.I.) when

$$\frac{d}{d\varepsilon} f_\varepsilon = \mathcal{F}_\varepsilon \circ f_\varepsilon$$

and $\mathcal{F}_\varepsilon$ is a globally Hamiltonian vector field. That is, we obtain f_ε as a time one map of a *globally* Hamiltonian vector field which also depends on time.

Clearly, a globally Hamiltonian isotopy is determined by the initial point and F_ε the Hamiltonian function of $\mathcal{F}_\varepsilon$. Conversely, a G.H.I. determines uniquely $\mathcal{F}_\varepsilon$ and this determines F_ε up to a constant, which we will fix by requiring $\int F_\varepsilon = 0$. We will call F_ε the Hamiltonian of f_ε.

As we will discuss at the end, the G.H.I.'s enjoy quite remarkable properties. For the moment, we will only work out some formal properties.

Notice that G.H.I.'s allow to express transformations by a simple function; in that respect, they are similar to generating functions, but, however, they are natural geometrical objects and, so, they can be defined in all manifolds; besides, the equations we will have to solve are simpler.

Lemma. Let f_ε, g_ε be G.H.I. of Hamiltonians F_ε, G_ε, then $f_\varepsilon \circ g_\varepsilon$ has Hamiltonian $F_\varepsilon + G_\varepsilon(f_\varepsilon^{-1})$.

The proof is very simple proceeding along the same lines as in the calculations of Eq. (5). We just have to observe that, since f_ε is a symplectic transformation, we can just transform the Hamiltonian.

Notice that, in particular, this Lemma shows that all the diffeomorphisms connected to the identity by a G.H.I. form a group.

If we decide to search for the g_ε among G.H.I., Eq. (5), — which is a necessary condition for Eq. (1) — becomes

$$F_\varepsilon = G_\varepsilon - G_\varepsilon(f_\varepsilon^{-1}) \ . \tag{6}$$

However, it turns out that, in spite of the fact that we used Eq. (1) twice to derive Eq. (5), and, therefore could fail to the sufficient on Eq. (6) is equivalent to Eq. (1).

If effect, calling

$$\Delta_\varepsilon = f_\varepsilon \circ g_\varepsilon \circ f_0^{-1} \circ g_\varepsilon$$

we can show that provided Eq. (6) holds

$$\frac{d}{d\varepsilon}\Delta_\varepsilon = 0 \quad \text{when} \quad \Delta_\varepsilon = \text{Id} \ .$$

Since clearly $\Delta_0 = \mathrm{Id}$ and everything is so smooth that uniqueness of O.D.E. holds, we have $\Delta_\varepsilon = \mathrm{Id}$.

3. Convergence

We now tackle the problem of actually constructing the solutions. More precisely, we want to prove the following.

Theorem. Let f_ε be an analytic globally Hamiltonian isotopy $f_0 = E_\alpha$, where α satisfies

$$|(\alpha, \mathbf{k}) - n|^{-1} \leq \exp \frac{A|\mathbf{k}|}{\log|\mathbf{k}| \cdot \log\log|\mathbf{k}| \times \cdots \times \underbrace{\log\ldots\log}_{\ell\ \text{times}}{}^{1+\nu}|\mathbf{k}|} + B$$

$$n \in \mathbf{Z} \quad , \quad \mathbf{k} \in \mathbf{Z}^{2d}$$

for some $A, B, \nu > 0$ ($|\mathbf{k}|$ means $\sum_i |k_i|$).

Then, for sufficiently small ε, there is an analytic globally Hamiltonian isotopy g_ε satisfying Eq. (1).

For convenience, we will call

$$\exp \frac{A|\mathbf{k}|}{\log|\mathbf{k}| \cdot \log\log|\mathbf{k}| \times \cdots \log\cdots\log^{1+\nu}|\mathbf{k}|} + B \equiv \Omega(|\mathbf{k}|)$$

following a notation due to Rüssmann. We will discuss later how abundant are the α for which these conditions are satisfied.

Remark. We will disregard what happens for $|\mathbf{k}|$ so small that one of the iterated log does not make sense. What is going to be relevant for the proof is that this inequality holds from a certain $|\mathbf{k}|$ on. If we were systematic about using that, we could do with the B, but it takes a bit of effort — a good exercise to check it — that the $|\mathbf{k}|$ big enough that we take can be uniform in all steps, so we still keep the B.

Given an analytic function in the torus, we will denote by

$$\|F\|_r = \sup_{|\operatorname{Im} z_i| \leq r} |F(\mathbf{z})|$$

and given an analytic family of analytic functions we will denote by

$$\|\!|F_\varepsilon|\!\|_r = \sup_{|\varepsilon| \in [0,1]} \|F_\varepsilon\|_r \ .$$

More precisely, what we are going to show is that, given r, u, there is some $S^*(r, u, \alpha)$ such that when $\|\!|F_\varepsilon|\!\|_r \leq S^*$ the conclusions of the theorem hold, when G_ε analytic in a slightly smaller domain. We remark that the loss of analyticity domains can be made as small as we please by imposing smallness conditions in $\|F_\varepsilon\|_r$. It is also important to remark that α only enters in S^* through Ω so that the proof holds at the same time for all the α with the same Ω.

Even if F_ε was defined in a smaller neighborhood in the variable, we can scale ε so that it is defined for $\varepsilon \in [0,1]$. Moreover, by scaling enough, we can adjust any smallness conditions in $\|F_\varepsilon\|$. So that the theorem can as well be enunciated for sufficiently small (1) can be written as

$$g_\varepsilon^{-1} \circ f_\varepsilon \circ g_\varepsilon = E_\alpha$$

and, if we apply repeatedly the formula for the Hamiltonian of a composition of G.H.I., we see that imposing that the Hamiltonian of the left hand side be equal to zero we obtain

$$F_\varepsilon \circ g_\varepsilon - G_\varepsilon \circ g_\varepsilon + G_\varepsilon \circ f_\varepsilon^{-1} \circ g_\varepsilon = 0 \tag{7}$$

(which, of course, is to hold in some appropriate complex extension of the torus).

Even if a direct solution of Eq. (7) is beyond hope, we can observe that there is a similar equation we can solve using Fourier coefficients.

$$F_\varepsilon - G_\varepsilon^0 + G_\varepsilon^0 \circ E_\alpha^{-1} = 0\ . \qquad (8)_0$$

Then, G_ε^0 will be commensurate, in some sense, to F_ε. Then, $g_\varepsilon^0 - \mathrm{Id}$ will also be controlled by F_ε and, finally, $E_\alpha^{-1} - f_\varepsilon^{-1}$ will also be controlled by F_ε.

If we substitute all this in the left hand side of Eq. (7), we obtain something which, even if not exactly zero, is "quadratic in F_ε."

The crucial fact is that Eq. (7) has some structure (usually called "group structure" [38]) that allows us to take advantage of approximate solutions.

In effect, if we could solve Eq. (7) with $(g_\varepsilon^0)^{-1} \circ f_\varepsilon \circ g_\varepsilon^0 = f^1$ in place of f_ε, we could solve the original problem. But, as we argued before, $(g_\varepsilon^0)^{-1} f_\varepsilon g_\varepsilon^0$ is much more favorable.

Of course, we still do not know how to solve the exact equation, so what we will do is to solve Eq. $(8)_1$ again and find another approximation g_ε^1 and keep repeating the process, hoping that the remainders will go to zero; $g = g_\varepsilon^0 \circ g_\varepsilon^1 \circ g_\varepsilon^2 \cdots$ will, then, be a solution of the problem.

The details of the proof of how this "quadratic convergence" is enough to beat the small denominators are quite fascinating and are the basis of a whole theory useful for many problems [38].

A very suggestive way of thinking about this technique is suggested in [14]. We could think of each one of our inductive steps as recognizing that a substantial part of the perturbation expansion can be absorbed by a change in the Hamiltonian, which is then closer to "free." This is quite analogous to the perturbative application of renormalization group and, indeed, this analogy can be exploited to obtain results in statistical mechanics of quantum field theory problems by methods reminiscent of K.A.M.. The theories involved have much more structure, their treatment has to include more technique. See [17] for a formulation of renormalization group close to this idea and for references about other points of view or results in this field.

The next four Lemmas will be a systematic control of the inductive step; then, we will show we can iterate indefinitely and that the iteration converges.

Lemma. Assume F is analytic in T^{2d}, $\int F = 0$ and α as before. Then, there is a unique G, analytic, $\int G = 0$, such that

$$F(\mathbf{x}) = G(\mathbf{x}) - G(\mathbf{x} + \alpha) \ .$$

Moreover,

$$\|G\|_{r-u} \leq e^{B+3n(u)} \|F\|_r$$

r, u sufficiently small where the smallness depends only on Ω ($n(u)$ to be defined later, is an explicit function, involving only Ω).

To avoid cluttering the proof with different constants, we will denote by C several constants which only depend on α and are independent of r, F, and all the other parameters of the proof.

Proof. The equation can be solved equating Fourier coefficients

$$G_{\mathbf{k}} = (e^{2\pi i(\mathbf{k},\alpha)} - 1)^{-1} F_k \qquad \mathbf{k} \neq 0$$
$$G_0 = 0 \ .$$

Now, we have

i)
$$|e^{2\pi i(\mathbf{k},\alpha)} - 1|^{-1} \leq \Omega(|\mathbf{k}|)$$

because

$$|e^{2\pi i x} - 1|^{-1} \leq C d(x, \mathbf{Z})^{-1}$$

($x = \frac{1}{2}$ gives the worst constant which is $C = 1$) also, because of Cauchy inequalities,

ii)
$$|F_{\mathbf{k}}| \leq e^{-2\pi r|\mathbf{k}|} \|F\|_r$$

and the triangle inequality gives

iii)
$$\|G\|_{r-u} \leq \sum |G_{\mathbf{k}}| e^{2\pi(r-u)|\mathbf{k}|} \ .$$

Hence,

$$\begin{aligned} \|G\|_{r-u} &\leq \|F\|_r \sum e^{-2\pi u|k|}(|k|) \\ &\leq \|F\|_r C \sum_{n \in \mathbf{N}} e^{-2\pi u n} \Omega(n) n^{2d-1} \end{aligned}$$

where C is the area of the unit sphere with respect to $|\_|$ in 2^d dimensions.

$$\sum_{n \in \mathbf{N}} e^{-2\pi u n} n^{2d-1} \Omega(n) \leq \left(\sum_{n \in \mathbf{N}} e^{-\pi u n} \right) \left(\max_{n \in \mathbf{N}} e^{-\pi u n} \Omega(n) n^{2d-1} \right) \ .$$

The sum can be done explicitly and bounded by Cu^{-1}. The maximum will be reached at the point that the log reaches the maximum and this can be computed taking derivatives

$$\begin{aligned}
0 &= \frac{d}{dn}\left[-\pi un + B + \frac{An}{\log n \log\log n \ldots \log \ldots \log^{1+\nu} n} + (2d-1)\log n\right] \\
&= -\pi u + \frac{A}{\log n \log\log n \ldots \log \ldots \log^{1+\nu} n} - \frac{A}{\log^2 n \log\log^2 n \ldots \log \ldots \log^{1+\nu} n} \\
&\quad - \frac{A}{\log^2 n \log\log^2 n \ldots \log \ldots \log^{1+\nu} n} \\
&\quad - \frac{A(1+\nu)}{\log^2 n \log\log^2 n \ldots \log^2 \ldots \log^{(1-\nu)2} n} + \frac{2d-1}{n} \\
&= -u + \frac{A}{\log n \ldots \log \ldots \log^{1+\nu} n}\left[1 - \frac{1}{\log n} - \frac{1}{\log\log n} - \cdots - \frac{1+\nu}{\log \ldots \log^{1+\nu} n}\right] \\
&\quad + \frac{2d-1}{n} .
\end{aligned}$$

So, we see that for u sufficiently small we have just one solution $n(u)$; the exact form of $n(u)$ will not be terribly important for the moment, we only have to observe that $n(u) \to \infty$ as $u \to 0$ $|\ell n(u)| = o(n(u))$. The maximum of the exponent will then be bounded by

$$\begin{aligned}
&-\pi un(u) + B + \frac{An(u)}{\log n(u) \ldots \log \ldots \log^{1+\nu} n(u)} \\
&= B + n(u)\left[\frac{A}{\log n(u) \ldots \log \ldots \log^{1+\nu} n(u)} - u\right] .
\end{aligned}$$

And using the equation satisfied by $n(u)$

$$\begin{aligned}
&= B + n(u)\frac{A}{\log n(u) \ldots \log \ldots \log^{1+\nu} n(u)}\Big(\frac{1}{\log n(u)} + \frac{1}{\log\log n(u)} + \cdots \\
&\quad + \frac{1+}{\log \ldots \log^{1+\nu} n(u)} \le B + 2n(u)
\end{aligned}$$

provided that $n(u)$ is large enough, that is, u small enough. We also observe that $e^{n(u)} \gg u^{-1}$ as $u \to 0$ so that the u^{-1} factor can be absorbed in adding a small coefficient of $e^{n(u)}$. Since everything was uniform with respect to extra parameters, the same result with the same bounds is true for families.

Lemma.

$$\|F_\epsilon - F_\epsilon \circ g_\epsilon\|_{r-u} \le Cu^{-2}\|F\|_r\|G_\epsilon\|_r$$

provided

$$\|G_\epsilon\|_r \le 3u^2C .$$

Proof. Using analyticity estimates, we have that $\|\nabla G_\epsilon\|_{r-u/2} \le Cu^{-1}\|G_\epsilon\|_r$. Since g_ϵ is obtained solving Hamilton's equations

$$\|\mathbf{x} - g_\epsilon(\mathbf{x})\|_{r-u} \le \|\nabla G_\epsilon\|_{r-u/2} \le Cu^{-1}\|G_\epsilon\|_r$$

where the first inequality assumes that $g_\epsilon(\mathbf{x})$ is always contained in $\{|\operatorname{Im}(\mathbf{z})| \leq r - u/2\}$ whenever $|\operatorname{Im}(\mathbf{x})| \leq r$. But this is true because of the smallness assumption on $\|G_\epsilon\|_r$ we included in the hypothesis of the lemma, which make the last bound $\leq u/3$. (We showed that provided $g_\epsilon(x)$ is contained in $\{\operatorname{Im}(z) \leq r - u/2\}$ it can not leave $\{\operatorname{Im}(\mathbf{z}) \leq r - u + u/3\}$.)

We also have

$$\|F_\epsilon - F_\epsilon \circ g_\epsilon\|_{r-u} \leq \|\nabla F_\epsilon\|_{r-u/2} \|\mathbf{x} - g_\epsilon(\mathbf{x})\|_{r-u}$$

and we can bound the first factor by analyticity estimates.

Applying the previous lemma twice and the formula of composition, we obtain

Lemma.

$$\|H_\epsilon(f_0^{-1}) - H_\epsilon(f_\epsilon^{-1} \circ g_\epsilon)\|_{r-u} \leq Cu^{-4} \|H_\epsilon\|_r (\|F_\epsilon\|_r + \|G_\epsilon\|_r)$$

provided

$$\|F_\epsilon\|_r \leq u^2 C$$
$$\|G_\epsilon\|_r \leq u^2 C \ .$$

Now, we can bound the left hand side of Eq. (7) provided that G_ϵ is obtained from F_ϵ by solving Eq. (8) (we will assume that the conditions to apply the lemma hold).

$$\begin{aligned} &\|F_\epsilon g_\epsilon - G_\epsilon g_\epsilon + G_\epsilon f_\epsilon^{-1} g_\epsilon\|_{r-u} \\ &= \|F_\epsilon g_\epsilon - G_\epsilon g_\epsilon + G_\epsilon f_\epsilon^{-1} g_\epsilon - F_\epsilon + G_\epsilon - G_\epsilon f^{-1}\|_{r-1} \\ &\leq \|F_\epsilon \circ g_\epsilon - F_\epsilon\|_{r-u} + \|G_\epsilon \circ g_\epsilon - G_\epsilon\|_{r-u} + \|G_\epsilon \circ f_\epsilon^{-1} \circ g_\epsilon - G_\epsilon \circ f^{-1}\|_{r-u} \leq \end{aligned}$$

Since we want, later, to bound G in terms of F, we will use only bounds of G in a slightly smaller domain. This amounts to taking $u/2$ in the previous lemma and, therefore, to multiplying some constants by 2^4

$$\leq Cu^{-2} \|F_\epsilon\|_r \|G_\epsilon\|_{r-u/2} + Cu^{-2} \|G_\epsilon\|_{r-u/2} + Cu^{-4} \|G_\epsilon\|_{r-u/2} \quad (\|F_\epsilon\|_r + \|G_\epsilon\|_{r-u/2}) \ .$$

Now (being incredibly wasteful), we bound all this by

$$Ce^{B+4n(u/2)} \|F_\epsilon\|_r^2 \ .$$

We have to assume $e^{n(u/2)} > u^{-4} u^{-2}$, which is certainly true for sufficiently small u.

Now, we have to show that this step can be iterated. (The only things to check are the smallness conditions we assumed for $\|G\|, \|F\|$) and that the iterates converge. Since each of the steps have a choice for u, we will have to show that there is a choice for the u's at each step so that indefinite iteration and convergence holds.

So, we pick once and for all a fixed sequence (u_k) in such a way that $n(u_k/2) = n_0(3/2)^k$ (n_0 to be adjusted later).

Notice that these choices imply (using the equation for $n(u)$) that

$$\sum_k u_k = \sum_k \frac{2}{\pi} \frac{A}{\log n_0(3/2)^k \ldots \log \ldots \log^{1+\nu} n_0(3/2)^k}$$

$$\left(1 - \frac{1}{\log n_0(3/2)^k} - \cdots - \frac{1+}{\log \ldots \log^{1+\nu} n_0(3/2)^k}\right)$$

which converges by comparison with

$$\sum \frac{1}{k \log k \ldots \log \ldots \log^{1+\nu} k}$$

which converges by the integral test. (The factor in brackets clearly tends to 1 when $k \to \infty$.)

Moreover, this sum is arbitrarily small provided n_0 is sufficiently large. Since $r - \sum u_k$ is going to be the measure of analyticity domain of G, we see that, if we can pick n_0 large, the analyticity loss will be a small factor of the original domain.

Calling $S_n = \|F_\epsilon\|_{r_n}$ after performing n inductive steps, we have shown that

$$S_{k+1} \le e^{B+3n\left(\frac{uk}{2}\right)} S_k^2 \equiv e^B e^{3n_0} \left(\left(\frac{3}{2}\right)^k\right) S_k^2$$

(assuming that the conditions to perform a step hold) so that

$$\begin{aligned} S_{k+1} &\le e^{B+3n_0(3/2)^k} e^{2B+2n_0(3/2)^{k-1}} S_{k-1}^{2^2} \\ &\le e^{B(1+2+\cdots+2^{k+1})} e^{2n_0((3/2)^k} \left(1+2\left(\frac{3}{2}\right)^{k-1} + \cdots + 2^k\right) S_0^{2^{k+1}} . \end{aligned}$$

Now,

$$\begin{aligned} 1+2+\cdots+2^{k+1} &\le 2^{k+1}\left(\frac{3}{2}\right)^k + 2\left(\frac{3}{2}\right)^k + \cdots + 2^{k+1} \\ &= 2^k \sum_{\ell=0}^{k} \left(\frac{3}{2}\right)^{k-\ell} \left(\frac{1}{2}\right)^{\ell-1} \\ &\le 2^k \sum_{\ell=0}^{\infty} \left(\frac{3}{4}\right)^{\ell} \le C 2^{k+1} \end{aligned}$$

so that

$$S_{k+1} \le \left(S_0 e^{B+Cn_0}\right)^{2^{k+1}} .$$

Reading the formula for u_k we obtain

$$u_k \le \frac{C(\nu)}{(\log(n_0) + (\log \frac{3}{2})k)^2}$$

where C depends only on ν.

The conditions to iterate that we found for G are implied then by

$$e^{-n(u_k/2)} S_k \le C_k^2$$

which are, in turn, implied by

$$e^{-(3/2)^k}\left(S_0 e^{B+Cn_0}\right)^{2k+1} \le \frac{C(\nu)}{[\log(n_0)+\log(3/2)k]^2}\,.$$

We can see that, given a certain n_0, we can find S_0 small enough so that the conditions hold for all k. Therefore, if they hold for the first step, the iteration keeps them valid.

Analogous, even easier, considerations hold for the other condition of the inductive step.

So, the inductive step can be iterated indefinitely. The remainder converges to zero and n_0 can be taken as big as we want provided S_0 is small enough.

Convergence of $g_\epsilon^0 \circ g_\epsilon^1 \cdots \circ g_\epsilon^k$ as $k \to \infty$ in the domain given by $r - \sum u_k$ can now be very easily established. At each step of the iteration, the same bounds we proved for the inductive hypothesis show that all the compositions make sense. We also have very good bounds on G_ϵ^k. Using them, the formula for the composition of G.H.I. and proceeding as in Lemma 6, we are done.

4. About the Nonresonance Conditions

The idea of introducing a function Ω to measure the non-resonance and try to determine the weakest conditions can be attributed to Russman [31].

Rather than calculate at each step of the process as he did, he left all the calculations for the end. The conditions he found are

$$\sum \frac{\Omega(2^n)}{2^{n+1}} < \infty$$

(plus monotonicity and logarithmic convexity).

The same conditions were derived by other methods by Brjuno [9]. These conditions are weaker than those of the proof we presented. For the proof we presented here, the left hand side of the inequality above becomes

$$\sum \frac{1}{n(\log n)(\log\log n)\dots(\log\dots\log^{1+\nu} n)}$$

which is not, of course, the slowest convergent series, but is probably the slowest which is comfortable to write.

In the next paragraph we will present some examples of α's failing to satisfy the condition and for which the theorem is false.

It is interesting to remark that there is a no man's land of numbers for which it is not known either how to prove the theorem or how to construct counterexamples. Certainly this no man's land is small in any sensible sense of the word "small" for uncountable sets. (It has zero measure, the complement is residual, has zero Hausdorff dimension, etc..) But nevertheless, its existence is vexing (it is quite similar to ruling out the existence of singular continuous spectrum whose presence prevents having a well behaved scattering theory). For a similar model problem, studying the conjugacion of analytic maps to a rotation in a vicinity of a fixed point with derivative of unit modulus, there are examples [39] that show that for a number failing to satisfy the contitions used in [9] to prove linearization, the conclusions of the theorem are false. See also [40] for some extensions of the methods.

What we will do now is to study the set of α's for which the conditions

$$\Omega(|k|) \leq |\alpha \mathbf{k}|^{-1}$$

are satisfied.

We want to show it is a big set (it has positive measure, as close to full as we want by taking $\Omega \to \lambda\Omega$, λ big) and also some geometric properties that will be useful in the proof of the full K.A.M. theorem.

We will introduce the set of "almost resonances of order $\mathbf{k}$"

$$R_{\mathbf{k}} = \left\{ \alpha \;\middle|\; |\alpha \mathbf{k}| \leq \frac{1}{\Omega(|k|)} \right\} .$$

Each one of those is a strip bounded by two planes perpendicular to $\mathbf{k}$ and separated by a distance

$$\frac{2}{\Omega(|k|)(\sum^{d} k_i^2)^{1/2}} .$$

The set in which the non-resonance conditions hold is the complement of $\bigcup_{\mathbf{k}\in \mathbf{Z}^d} R_{\mathbf{k}}$.

If we restrict ourselves to a ball in α space,

$$\begin{aligned} \mathrm{Vol}(R_{\mathbf{k}}) &\leq \mathrm{diam(Ball)} \frac{2}{\Omega(|k|)(\sum_{i=1}^{d} k_i^2)^{1/2}} \\ &\leq \frac{\mathrm{diam(Ball)} 2\sqrt{d}}{\Omega(|k|)|\mathbf{k}|} . \end{aligned}$$

(By Cauchy Schwartz we have $|\mathbf{k}| \leq \sqrt{d}(\sum_{i=1}^{d} |k_i|^2)^{1/2}$). We then see that

$$\mathrm{Vol}\left(\bigcup_{\mathbf{k}} R_{\mathbf{k}}\right) \leq \sum_{\mathbf{k}\in \mathbf{Z}^d} \frac{\mathrm{diam(Ball)} 2\sqrt{d}}{\Omega(|k|)|k|}$$

which, for our choice of $\Omega(|k|)$ certainly is finite. Moreover, by choosing big enough constants we can make this measure of the excluded set as small as we wish.

But remember that the proof could go through for arbitrarily large constants in Ω provided that we took small enough perturbations.

This is one of the key ingredients in the proof that, for the full K.A.M. theorem, the proportion of phase space occupied by invariant tori is close to one, if the perturbation is sufficiently small.

The actual proof is slightly more complicated and involves other slightly stronger statements about how big the set is; something more is necessary to control the changes introduced by the partial conjugacies. The "bad" set is either those of resonances or the things that can be moved into resonances by a movement of the size of the perturbations at the corresponding stage of the iterative process.

We call

$$\widetilde{R}_{\mathbf{k},u} = \{\alpha \mid \mathrm{dist}(\alpha, R_k) \leq u\} .$$

In the ℓth step of the proof the "bad" set will turn out to be

$$\bigcup_{|\mathbf{k}| \leq n(u_\ell)} \widetilde{R}_{\mathbf{k},\varepsilon_\ell} \quad \text{where} \quad (\varepsilon_\ell = \|F^\ell\|_{r_\ell}) .$$

These sets still have small measure when restricted to a ball because $\widetilde{R}_{\mathbf{k},u_\ell}$ is still bounded by two planes perpendicular to $\mathbf{k}$ and separated by a distance bounded above by $(\frac{2\sqrt{d}}{\Omega(|k|)|\mathbf{k}|} + 2\varepsilon_\ell)$ and the sum of the measure of the bad sets is the same as before plus

$$\sum_\ell \sum_{|j| \leq |k| \leq n_{\ell+1}} u_\ell \leq C \sum_\ell n_{\ell+1}^{d-1} \varepsilon_\ell$$

which also converges and is as small as we wish provided ε_0 is small.

Notice that, given a Cantor set even of large measure, it is by no means obvious that we can enlarge the complement at each of the stages of the construction and still obtain something non-empty.

5. A Counterexample

Here we construct a G.H.I. f_ε in the torus T^2 that satisfies all the conditions of the previous theorem except for non-resonance and, such that it is not topologically conjugate for any ε to f_0.

We can find an $a \in \mathbf{R}$ such that $\limsup_{k\to\infty} \exp(-Ak) \min_{n\in\mathbf{Z}} |ak - n|^{-1} \to \infty$. An explicit construction of such an a can be achieved by, e.g., taking a number with a decimal expression $0.1\overbrace{0\ldots0}^{n_1}\ldots01\overbrace{0\ldots0}^{n_2}1\ldots$ we have

$$[10^{n_1+\cdots+n_j+j}a - 1]^{-1} = 10^{(n_1+\cdots+n_j-n_{j+1}+j+1)}$$

we can recursively choose the n_{j+1} so that the limit of the resonances along this sequence is ∞. A more elegant way could be using continued fractions.

Notice that the construction can be modified to accommodate any arbitrary finite sequence of numbers before starting with the 0's and 1's. So, these counterexamples are dense (although we showed before that they are of zero Lebesgue measure).

We will be considering families

$$f_\varepsilon = (x + a, y + b + \varepsilon\psi(x))$$

which are G.H.I.'s whenever $\int_0^1 \psi(x)\,dx = 0$. b will not play any role in what follows and we can choose it in any way we like. In order to get a counterexample as strong as possible we will choose it in such a way that the resonances of the pair (a, b) are as weak as possible. (It turns out that we can make them as weak as those of a.)

We will call $\mathcal{H}_A$ the set of functions analytic in S^1 integrating to zero, and with finite norm $||\psi||_A = \sup_{k\in Z^2} e^{kA}|\hat{\psi}_k|$. ($\hat{\psi}_k$ is the kth Fourier coefficient)

Proposition. There is a residual set in $\mathcal{H}_A$ such that when ψ belongs to this set f_ϵ is not topologically conjugate to f_0 for any ε.

Proof. We will just establish disconjugacy of $(x+\alpha, y+\beta+\psi(x))$ to $(x+\alpha, y+\beta)$ in a residual set which will also be invariant by multiplication by scalars.

We observe that the set $\{f_0^k\}$ is equicontinuous and, if f_1 were topologically conjugate to f_0 so would be $\{f_1^k\}$. We show that this is not the case.

$$f_1^k(x,y) = \left(x + k\alpha, y + k\beta + \sum_{\ell=1}^{k} \psi(x+\ell\alpha)\right)$$

and it suffices to show that $\{C_k\psi(x) \equiv \sum_{\ell=1}^{k} \psi(x+\ell\alpha)\}$ are not equicontinuous for a residual set of ψ.

To establish the theorem it suffices to show that for the operators C_k, $||C_k||_{\mathcal{H}_A\to C^0}$ is unbounded, and invoke Baire Category theorem. (See e.g., [31], p.44.)

We have

$$\begin{aligned}
C_k e^{2\pi i n x} &= e^{2\pi i n x} \sum_{\ell=0}^{k-1} e^{2\pi i \ell n \alpha} \\
&= e^{2\pi i n x} \frac{e^{2\pi i k n} - 1}{e^{2\pi i n \alpha} - 1} \\
\sup_k ||C_k||_{\mathcal{H}_A\to C^0} &\geq \sup_k \sup_n e^{-An} \frac{|e^{2\pi i k n} - 1|}{|e^{2\pi i n} - 1|} \\
&= \sup_n \frac{e^{-An}}{|e^{2\pi i n \alpha} - 1|} \sup_k |e^{2\pi i k n \alpha} - 1| \\
&= \infty\ .
\end{aligned}$$

It is quite easy to modify the construction of the example so that it shows disconjugacy generically in C^r when the growth of resonances is faster than k^r.

Notice also that the conjugacy would be provided by function $g(x,y) = (x, y + \eta(y))$ where $\eta(y)$ satisfies $\eta(y) - \eta(y+\alpha) = \psi(x)$ so that one can also play the game of finding resonance conditions so that there is conjugacy in one regularity class and not in another.

Some partial converses of this counterexample are possible.

Theorem. If $f : T^d \to T^d$ is an area preserving C^k diffeomorphisms $k > 2$ and $||f^{n(k}||_{C^0}$ is bounded independently of n, then there exists a C^{k-1} area preserving $h^* : T^d \to T^d$ satisfying

$$h^*(f(x)) = h^*(x) + \mathbf{a}$$

where $\mathbf{a} = \int_{T^d} f(\mathbf{x}) - \mathbf{x}\,dx$. (The integral is to be interpreted as the functions f and the vectors taking values in $\mathbf{R}^d$ — the universal cover of T^d and the integral ranging over a fundamental domain. $f^{(k}$ means now k^{th} derivative.)

We start by observing that since

$$f(\mathbf{x}) = x + \mathbf{a} + \varphi(\mathbf{x}) \ , \quad \text{where} \quad \int_{T^d} \varphi(x) = 0$$

we have

$$f^2(x) = x + 2\mathbf{a} + \varphi(f(x)) + \varphi(x)$$

in general

$$f^n(x) = x + n\mathbf{a} + \sum_{\ell=0}^{n-1} \varphi(f^\ell(x))$$

since f preserves area

$$\int_{T^d} \varphi(f^\ell(x)) = \int_{T^d} \varphi(x) = 0 \ .$$

Hence

$$\int_{T^{2d}} f^n(\mathbf{x}) - \mathbf{x} = n\mathbf{a} \ .$$

Now, if we form

$$h_n(\mathbf{x}) = \frac{1}{n} \sum_{\ell=0}^{n-1} (f^\ell(\mathbf{x}) - \ell\mathbf{a})$$

we see that $\|h_n^{(k}\|_{C^0}$ is also bounded independently of n. Moreover, since

$$\int_{T^d} h_n(\mathbf{x}) - \mathbf{x} = 0$$

it also follows that

$$\sup_{x \in T^d} |h_n(\mathbf{x}) - n\mathbf{a}|$$

is also uniformly bounded.

Then, we have a set of equicontinuous equibounded h_n and we can select a convergent subsequence

$$h_{n_i} \to h^*$$

but

$$h_n(f(x)) = h_n(x) + a + \frac{1}{n}[f^n(x) - n\mathbf{a} - x] \ .$$

Again $f^n(\mathbf{x}) - n\mathbf{a} - \mathbf{x}$ has k derivative bounded independently of n and integrate to zero.

Hence, the bracket converges to zero pointwise so that h^* satisfies the equation of the theorem.

The fact that h^* preserves area follows from the fact that h_n does, and the convergence $h_{n_i} \to h^*$ is uniform C^{k-1} convergence so that the property of conserving the area is preserved under those limits.

This argument is very useful in many contexts. It shows that in order to produce smooth solutions of conjugacy equations, it suffices to obtain uniform bounds for the iterates in a smooth norm.

(Similar arguments appear very often in scattering theory, and are very similar to the Krylov–Bogoliubov argument [33].)

Notice however that we had to use here many properties of the space, like the fact that the cover is $\mathbf{R}^d$ so that we can add. In particular, I do not know how to prove by the same method existence of solutions to

$$f(h^{**}(\mathbf{x})) = h^{**}(\mathbf{x} + \mathbf{a}) \; .$$

(This, however follows if we can guarantee h^* is invertible which can be established assuming that $\|D(f^n) - \mathrm{Id}\|_{C^0}$ is small.)

Remark 1. We can think about the previous proof as an implicit function theorem. Eq. (7) is a particular case of a relation between F and G (which we write $\varphi(F, G) = 0$). F and G range on a linear space and φ, even if it is very complicated, has the advantage that we know that $F = 0$, $G = 0$ is a solution and that we know how to make first order perturbation theory around it. That is, the relation φ is differentiable in some sense and we know how to invert $D_2(0, 0)$.

The standard way of proceeding is to produce a scheme of successive approximations and show it converges when all the differentials, etc. are taken in the appropriate sense.

For example if we took the scheme

$$\begin{aligned} G_{n+1} &= -D_2^{-1}(0,0)\;\left[\varphi(F, G_n)\right] + G_n \\ G_0 &= 0 \; , \end{aligned}$$

Provided that F and G range in a Banach space, that φ is differentiable in the strong sense, that the differential is continuous and that existence of the inverse is supposed to mean existence of a bounded inverse, it is possible to show that this scheme is a contraction (see e.g., [11]).

This is very similar to what we did, but such a simple framework does not suffice. For our problem $D_2^{-1}(0, 0)$ is not bounded in any space of functions, and moreover, the derivative is not continuous in almost any sense.

So, we have to use some extra structure of our problem. We had a natural scale of Banach spaces, the spaces of analytic functions on strips, and the operator $D_2\varphi(0, 0)^{-1}$ was bounded from one of those spaces to the others with smaller domain; the norm of those operators gets unboundedly large if the loss of domain gets small (as it should if we are to be able to iterate indefinitely).

The saving grace for our scheme was the "quadratic convergence": The new remainder was obtained from the old by multiplying by a factor that contains not only the bad factor due to the analyticity loss, but also a term proportional to the error. This combination is less than one, and we keep on improving.

Notice that the scheme we proposed for the easy implicit function theorem does not have this quadratic convergence in Banach spaces. The new error

$\|\varphi(F, G_{n+1})\|$ can be bounded only by $\|D_2^{-1}\varphi(0,0)\|\ \|D^2\varphi\|\ \|\varphi(F, G_n)\|$ which, even if a contraction in Banach space for small F, is not quadratic.

The best known scheme with quadratic convergence is Newton's method

$$G_{n+1} = -D_2^{-1}\varphi(F, G_n)\varphi(F, G_n) + G_n \ .$$

However, this was not the one we used and, indeed, is unsuitable for our problem, because it requires to invert the linearized equation away from $(0,0)$, a task that we could not accomplish by using Fourier analysis.

What we did was to use the scheme involving only the solution of $D_2\varphi(0,0)$ but we used this knowledge to improve F (make F closer to zero) so that we can get quadratic convergence even if not using Newton's scheme.

This is something that cannot be done for all equations, only for those that have some "group structure."

To summarize: we can prove the implicit function theorem in a more general framework that in Banach space, but we need some structure on the equation: either solvability in an open neighborhood or group structure. We also need some structure on the framework in which we set up the problem (the scale of Banach spaces).

Since non-linear functional equations are pervasive in applications and the implicit function theorem is a very natural tool, there has been lots of thought given to the problem of finding more and more general settings. For Dynamical Systems, probably the most useful setting is that of Zehnder [38] but there are others, probably more useful in geometry [19]. There are still other approaches (e.g., [18]).

There is however lots of room for improvement. In particular the results of [26] which are a kind of implicit function theorem do not fit in any of the schemes above.

Besides the example we already discussed it is worth keeping the following counterexample in mind. In comes from [2].

Consider the Frechet space of functions analytic in the unit disk endowed with the usual topology of uniform convergence on compact sets.

If $f(z) = \sum_{k\geq 0} f_k z^k$ define φ by $(\varphi(f))(z) = \sum_{k\geq 0} f_k^2 z^k$ (which is again a function of radius of convergence at least one).

The function $f^* = 1/1 - z$ is a fixed point of φ and, moreover, $D\varphi(f^*) = 2$.

However, in any neighborhood, there are functions $\tilde{f}$ different from f^* so that $\varphi(\tilde{f}) = f^*$ (just take $\tilde{f}_k = 1$ if $k \neq N$, $\tilde{f}_N = 1$), one can produce even an uncountable number of them in any neighborhood (so there is no inverse function; the example can be modified to be a counterexample of the implicit function theorem).

Remark 2. Giovanni Gallavotti has repeatedly drawn attention to the analogy between this treatment and the renormalization group approach to the obtention of uniform estimates in quantum field theory. There one also uses the group structure of the problem. The perturbation theory is used to guess a transformation that substitutes our problem by one closer to "free."

Remark 9. Observe that the quadratic convergence does not get spoiled if instead of using the full solution of Eq. (8) we only use an approximate one, solving Eq. (8) up to an error comparable to the square of $||F||$.

In the full K.A.M. theorem, or in the standard proof of the toy theorem discussed here with a finite number of derivatives, it becomes quite useful to use a truncation

$$G^{[\leq N]} = \sum_{|\mathbf{k}|\leq N} e^{2\pi i \mathbf{k}\mathbf{x}} G_{\mathbf{k}} \ .$$

Using the triangle inequality

$$||G - G^{[\leq N]}||_{A-\delta} \leq \sum_{|\mathbf{k}|>N} |G_k| e^{(A-\delta)|\mathbf{k}|} \ .$$

By Cauchy estimates

$$\leq \sum_{|\mathbf{k}|>N} e^{-\delta|\mathbf{k}|} ||G||_A \leq$$

$$\leq ||G||_A e^{-\delta N} C \sum_{\mathbf{Z}} \ell^{d-1} e^{-\delta\ell}$$

$$\leq ||G||_A C e^{-\delta N} \delta^{-(d-1)} \ .$$

We can, at each step, accommodate an extra loss of domain and an N in such a way that

$$e^{-\delta N} \leq ||F||_A \ .$$

If we fix that N grows exponentially with the number of steps this makes a δ which is comparable to the other losses of domains and the proof of infinite iteration and convergence needs no modification.

6. The Full K.A.M. Theorem

The most famous of the applications of these techniques is the theorem of persistence of quasi-periodic motions in perturbations of integrable systems, that is, systems that have coordinates $\mathbf{A}$, φ (φ are angles) such that the Hamiltonian is only a function of $\mathbf{A}$. The equations of motion for an integrable system are

$$\dot{\varphi}_i = \frac{\partial H}{\partial A_i} = \omega(\mathbf{A})$$

$$\dot{A}_i = -\frac{\partial H}{\partial \varphi_i} = 0$$

so that the solutions are

$$\varphi = \varphi(0) + \omega(\mathbf{A})\mathbf{t}$$

$$\mathbf{A} = \mathbf{A}(0)$$

and all of them are quasi-periodic.

The theorem states that, provided that ω appears in the range of $\omega(\mathbf{A})$ (and some other conditions that we will discuss later) and satisfies irrationally conditions, any sufficiently small perturbation H will have an invariant torus of dimension $2d$. The motion on this torus will be conjugate to the irrational flow of angular frequencies ω.

The conditions we need can be motivated by the following example

$$H_\varepsilon = (\omega_0 + \varepsilon\omega_1)\mathbf{A} \ .$$

Hamilton's equation for this Hamiltonian can be readily solved but there are no motions with frequency ω_0. This shows that we need some conditions that "clamp" the frequency forcing it to appear in the motions of H_ε.

Conditions which are sufficient are that $\det(\frac{\partial^2 H_0}{\partial\mathbf{A}\partial\mathbf{A}}) \neq 0$. Heuristically, the range of $\omega(\mathbf{A}) \equiv \nabla_{\mathbf{A}}\mathbf{H}$ is persistent and those are the conditions under which the theorem is proved most often.

Unfortunately these hypotheses do not apply to the most interesting case, when H_0 is the Hamiltonian of the solar system with the interactions between the planets turned off. (As is very well known the frequency of orbit does not depend on the angular momentum so that the Hessian has less rank than the dimension.) However, there are weaker hypotheses that apply to this case [4] and allow us to prove the theorem.

There are basically two strategies to prove this theorem:

The first — we will call it Kolmogorov's strategy — starts with the observation that Hamiltonians of the form

$$H(\mathbf{A},\varphi) = \omega\mathbf{A} + 0(|\mathbf{A}|^2)$$

admit quasi-periodic solutions of frequency ω

$$\begin{aligned} \varphi(t) &= \varphi(0) + \omega\mathbf{t} \\ \dot{\mathbf{A}}(t) &= 0 \ . \end{aligned}$$

However, if we take coordinates in which $\omega(0) = \omega_0$, the most general perturbation of our integrable system is

$$H(\mathbf{A},\varphi) = \mathbf{F}(\varphi) + \omega_0\mathbf{A} + \mathbf{G}(\varphi)(\mathbf{A} + o(|\mathbf{A}|^2))$$

(as can be readily seen by expanding in Fourier series) where F, G are assumed to be small. So, we try to make a change of variables in which the terms F, G are absent.

If we use the variables A', φ' given by $\exp\mathcal{L}$, where $\mathcal{L}$ is a globally Hamiltonian vector field of Hamiltonian L, we see that "up to higher order terms in L, F, G," the Hamiltonian in the new variables is

$$H(A',\varphi') = F(\varphi') + \omega_0\mathbf{A}' + \mathbf{G}(\varphi')\mathbf{A}' + \{\omega_0\mathbf{A}', \mathbf{L}(\mathbf{A}'\varphi')\}$$

where { } means Poisson Bracket.

So, we try to determine L in such a way that it cancels as many as possible undesirable terms in first order.

The equations that imply this cancellation can be studied because the Poisson bracket means either derivative $\omega\,\mathbf{A}'$ along the Hamiltonian field of L or minus derivative of L along the field of Hamiltonian $\omega\,\mathbf{A}'$. In this later interpretation, we can solve by taking Fourier coefficients eliminating all Fourier coefficients except the constants.

That is, we can find L so that

$$H(A',\varphi') = \left(\int F(\varphi')\mathrm{d}^{\mathbf{d}}\varphi'\right) + \omega\mathbf{A}' + \left(\int \mathbf{G}(\varphi')\mathrm{d}^{\mathbf{d}}\varphi'\right)\mathbf{A} + o(|\mathbf{A}|^2)$$

(up to quadratic in F, G) .

We can omit the constant $\int F(\varphi')$ because adding a constant to the Hamiltonian does not change the dynamics.

As it stands now, the step is impossible to iterate because $(\omega + \int \mathbf{G})$ could fail to satisfy the irrationality conditions.

But we now try to make a change of coordinates $A'' = A' + \mathbf{Y}(\varphi')$ (which is clearly symplectic) and that restores the form we started with

$$H(\mathbf{A}'',\varphi'') = \tilde{\mathbf{F}}(\varphi'') + \omega\mathbf{A} + \tilde{\mathbf{G}}(\varphi'')\mathbf{A}'' + o(\mathbf{A}^2)$$

but with $\tilde{F}$, $\tilde{G}$ which are "quadratic" in F, G.

We can indeed find such a $\mathbf{Y}(\varphi)$ because of the hypothesis on the Hessian. Moreover, the size of $\mathbf{Y}$ will be commensurate with that of $\int \mathbf{G}$. (This only requires the standard implicit function theorem in $\mathbf{R}^n$.)

To prove convergence, we will proceed as before, but with the extra induction hypothesis that the Hessian is uniformly bounded away from zero; again, this is no problem because controlling the change of the Hessian involves controlling only three derivatives which, in the previous notation means that controlling $\sum u_k^{-3} S_k$ which follows from estimates we had already done.

Notice that since the hypothesis of the theorem include that $\omega(\mathbf{A})$ covers an open set when $\mathbf{A}$ ranges in an open set, there will be, for sufficiently small perturbations, many invariant tori. However, the proof we gave requires a different transformation for each one of them. This makes it very hard to answer questions about the nature of the sets of invariant tori.

For example, it is known that there is a set of smooth invariant tori which can be embedded in a differentiable family of tori and which occupy a fraction of phase space arbitrarily close to 1 for sufficiently small perturbations. (It is, in principle, possible that there are some invariant tori, corresponding to frequencies closely approximated by rationals, which do not fit in this smooth family, or which are not smooth, or such that the motion of them is not conjugate, to a linear flow. I will be very curious to know whether the same techniques in the counterexample provided here can be used to provide counterexamples for flows.)

To prove these results about families of tori, it is necessary to use another strategy, due to Arnold [3]. We try to reduce the Hamiltonian flow to linear simultaneously in several places of the phase space. Again, we proceed inductively.

At stage ℓ, we try to construct the conjugacy in a domain of points that satisfy

$$|\mathbf{k}\omega(\mathbf{A})|^{-1} \leq \Omega(|\mathbf{k}|) \qquad 0 \leq |\mathbf{k}| < (3/2)^{\ell} \ ,$$

Then, we solve the linearized conjugacy equation in Fourier coefficients up to order $(3/2)^k$. As explained before, this does not affect the quadratic convergence. However, the structure of the set where we are solving is very complicated and we should worry that each of the conjugacy does not take us out of the space for the next. The idea is to use the anisochromy condition to argue that the geometry of the resonant set, in $\mathbf{A}$, is very similar to that of the resonant set in ω which we discussed. This produces the estimates in the fraction of the volume preserved and, taking care of keeping all the estimates uniform, also the smoothness of the family interpolating the invariant tori. For the details and quantitative statements, the reader should look at [14,29].

Besides the above statements about the global appearance of a large set of invariant tori the previous argument can be modified to yield other results:

If we stop the elimination procedure at a certain stage, we can still show that in non-resonant regions (not just in tori), the motion is up to a change of coordinates, extremely similar to a linear flow. In particular, it does not move off for an extremely long time.

This can be combined with an analysis of the motion in the resonant region to show that the motion is essentially perpendicular to the resonance line to conclude that small perturbations of integrable systems do no wander off for extremely long times, much longer than what a naive dimensional argument would suggest. This is the content of a famous theorem proved by Nekhoroshev [27]. (See simpler proofs in [7].)

The fact that for even longer times, this recurrence disappears and points eventually do wander off was established in some examples [5,1], conjecturing this type of behavior is the rule. Some people claim that they have observed and measured this Arnold diffusion in some systems. Nevertheless, the effect to be measured is several orders of magnitude smaller than the truncation error due to the numerical methods and quite comparable to round-off and the results are not quite reproducible.

7. Computational Aspects

Most of the proofs based on the previous strategy suffer the shortcoming that the smallness conditions required are much stronger than those that are relevant for practical applications (we recommend [41] for a discussion of the applicability of K.A.M. theory specifically to the solar system).

Of course, one could argue that it is possible that the method of proof is not optimized to produce larger constants. Presumably, the figure of merit that has been optimized in many of the proofs is the number of derivatives. Since the K.A.M. theorem makes statements for whole classes of functions and this is an infinite dimensional space it is not even clear what does it mean to optimize the constants in general because there are many different ways to measure distances and sizes of functions.

There are two basic numerical methods to study directly the existence of quasi-periodic orbits lying on smooth tori. One of them is to write down the equation for conjugacy and then apply a non-linear equation solver (e.g., a Newton or continuation method) and the other is to implement numerically the perturbation expansions that we discussed in the previous section.

The numerical literature about implementation of perturbation series expansions is quite extended. Let us mention, among others[42] for an implementation of perturbation series for the quasiperiodic orbit (sometimes called Lindstedt series) or [43] (and references there), [44] for a discussion of the calculation of the approximate chage of variables that integrates the whole system approximately. For numerical implementation of direct solutions of the equation, and a very lucid discussion of the difficulties, let us mention [44].

The validity of these numerical methods has been very controversial. Of course, all numerical methods are subject to round-off and truncation so that, strictly speaking, all numerical results are not mathematically conclusive and, if not accompanied by an error analysis, suspect.

For the computation of quasi-periodic orbits this has been a little bit more so because all the counterexamples point to the effect that even very small values of derivatives of relatively high order can make a difference. It is well known that many numerical discretizations of functions — specifically of several variables — do not allow numerically stable evaluation of derivatives. (The only method we know that allows stable discretization of derivatives is the storage of Fourier coefficients.) Also the examples point out that the irrationality properties of the frequencies play a significant role. Of course, all the numbers stored in a computer are rational and it is not obvious how one could attempt to discretize such problems.

In practice, the methods based on the perturbation series expansions are somewhat easier to implement since the recursion expansion is always done by solving the first order expansion around the unperturbed case. The Newton method requires one to solve a system of linear equations of dimension the number of discrete variables. Even if with todays computers solving a system of equations with 1000 variables is quite feasible, it is a computation of the order of one hour in a typical workstation. Even if shortcuts are possible, this remains a calculation much slower than one step of a recursion. An important reason why these methods based on the solving of functional equations are feasible is that there is a natural variational formulation of the problem.

It is well known that numerically, variational methods — if the variation principle is reasonable — are more reliable and faster than direct solution of the variational equations. Moreover, expansions give you information for several values of the parameter.

On the other hand perturbation expansions in powers of the parameter have several shortcomings. First notice that power series expansions always converge in circles and diverge outside of them. So that if the domain of analyticity of the function we are studying is not a circle, there are values of the parameter for which the result is true but which cannot be studied by the perturbation expansion. It could very well happen that some of these inaccessible values are real — i.e., physical — because there are some unphysical singularities for complex values very close to the origin.

Also, for many of the perturbation expansions of periodic orbits the coefficients differ greatly in size and involve large amounts of cancellations. Especially close to the domain of convergence, evaluation may be quite unstable.

The methods of successive changes of variables used frequently in the mathematical literature are not very convenient for numerical implementations. The main reason is that the difficulty of discretizing a function increases very rapidly with the number of variables. Since the tori are half the dimension of phase space, the embedding of the torus in the phase space involves many less variables than studying functions in the phase space.

From the point of view of numerical applications, one of the most successful methods to compute invariant tori has been its approximation by periodic orbits. This indirect method, even if it has very good empirical justification [45] and can be justified from the point of view of renormalization [46] is hard to justify mathematically (see [47], [48] for some partial results).

Recently, there has been considerable progress in the study of the validity of numerical computations of K.A.M. tori.

At the root is the use of interval arithmetic and its extension to function spaces [49]. The basic idea of interval arithmetic is very simple. Rather than represent a real quantity by a machine number that we hope approximates its value, we represent it by two machine numbers that are upper and lower bounds for its value.

One can write operations among pairs of representable numbers that bound the usual arithmetic operations. That is, the sum of two intervals is another interval that is guaranteed to contain the possible values of the sums of two numbers, whenever these two numbers belong to the summand intervals. (If it is impossible to find a pair of representable numbers with this property, the operation should just raise an exception.)

Then, given an algebraic expression, by translating these algebraic operations into calls to the corresponding interval routines one obtains a routine that produces rigorous bounds for the values of algebraic expression given bounds on the variables.

Similarly, for all the functions for which one knows rational approximations and explicit bounds on the error, one can also obtain routines that bound them by enlarging by a convenient amount the result returned by the routines that bound the polynomial approximation.

A similar circle of ideas can be extended to function spaces. We need to identify families of sets that can be finitely specified and for which it is nevertheless easy to implement operations that bound the usual operations of analysis.

There are many such families of sets that one could consider. One, introduced in [50] that has been found to be particularly useful in the type of questions discussed in these lectures is the family of sets:

$$\begin{aligned} B_{I_{-N}\cdots I_N;\epsilon} = \Big\{ f : T^1 \to \mathbb{C} \mid\ & f = f_p + f_e \\ & f_p(\theta) = \sum_{k=-N}^{N} \hat{f}_{p,k} e^{2\pi i k\theta} \ , \ \hat{f}_{p,k} \in I_i \\ & f_e(\theta) = \sum_{k=-\infty}^{\infty} \hat{f}_{e,k} e^{2\pi i k\theta} \ , \ \sum |\hat{f}_{e,k}| e^{-\delta|k|} \le \epsilon \Big\} \end{aligned}$$

Roughly speaking, we are fixing the main Fourier coefficients to lie in some intervals but allowing some error.

It is quite possible to bound many of the operations of analysis — including, of course, the elementary arithmetic ones — in this class of functions. Roughly, one performs the truncated operations in the Fourier coefficients and then, estimates the errors incurred either because of the uncertainties in the original data or because of the truncation process.

Notice that the norm in this particular example satisfies the Banach algebra property $\|fg\| \le \|f\|\,\|g\|$. Another important property of such discretizations is that they contain information about all derivatives and allow for reliable evaluation of them.

There are many variations that improve several practical considerations. For example, one can keep several error terms, some of which enjoy special properties (e.g., having only coefficients of high order). Or, one can keep more detailed information about the error terms than that afforded by just one norm. For example, keeping estimates both $\sum |\hat{f}_k| e^{\delta k}$ and $\sum |\hat{f}_k|\,|k| e^{\delta k}$ leads to a more efficient proof of the K.A.M. theorem (see [51]) and hence, it is a good idea to incorporate it in the algorithms to compute how good teh approximate solutions are.

There are other discretizations of spaces of functions in the literature. For example [52] discusses in detail discretizations based on pointwise bounds of the functions. These methods are much more efficient than the one above in some problems such as integral equations. They are not so useful when one requires derivatives.

Once one has a powerful enough interval arithmetic package it is possible to show that a specific trigonometric polynomial is an approximate solution of (1). One can verify that the right and left hand sides differ by a small amount and one can also obtain estimates of the properties of an analytic extension.

It is possible to prove a theorem — more or less along the same lines as the one discussed in the text — that shows that if a function satisfies (1) with an error sufficiently small with respect to other analyticity of properties, then a Newton-like method started on it will converge to a true solution.

Hence, one can verify that there is a true solution of (1) and that it is close to our initial guess.

Notice that the way that the solution is produced is irrelevant for the verification that a Newton method started on it converges. In practice, the solution is produced using a numerical method that may include as many shortcuts as are considered expedient. The verification step, guarantees that this is a solution.

The relevant theorems and implementation for some cases is described in [Ra], [53] where it is shown that this way of proceeding — given enough time — can explore as much as desired of the region of validity of the K.A.M. theorem. The actual implementation was run in some concrete problems and got verified the results for some values of ε bigger than 95% of values for which it is known that the conclusions of the theorem are false.

Another different method to take advantage of interval arithmetic is considered in [54]. There, interval arithmetic is used to obtain bounds of each of the coefficients in the perturbative expansions and a theorem is proved stating that certain properties of the perturbative expansion imply lower bounds for the radius of convergence.

As mentioned before, there are reasons why power series expansions cannot cover the whole domain of analyticity. Nevertheless, they provide information for whole ranges of parameters, which may be preferable in some practical situations where the relevant values of parameters are covered.

One should also point out that the use of interval arithmetic prevents the use of the cancellations that provide the convergence of this series so that one should expect that the method reaches smaller values than the true ones.

One improvement of the method in which the coefficients of the power series expansions are computed without using interval arithmetic is [55].

The validity of the K.A.M. theorem can be delineated with the help of counterexamples to the conclusions. For twist maps (two dimensional) there are several counterexamples to the differentiability and irrationality properties [56], [57], [58]. For the numerical values allowed in the perturbation [59], [60], [61]. The latter two involve computer assisted calculations. The situation in higher dimensions is more confusing since tori can be notably more complicated [62]. Computer assisted counterexamples to several possible statements for sufficiently large values of the parameter can be found in [Ma].

For the model problem of iteration of analytic functions near a fixed point with derivative of modulus 1, a very satisfactory description of the boundaries of validity exists [39], [40].

References

1. V. A. Arnold and A. Avez, *Ergodic Problems of Classical Mechanics*, (Benjamin, Reading, 1968).
2. R. Abraham and J. Marsden, *Foundations of Mechanics*, (Addison-Wesley, New York, 1978).
3. V. A. Arnold, Proof of A. N. Kolmogorov's theorem on the preservation of quasi-periodic motions under small perturbations of the hamiltonian, *Russ. Mat. Surv.* **18** (1965).

4. V. A. Arnold, Small divisor problems in classical and celestial mechanics, *Russ. Mat. Surv.* **18** (1965).
5. V. A. Arnold, Instability of dynamical systems with several degrees of freedom, *Dok. Aad Nauk* **156** (1964).
6. A. Banyaga, Sur le structure du groupe de diffeomorphism qui preservent une forme symplectique, *Comm. Mat. Helv.* **53** (1978).
7. G. Benettin, L. Galgani and A. Giogilli, A proof of Nekhoroshev's theorem for the stability times in nearly integrable systems, preprint.
8. N. Bogoliubov and A. Mitropolskii, *Asymptotic Methods in the Theory of Non-Linear Oscillations*, (Gordon and Breach, New York, 1961).
9. A. D. Brjuno, Analytic form of differential equations, I. *Trans. Moscow Math. Soc.* **25** (1971); II. *Trans. Moscow Math. Soc.* **26** (1972).
10. L. Chierchia and G. Gallavotti, Smooth prime integrals for quasi-integrable Hamiltonian systems, *Nuovo Cimento* **67B** (1982).
11. J. Dieudonne, *Foundations of Modern Analysis*, (Academic Press, New York, 1960).
12. E. Dynkin, The structure of semi-simple Lie algebras, *Trans. A.M.S.* Ser.1 **9** (1950).
13. D. B. A. Epstein, The simplicity of certain groups of homeomorphisms, *Compositio Matematica* **22** (1970).
14. G. Gallavotti, in *Scaling and Self-Similarity in Physics*, ed. J. Fröhlich (Birkhauser, Boston, 1983).
15. G. Gallavotti, in *Regular and Chaotic Motions in Dynamic Systems*, eds. G. Velo, A. S. Wightan (Plenum, New York, 1985).
16. G. Gallavotti, A criterion of integrability for perturbed non-resonant harmonic oscillators. "Wick ordering" of the perturbations in classical mechanics and invariance of the frequency spectrum, *Comm. Math. Phys.* **87** (1982).
17. G. Gallavotti, Renormalization theory and ultraviolet stability for scalar fields via renormalization group methods, *Rev. Mod. Phys.* **57** (1985).
18. R. Graff, Elements of non-linear functional analysis, *Memoirs A.M.S.* **206** (1978).
19. D. Hamilton, The implicit function theorems of Nash and Moser, *Bull. A.M.S.* **7** (1982).
20. M. Herman, Sur les diffeomorphisms du tore, *Ann. Inst. Fourier* **23** (1974).
21. R. de la Llave, J. M. Marco and R. Moriyon, Canonical perturbation theory for Anosov diffeomorphisms and regularity results for Livsic cohomology equation, *Ann. of Math.* in print.
22. J. Moser, *Stable and Random Motions in Dynamical Systems With Special Emphasis on Celestial Mechanics* (Princeton Univ. Press, Princeton, 1973).
23. J. Mather, Commutators of diffeomorphisms, I. *Comm. Math. Helv.* **49** (1974).
24. J. Mather, Commutators of diffeomorphisms, II. *Comm. Math. Helv.* **50** (1975).
25. J. Mather, Commutators of diffeomorphisms, III. *Comm. Math. Helv.* **60** (1985).
26. J. Mather, Stability of C^∞ mappings, II, Infinitesimal stability implies stability, *Ann. of Math.* **89** (1969).
27. N. N. Nekhoroshev, An exponential estimate of the time of stability of nearly integrable systems, *Russ. Math. Surv.* **32** (1977).
28. H. Poincaré, *Les Methodes Nouvelles de la Mecanique Celeste* (Gauthier Villars, Paris, 1892).

29. J. Poschel, Integrability of Hamiltonian systems on Cantor sets, *Comm. Pure and Appl. Math.* **35** (1982).

30. R. D. Ritchmyer, *Principles of Advanced Mathematical Physics, II* (Springer, New York, 1981).

31. W. Rudin, *Functional Analysis* (McGraw-Hill, New York, 1973).

32. H. Rüsmann, On the one dimensional Schrödinger equation with a quasiperiodic potential, *Ann. New York Acad. Sci.* **357** (1980).

33. Y. G. Sinai, *Introduction to Ergodic Theory* (Princeton Univ. Press, Princeton, 1976).

34. S. Sternberg, Infinite Lie groups and the formal aspects of mechanics, *Jour. Math. Mech.* **10** (1961).

35. W. Thirring, *A Course in Mathematical Physics, Vol. 1, Classical Dynamical Systems* (Springer Verlag, New York, 1978).

36. A. Weinstein, Lectures on symplectic manifolds, *C.B.M.S. Conf. Series A.M.S.* **29** (1977).

37. A. Wintner, *The Analytical Foundations of Celestial Mechanics* (Princeton Univ. Press, Princeton, 1941).

38. E. Zehnder, Generalized implicit function theorems with applications to some small divisor problems, I. *Comm. Pure and Appl. Math.* **28** (1975); II. *Comm. Pure and Appl. Math.* **29** (1975).

39. J. C. Yoccoz, Theoreme de Siegel, polynomes quadratiques et nombres de Brjuno, preprint (1988).

40. R. Perez-Marco, Sur la structure des germes holomorphes non-linearizables, *C. R. Acad. Sci. Paris* **312** (1991).

41. J. Moser, Is the solar system stable?, *Math Intelligencer* **1** (1978).

42. W. H. Pressler, R. Broucke, Computerized formal solutions of dynamica systems with two degrees of freedom and an application to the Contopoulos Potential, I,II *Comp. & Maths. with Appl.* **7** (1981).

43. S. Coffey, A. Deprit, E. Deprit, L. Healy, B. Miller A toolbox for non-linear dynamics. *Computer aided Proof in Analysis, K.R. Meyer, D. S. Schmidt ed.* Springer (1991).

44. D. Braess, E. Zehnder, On the numerical treatment of a small divisor problem, *Numer. Math.* (1982).

44. A.J. Dragt, J. Finn, Lie Series and invariant functions for analytic symplectic maps *Jour. Math. Phys.* **17** (1976).

45. J. Greene, A method for determining a stochastic transition *J. Math. Phys.***20** (1979).

46. R. S. MkKay, Renormalization in area preserving maps. *Princeton U. thesis* (1982).

47. C. Falcolini, R. de la Llave, A rigorous partial justification of Greene's criterion *Jour. Stat. Phys* **67** (1992).

48. R. S. McKay, On Greene's residue criterion *Nonlinearity* **5** (1992).

49. R. Moore, *Methods and applications of interval analysis* Siam, Philadelphia (1979).

50. O. E. Lanford, Computer assisted proofs in analysis *Physica* **124A** (1984).

51. D. Rana, Proof of accurate upper and lower bounds to stability domains in small denominator problems *Princeton thesis* (1987).

52. E. W. Kaucher, W. Miranker, Self-Validating numerics for function space problems *Academic Press* (1984).

53. R. de la Llave, D. Rana, Accurate stategies for K. A. M. bounds and their implementation. *In Computer Aided Proofs in Analysis, K. Meyer, D. Schmidt eds. Springer* (1991).
54. A. Celletti, L. Chierchia, Construction of analytic K. A. M. surfaces and effective stability bounds *Comm. Math. Phys.* **118** (1988).
55. A. Celletti, L. Chierchia, Invariant curves for area-preserving twist maps far from integrable *Jour. Stat. Phys.* **65** (1991).
56. J. Mather, Destruction of invariant circles *Erg. Theo. & Dyn. Syst.* **8** (1998).
57. M. R. Herman, Sur les courbes invarinates par les difféomorphismes de l'anneau vol 2 *Asterisque* **144** (1986).
58. G. Forni, Analytic destruction of invariant circles *Princeton Univ. Preprint* (1992).
59. J. Mather, Non-existence of invariant circles, *Erg. Theo. & Dyn. Syst.* **4** (1984).
60. R. S. McKay, I. Percival, Converse K. A. M.: theory and practice *Comm. Math. Phys.* **98** (1985)
61. I. Jungreis, A method for proving that monotone twist maps have no invariant circles *Erg. Theo. & Dyn. Syst.* **11** (1991).
62. M. R. Herman, Dynamics connected with indefinite torsion *Ecole Polytechnique preprint* (1991).
63. M. Muldoon, Ghosts of order on the frontier of chaos *Caltech thesis* (1988).

STABILITY AND CHAOS
IN CLASSICAL AND QUANTUM HAMILTONIAN SYSTEMS

H.R. JAUSLIN
Départememt de Physique, Université de Bourgogne
6,bd. Gabriel, 21100 Dijon, France

ABSTRACT

We discuss regular and chaotic aspects of the dynamics of Hamiltonian systems, from the classical as well as from the quantum mechanical point of view. The contents of these lectures are organized as follows:

1. Introduction

The complete information on the classical dynamics of a conservative system is in principle contained in a single function – the Hamiltonian – and in the initial conditions. It is however a highly non-trivial task to extract this information. Since the pioneering work of Poincaré it has been established that the dynamics defined by Hamiltonians of even very simple forms can be extremely complex. Furthermore, the dependence on the initial conditions can be very sensitive, in the sense that small differences are strongly amplified. This puts fundamental limitations on the ability to predict the dynamical behavior for long times. The advent of modern computers with powerful visualization capabilities has given a big boost to the study of these complex phenomena. There have also been in recent years important analytical advances, that allow a better interpretation and understanding of the results obtained by numerical computation. They allow also to devise better adapted and more precise algorithms to probe these structures.

The numerical solution of Hamilton's evolution equations involves a discretization of the time variable. The evolution is then implemented as the iteration of a map. In order to be able to get reliable information on the long time behavior of the system, the approximating maps should be compatible with the Hamiltonian structure of the problem. The maps should, among other things, conserve volume in phase space, as required by Liouville's theorem. Many new algorithms that satisfy this requirement have been developed in the last couple of years. The maps involved are interesting dynamical systems on their own right. In fact some of them have traditionally played a fundamental role in the development of the understanding of Hamiltonian systems.

In quantum mechanics the complete information about a system is also contained in the Hamiltonian, which in this case is an operator acting on a Hilbert space, and in the initial state. Again, the simplicity of a Hamiltonian by no means implies that one can easily extract the relevant static or dynamic information from it. A very natural question, on which a lot of effort has been focused in recent years, is whether the rich and complex structures observed in classical mechanics have some counterparts in the corresponding quantum models.

In this article we will discuss some aspects of classical Hamiltonian dynamics, its numerical implementation, and the properties of the corresponding quantum systems. The general focus will be on Hamiltonian systems driven by time dependent external forces. We will emphasize connections between different concepts and phenomena, both on a physical and on a mathematical level. The physical connections rely on the fact that we are discussing basically the same system under different points of view: its exact continuous-time classical evolution, a discretized approximation suitable for numerical implementation, and the quantum version of the model. There are other connections of methodological nature, based on the similarity of mathematical structures: In order to devise good algorithms one uses the knowledge on the symplectic structure of the equations. Furthermore, in order to analyze the validity of the approximations involved in the algorithms, one uses the qualitative information about generic aspects of chaotic Hamiltonian evolution. For instance, although the exact evolution might be integrable, i.e. completely regular, the necessary discretization of time can introduce a spurious non-integrability and chaotic behavior.

We will also see that methods like the superconvergent iterations of the KAM theory, devised to prove stability of classical systems, furnishes the right tools to control the perturbation theory of quasienergy eigenvalues of time dependent quantum systems. Finally, the concept of complete integrability of classical Hamiltonian systems, which is central to their qualitative understanding, appears in a somewhat unexpected way in the study of the dependence of eigenvalues on the amplitude of the perturbation.

CLASSICAL SYSTEMS

2. Canonical Transformations, Symplectic Structure

The canonical equations of motion associated to a general, possibly time dependent Hamiltonian $H(p,q,t), \quad p,q \in R^N$ are

$$\frac{dp}{dt} = -\frac{\partial H}{\partial q}, \qquad \frac{dq}{dt} = \frac{\partial H}{\partial p} \tag{2.1}$$

A coordinate transformation in phase space $(p,q) \mapsto (P,Q) = C(p,q,t)$ is called *completely canonical* [1] if the equations of motion for the new variables (obtained by insertion into (2.1)) have the same form as (2.1), with a new Hamiltonian $\hat{H}$:

$$\frac{dP}{dt} = -\frac{\partial \hat{H}}{\partial Q} \qquad \frac{dQ}{dt} = \frac{\partial \hat{H}}{\partial P} \tag{2.2}$$

Remark: This definition requires that the transformation C satisfies this property for all possible Hamiltonians, i.e. it excludes transformations that satisfy it for one special H but not for others [1].

In order to give a characterization of canonical transformations that is more convenient to use, we first write Hamilton's equations (2.1) in a more compact form

$$\frac{d\eta}{dt} = J\frac{\partial \hat{H}}{\partial \eta} \qquad \frac{d\xi}{dt} = J\frac{\partial \hat{H}}{\partial \xi} \tag{2.3}$$

where we have introduced the vectors $\eta = (p_1, q_1 \ldots, p_N, q_N)$, $\xi = (P_1, Q_1, \ldots \ldots, P_N, Q_N)$ and the matrix

$$J = \begin{pmatrix} 0 & -1 & & & & 0 \\ 1 & 0 & & & & \\ & & \cdot & & & \\ & & & \cdot & & \\ & & & & 0 & -1 \\ 0 & & & & 1 & 0 \end{pmatrix}$$

which satisfies $J^T = J^{-1} = -J$, and $J^2 = -1$.

A transformation is called *symplectic* if the corresponding Jacobi matrix $M_{i,j} \equiv \partial\xi_i/\partial\eta_j$ is symplectic, i.e. satisfies the relation [1-3]

$$M^T J M = J. \tag{2.4}$$

We remark that if M is symplectic, then also M^{-1} and M^T.

A transformation is called *generalized symplectic* if there is a constant λ such that $\lambda^{-1}M$ is symplectic, i.e.

$$M^T J M = \lambda^2 J. \tag{2.5}$$

An example for this last type is the change of scale $q \mapsto \nu q \equiv Q$, $p \mapsto \mu p \equiv P$. It is not symplectic since $M = \begin{pmatrix} \nu & 0 \\ 0 & \mu \end{pmatrix}$ and $M^T J M = \nu\mu J$. The equations of motion in the new variables are

$$\begin{aligned} \dot{P} &= \mu\dot{p} = -\mu\frac{\partial H}{\partial q} = -\nu\mu\frac{\partial H}{\partial Q} \\ \dot{Q} &= \nu\dot{q} = \nu\frac{\partial H}{\partial p} = \nu\mu\frac{\partial H}{\partial P} \end{aligned}$$

The new Hamiltonian is thus $\hat{H} = \nu\mu H$. The factor $\mu\nu$ could be absorbed by rescaling the time variable $t \mapsto t' = \nu\mu t$.

The interest of this definition is that the generalized symplectic transformations are precisely the ones that leave the structure of Hamilton's equations invariant:

Theorem 2.1: Let $C(\eta,t)$ be a family of invertible transformations, such that C and C^{-1} are twice continuously differentiable. Then C is a completely canonical transformation if and only if it is generalized symplectic.

The proof of this theorem is quite simple. We describe it in an appendix at the end of this section.

As a consequence of the particular structure of Hamilton's equations (2.1), the flow in phase space, $\Phi_{t,t_0} : (p(t_0), q(t_0)) \mapsto (p(t), q(t))$, has some special properties:

1) The flow Φ_{t,t_0} for any fixed initial and final times, t_0 and t, is a canonical transformation, i.e. if we identify $\eta = (p_1(t_0), q_1(t_0) \ldots, p_N(t_0), q_N(t_0))$ and $\xi = (p_1(t), q_1(t) \ldots \ldots, p_N(t), q_N(t))$ then the Jacobi matrix $M \equiv D\Phi_{t,t_0}$ is symplectic. This can be seen from the following simple calculation: Since $\xi(t) = \Phi_{t,t_0}\eta$ satisfies the equations of motion for arbitrary initial conditions η, we can write the equation of motion for the time evolution operator as

$$\frac{\partial \Phi_{t,t_0}}{\partial t} = J\frac{\partial H}{\partial \xi} \circ \Phi_{t,t_0} \tag{2.6}$$

where the sign $\circ$ denotes composition of operators. Applying the differential operator D on Eq. (2.6) and using the chain rule $D(G \circ H)(y) = DG(H(y))\ DH(y)$ we obtain

$$-J\frac{\partial}{\partial t}D\Phi_{t,t_0}(\eta) = (D\frac{\partial H}{\partial \xi})(\xi)\ D\Phi_{t,t_0}(\eta).$$

Remark that

$$\left[D\frac{\partial H}{\partial \xi}\right]_{ij} = \frac{\partial^2 H}{\partial \xi_i \partial \xi_j}.$$

To prove that $D\Phi_{t,t_0}$ satisfies (2.4) we first show that the matrix $[D\Phi_{t,t_0}]^T\ J\ D\Phi_{t,t_0}$ does not depend on t:

$$\frac{\partial}{\partial t}\Big\{[D\Phi_{t,t_0}]^T\ J\ D\Phi_{t,t_0}\Big\} = [D\Phi_{t,t_0}]^T\Big([\frac{\partial^2 H}{\partial \xi \partial \xi}]^T - \frac{\partial^2 H}{\partial \xi \partial \xi}\Big)D\Phi_{t,t_0} = 0$$

since the matrix $\partial^2 H/\partial\xi\partial\xi$ is symmetric. If we now take $t = t_0$ we have $\Phi_{t_0 t_0} = 1$, which implies

$$[D\Phi_{t,t_0}]^T\ J\ D\Phi_{t,t_0} = J \qquad \forall t, t_0 \tag{2.7}$$

2) *Liouville theorem and Poincaré invariants.* The fact that the Jacobi matrix $D\Phi_{t,t_0}$ (which represents the linear part of the flow) is symplectic, i.e. satisfies (2.7) implies that

$$|\det D\Phi_{t,t_0}| = 1 \tag{2.8}$$

which means that the canonical flow preserves volume in phase space (Liouville's theorem): The evolution of the volume $V(t)$ of a set of points in phase space contained initially in a volume $V(t_0)$ is given by

$$\int_{V(t)} d^N p\ d^N q = \int_{V(t_0)} d^N p_0\ d^N q_0\ |\det D\Phi_{t,t_0}| = \int_{V(t_0)} d^N p\ d^N q$$

The volume is only one among the hierarchy of N Poincaré invariants, which are integrals (over surfaces of the appropriate dimension) of the following N differential forms

$$\omega^{(2)} = \sum_i dp_i \wedge dq_i$$
$$\omega^{(4)} = \sum_{i<j} dp_i \wedge dp_j \wedge dq_i \wedge dq_j \tag{2.9}$$
$$\omega^{(2k)} = \sum_{i_1<\ldots<i_k} dp_{i_1} \wedge \ldots \wedge dp_{i_k} \wedge dq_{i_1} \wedge \ldots \wedge dq_{i_k}, \qquad k = 1 \ldots N$$

Remarks:

a) The last one, $\omega^{(N)}$, is proportional to the volume.

b) The invariance of all the differential forms (2.9) can be deduced from the invariance of $\omega^{(2)}$, since $\omega^{(2k)} = \omega^{(2)} \wedge \omega^{(2)} \wedge \ldots \wedge \omega^{(2)}$. In fact, canonical transformations can be equivalently defined as the maps that leave the two-form $\omega^{(2)}$ invariant [2, 4].

c) The integral of $\omega^{(2)}$ over a two dimensional surface S in phase space

$$\int_S \sum_i dp_i \wedge dq_i$$

can be interpreted geometrically as the sum of the areas of the projections of the surface S on the planes defined by (p_i, q_i).

d) The property (2.8) implies that canonical transformations are always locally invertible. From the symplectic condition (2.7) one sees further that the inverse of a canonical transformation is also canonical.

2.1. Generating Functions

There is a special type of canonical transformations, called contact transformations, that can be constructed from a generating function. The fact that the transformation can be characterized by a single function (instead of $2N$) makes them very convenient in many contexts. They can be introduced as follows: The trajectories $p(t), q(t)$ of a Hamiltonian system can be characterized as the extremals of the variational principle

$$\delta \int_{t_1}^{t_2} dt\ (p\dot{q} - H(p,q,t)) = 0$$

This principle is valid in any canonical coordinates, i.e. if $(p,q) \mapsto (P,Q)$ then

$$\delta \int_{t_1}^{t_2} dt\ (P\dot{Q} - \hat{H}(P,Q,t)) = 0$$

A *sufficient* condition for two Lagrangeans to give identical extremals is the existence of a function $F(p,q,t)$ such that

$$p\dot{q} - H(p,q,t) = P\dot{Q} - \hat{H}(P,Q,t) + \frac{dF}{dt} \tag{2.10}$$

(since $\delta \int_{t_1}^{t_2} dt\ dF/dt = 0$). Because of the functional relation between p, q, P, Q, F can be considered as a function of any pair of those variables. Equation (2.10) can be written in differential form as

$$pdq - H(p,q,t)\ dt = PdQ - \hat{H}(P,Q,t)\ dt + dF \tag{2.11}$$

Since the independent differentials that are explicit in this relation are dq, dQ and dt the simplest choice is $F = F_1(q,Q,t)$, which leads to

$$dF_1 = \frac{\partial F_1}{\partial q}\ dq + \frac{\partial F_1}{\partial Q}\ dQ + \frac{\partial F_1}{\partial t}\ dt$$

Comparing the coefficients of dq, dQ and dt we obtain

$$p = \frac{\partial F_1}{\partial q}(q,Q,t) \tag{2.12}$$

$$P = -\frac{\partial F_1}{\partial Q}(q,Q,t) \tag{2.13}$$

$$\textit{and} \qquad \hat{H}(P,Q,t) = H(p,q,t) + \frac{\partial F_1}{\partial t}(q,Q,t). \tag{2.14}$$

The explicit form of the transformation, e.g. expressing the old variables as functions of the new ones, is obtained by inverting Eq. (2.13) to obtain $q = G(P,Q,t)$ and inserting it into Eq. (2.12).

For some applications it is more convenient to work with generating functions that depend on other pairs of variables. They can be constructed by applying Legendre transformations on F_1. For instance we will use a generating function depending on P and q, which is defined as

$$F_2(q,P,t) = F_1 - Q\frac{\partial F_1}{\partial Q} = F_1 + QP$$

The differentials are related by

$$dF_2 = dF_1 + QdP + PdQ.$$

The purpose of the Legendre transformation is to substitute a dQ term by a dP term. Inserting into (2.11) we obtain

$$pdq - H(p,q,t)dt = -QdP - \hat{H}(P,Q,t) + dF_2$$

$$dF_2 = \frac{\partial F_2}{\partial q}dq + \frac{\partial F_2}{\partial P}dP + \frac{\partial F_2}{\partial t}dt$$

and comparing coefficients we obtain

$$p = \frac{\partial F_2}{\partial q}(q,P,t), \qquad Q = \frac{\partial F_2}{\partial P}(q,P,t), \qquad \hat{H}(P,Q,t) = H(p,q,t) + \frac{\partial F_2}{\partial t}(q,P,t) \tag{2.15}$$

A third type of generating function that we will use depends on Q and p and is related to F_1 by $F_3(p,Q,t) = F_1(q,Q,t) - qp$, which leads to the relations

$$q = -\frac{\partial F_3}{\partial p}(p,Q,t), \qquad P = \frac{\partial F_3}{\partial Q}(p,Q,t), \qquad \hat{H}(P,Q,t) = H(p,q,t) + \frac{\partial F_3}{\partial t}(p,Q,t) \tag{2.16}$$

Remark: It is not possible in general to transform the differential relation (2.3) into one which involves e.g. only dp, dq or only $dPdQ$ (without prior knowledge of the function C). As a consequence the canonical transformations constructed from a generating function are always in implicit form. To obtain the explicit transformation one has to invert a function, which can only rarely be done analytically in closed form.

2.2. The Hamilton-Jacobi Equation

Since, as we have seen, the time evolution is at each fixed time a canonical transformation, we can try to construct the corresponding generating function, which we will denote S_2. We chose it of the type F_2, i.e. $S_2 = S_2(q,P,t;t_0)$. The fourth variable indicates the parametric dependence on the initial time t_0. If one can determine S_2 one gets the general solution of the equations of motion, since all trajectories can be obtained from it. As we will see, this generating function plays an important role in the construction of numerical algorithms to integrate Hamilton's equations. The idea is to identify the new variables with the initial conditions $Q \equiv q(t_0)$, $P \equiv p(t_0)$, and the old ones with the trajectory at a given time t: $q \equiv q(t)$, $p \equiv p(t)$. The new equations of motion will be

$$\begin{aligned} \dot{P} &= -\frac{\partial \hat{H}(P,Q,t)}{\partial Q} = 0 \\ \dot{Q} &= \frac{\partial \hat{H}(P,Q,t)}{\partial P} = 0 \end{aligned} \tag{2.17}$$

since the initial conditions don't change. Eq. (2.17) implies that the new Hamiltonian will be independent of P and Q, i.e. it can at most depend on t: $\hat{H} = \hat{H}(t)$. From Eqs. (2.15) it follows that the generating function S_2 satisfies the equation

$$\frac{\partial S_2}{\partial t} + H\left(\frac{\partial S_2}{\partial q}, q, t\right) = \hat{H}(t)$$

We can absorb $\hat{H}(t)$ by defining another function

$$S(q,P,t;t_0) \equiv S_2(q,P,t;t_0) - \int_{t_0}^{t} dt' \; \hat{H}(t') \tag{2.18}$$

which satisfies

$$\frac{\partial S}{\partial t} + H\left(\frac{\partial S}{\partial q}, q, t\right) = 0 \tag{2.19}$$

and generates the same trajectories as S_2 since

$$\frac{\partial S}{\partial P}=\frac{\partial S_2}{\partial P}, \qquad \frac{\partial S}{\partial q}=\frac{\partial S_2}{\partial q}.$$

Eq. (2.19) is the Hamilton-Jacobi equation.

2.3. Appendix: Proof of Theorem 2.1

We first show that, independently of the nature of the transformation C, if the old variables satisfy Hamilton's equations

$$\dot{\eta}=J\frac{\partial H}{\partial \eta} \tag{2.20}$$

then the new variables $\xi=C(\eta,t)$ satisfy an evolution equation that can be written in the form

$$\dot{\xi}=J\ [L\frac{\partial h}{\partial \xi}+JM\frac{\partial C^{-1}}{\partial t}] \tag{2.21}$$

where

$$\begin{aligned} h(\xi,t)&=H(C^{-1}(\xi,t),t) \\ L&=-JMJM^T \end{aligned} \tag{2.22}$$

and M is the Jacobi matrix $M_{jk}=\partial\xi_j/\partial\eta_k$.

This equation can be obtained as follows. Taking the time derivative of $\eta=C^{-1}(\xi,t)$, we obtain

$$\dot{\eta}=M^{-1}\dot{\xi}+\frac{\partial C^{-1}}{\partial t} \tag{2.23}$$

The function h defined in (2.22) satisfies

$$\frac{\partial h}{\partial \xi_j}=\sum_k \frac{\partial H}{\partial \eta_k}\frac{\partial \eta_k}{\partial \xi_j}=\sum_k (M^{-1})_{kj}\frac{\partial H}{\partial \eta_k}$$

i.e.

$$\frac{\partial h}{\partial \xi}=(M^{-1})^T\frac{\partial H}{\partial \eta} \qquad or \qquad \frac{\partial H}{\partial \eta}=(M^T)\frac{\partial h}{\partial \xi} \tag{2.24}$$

Inserting into the inverse of (2.23) and using (2.20)(2.24) we obtain

$$\begin{aligned} \dot{\xi}&=M\dot{\eta}-M\frac{\partial C^{-1}}{\partial t} \\ &=MJM^T\frac{\partial h}{\partial \xi}-M\frac{\partial C^{-1}}{\partial t} \end{aligned} \tag{2.25}$$

which is (2.21). With this preparation we can now prove the equivalence of completely canonical and generalized symplectic transformations.

(i) We first prove that generalized symplectic implies completely canonical. This amounts to show that the expression inside the bracket in Eq. (2.21) is the gradient of some function. If C is generalized symplectic ($M^TJM=\lambda^2J$) then

$$L\equiv -JMJM^T=\lambda^2. \tag{2.26}$$

Thus the first term of (2.21) is a gradient, and it is left to show that it is also the case for $JM\partial C^{-1}/\partial t$. Using (2.26) we can write

$$JM\frac{\partial C^{-1}}{\partial t} = \lambda^2 (M^T)^{-1} J \frac{\partial C^{-1}}{\partial t}. \tag{2.27}$$

This expression will be the gradient of a function provided that

$$\frac{\partial}{\partial \xi_k}\Big((M^T)^{-1} J \frac{\partial C^{-1}}{\partial t}\Big)_j = \frac{\partial}{\partial \xi_j}\Big((M^T)^{-1} J \frac{\partial C^{-1}}{\partial t}\Big)_k. \tag{2.28}$$

The left-hand side can be rewritten as

$$\frac{\partial}{\partial \xi_k}\sum_n \frac{\partial \eta_n}{\partial \xi_j}\Big(J\frac{\partial C^{-1}}{\partial t}\Big)_n = \sum_n \frac{\partial^2 \eta_n}{\partial \xi_j \partial \xi_k}\Big(J\frac{\partial C^{-1}}{\partial t}\Big)_n + \sum_n \frac{\partial \eta_n}{\partial \xi_j}\frac{\partial}{\partial \xi_k}\Big(J\frac{\partial C^{-1}}{\partial t}\Big)_n$$

We see that the first term is symmetric in $j \leftrightarrow k$. In order to check it for the second term we rewrite it as

$$\sum_n \frac{\partial \eta_n}{\partial \xi_j}\frac{\partial}{\partial \xi_k}\Big(J\frac{\partial C^{-1}}{\partial t}\Big)_n = \sum_{n,m} \frac{\partial \eta_n}{\partial \xi_j} J_{nm} \frac{\partial}{\partial t}\frac{\partial \eta_m}{\partial \xi_k}$$

Doing the same with $j \leftrightarrow k$, the condition (2.28) can be expressed as

$$(M^{-1})^T J \frac{\partial M^{-1}}{\partial t} = \Big(\frac{\partial M^{-1}}{\partial t}\Big)^T J^T M^{-1}$$

which is satisfied, since

$$(M^{-1})^T J \frac{\partial M^{-1}}{\partial t} M^{-1} + \Big(\frac{\partial M^{-1}}{\partial t}\Big)^T J M^{-1} = \frac{\partial}{\partial t}\Big((M^{-1})^T J M^{-1}\Big) = \frac{\partial}{\partial t}(\lambda^{-2} J) = 0 \tag{2.29}$$

(ii) We now prove the implication in the opposite direction. Since completely canonical means that Hamilton's equations are invariant for all choices of Hamiltonians, we will take three different ones with special forms, in order to deduce the necessity of the generalized symplectic condition. The claim is that the condition

$$L\frac{\partial h}{\partial \xi} + JM\frac{\partial C^{-1}}{\partial t} \qquad \textit{is a gradient for all choices of } h \tag{2.30}$$

implies that L is equal to a constant times the identity. We proceed as follows:

1) We start with the choice $h(\xi, t) = h_0 = constant$. Then the first term in (2.30) vanishes, and therefore $JM\partial C^{-1}/\partial t$ must necessarily be a gradient.

2) As a consequence, $L\partial h/\partial \xi$ must also be a gradient for arbitrary h. We now choose $h(\xi,t) = h_1(\xi) = \sum_j b_j \xi_j$, with contants b_j. Then the condition implies that $L\underline{b}$ is a gradient, and therefore

$$\frac{\partial}{\partial \xi_k}\sum_m L_{jm} b_m = \frac{\partial}{\partial \xi_j}\sum_m L_{km} b_m$$

If we now make a series of specific choices $b_m = \delta_{m,r}$, $r = 1,\dots 2N$, we obtain

$$\frac{\partial L_{jr}}{\partial \xi_k} = \frac{\partial L_{kr}}{\partial \xi_j}, \qquad \forall j,k,r. \tag{2.31}$$

3) Next, we consider a choice of the form $h(\xi,t) = h_2(\xi) = \sum_{j,k} S_{jk}\xi_j\xi_k$, where S is a constant symmetric matrix. The condition implies here that $LS\xi$ is a gradient. Therefore

$$\frac{\partial}{\partial \xi_k}\sum_{l,m} L_{jl}S_{lm}\xi_m = \frac{\partial}{\partial \xi_j}\sum_{l,m} L_{kl}S_{lm}\xi_m$$

or

$$\sum_{l,m}\Big(\frac{\partial L_{jl}}{\partial \xi_k}S_{lm}\xi_m + L_{jl}S_{lm}\delta_{k,m}\Big) = \sum_{l,m}\Big(\frac{\partial L_{kl}}{\partial \xi_j}S_{lm}\xi_m + L_{kl}S_{lm}\delta_{j,m}\Big)$$

The first terms of the two sides are equal because of Eq. (2.31), and therefore

$$\sum_l L_{jl}S_{lk} = \sum_l L_{kl}S_{lj} \tag{2.32}$$

i.e. $LS = (LS)^T$. If in particular we take $S = 1$, we obtain $L = L^T$. On the other hand, if we take the series of choices $S_{ij} = \delta_{i,j}\delta_{j,r}$, $r = 1,\dots 2N$, we obtain

$$L_{jk}\delta_{k,r} = L_{kj}\delta_{j,r}$$

For $r = j \neq k$ this implies $L_{kj} = 0$, and thus L is diagonal.

With this information Eq. (2.32) becomes $L_{jj}S_{jk} = L_{kk}S_{kj}$ and since $S_{jk} = S_{kj}$, we obtain $L_{jj} = L_{kk}$, i.e. L can be written as

$$L \equiv -JMJM^T = a(\xi,t)1 \tag{2.33}$$

where a could in principle still depend on ξ and on t. But Eq. (2.31) can now be rewritten as

$$\frac{\partial a}{\partial \xi_k}\delta_{jr} = \frac{\partial a}{\partial \xi_j}\delta_{kr}$$

and thus for $j = r \neq k$ we obtain $\partial a/\partial \xi_k = 0$, i.e. a is independent of ξ.

4) We still have to show that a must be independent of t. We first show that a(t) would not have any zeros and therefore $1/a$ is well defined: From (2.33) it follows that $det\, L = -(det\, M)^2 = (a(t))^{2N}$. Since by hypothesis $C(\eta,t)$ is invertible for all t, $det\, M \neq 0$, and thus $a(t) \neq 0$, $\forall t$.

We know from 1) that $JM\partial C^{-1}/\partial t$ must be a gradient. Through the calculation (2.27)–(2.29), with $a(t)$ in place of λ^2 this implies that

$$\frac{\partial}{\partial t}((M^{-1})^T JM^{-1}) = 0$$

Taking the inverse of both sides of (2.33) we can write $(M^{-1})^T JM^{-1} = a(t)^{-1}J$. Therefore $\partial a^{-1}/\partial t = 0$, i.e. a is independent of t. This completes the proof of the theorem.

3. Symplectic Integration of Hamiltonian's Equations

The numerical solution of differential equations involves a discretization of the time variable. The integration is then performed by the iteration of a map. In the simplest case the iteration maps one point of the (approximate) trajectory to the next. More elaborate schemes use several points or several intermediate steps to calculate each new point (like the fourth order Runge-Kutta algorithm). One problem encountered when these methods are applied to Hamiltonian systems is that the discretized map of the usual algorithms does not preserve volume in phase space, i.e. the map produces a spurious dissipation or antidissipation. If one is interested in short time properties this effect can be negligible if one takes small enough time steps. However, if one is interested in the qualitative properties of long time behavior, as is often the case in the study of chaotic dynamics, it is important to use approximating maps that conserve volume in phase space. In fact, as we have seen the Hamiltonian evolution should conserve not only volume but also the other Poincaré invariants. This can be summarized by saying that the discretized map should be a canonical transformation. As we have discussed, a systematic way to construct canonical transformations is by deducing them from a generating function. In recent years a variety of schemes have been proposed [5 - 11]. There is no single algorithm that is best for a general Hamiltonian. Here we will discuss some algorithms that are applicable when the Hamiltonian is a sum of a kinetic term depending only on the momenta p and a potential term depending only on the position q. The main difficulty in more general Hamiltonians with mixed dependence on p and q is that in the canonical transformation constructed from a generating function, a function must be inverted in order to get the old variables explicitly as functions of the new ones. While this inversion can be performed analytically in the case with separate kinetic and potential terms, it must be performed numerically in the general case. The algorithms become then much more involved, in particular if they are required to be of higher order in the time step. The more complicated cases are discussed [10, 9]. There are also algorithms whose construction doesn't involve explicit use of generating functions. One of these methods is based on the traditional Runge-Kutta construction, with a set of free parameters that are chosen such that the final map is symplectic. This approach is reviewed in Ref.11.

3.1. Infinitesimal Generator of the Time Evolution, First Order Algorithms.

We start by discussing a first order algorithm that can be easily constructed following our discussion of Section 2. We develop the generating function (2.18) as

$$S(q,P,t;t_0) = S^{(0)}(q,P) + (t-t_0)S^{(1)}(q,P;t_0) + \mathcal{O}((t-t_0)^2). \tag{3.1}$$

$S^{(0)}$ must generate the identity; therefore $S^{(0)} = Pq$. The first order term is given by

$$S^{(1)}(q,P;t_0) = -H(P,q,t_0) \tag{3.2}$$

i.e. the original Hamiltonian evaluated at $q = q(t)$, $p = p(t_0) \equiv P$ and $t = t_0$. The result (3.2) can be obtained as follows: Inserting the series (3.1) into (2.15) we obtain

$$\begin{aligned} p &= \frac{\partial S}{\partial q} = P + (t - t_0)\frac{\partial S^{(1)}}{\partial q} + \mathcal{O}((t - t_0)^2) \\ Q &= \frac{\partial S}{\partial P} = q + (t - t_0)\frac{\partial S^{(1)}}{\partial P} + \mathcal{O}((t - t_0)^2). \end{aligned} \tag{3.3}$$

Using this expression we compute

$$\begin{aligned} \dot{p}(t_0) &= \lim_{t\to t_0} \frac{p - P}{t - t_0} = \frac{\partial S^{(1)}}{\partial q}(q, P; t_0) \\ \dot{q}(t_0) &= \lim_{t\to t_0} \frac{q - Q}{t - t_0} = -\frac{\partial S^{(1)}}{\partial P}(q, P; t_0). \end{aligned} \tag{3.4}$$

On the other hand, Hamilton's equations are

$$\begin{aligned} \dot{p}(t_0) &= -\frac{\partial H}{\partial q}(p, q, t_0) \\ \dot{q}(t_0) &= \frac{\partial H}{\partial p}(p, q, t_0). \end{aligned} \tag{3.5}$$

Comparison of (3.4) and (3.5) yields the result (3.2). The function

$$\tilde{S}(q, P, t; t_0) \equiv Pq - (t - t_0)H(P, q, t_0)$$

satisfies the Hamilton-Jacobi equation to first order, i.e. the error consists of terms of order $\mathcal{O}((t - t_0)^2)$. The canonical transformation generated by $\tilde{S}$ provides a symplectic integration algorithm that is of first order in δt. We obtain the explicit expression by inserting into (3.3)

$$p = P - (t - t_0)\frac{\partial H}{\partial q}(P, q, t_0) \tag{3.6}$$

$$Q = q - (t - t_0)\frac{\partial H}{\partial P}(P, q, t_0) \tag{3.7}$$

In order to obtain the explicit map

$$\begin{aligned} Q &\equiv q(t_0) \mapsto q(t) \\ P &\equiv p(t_0) \mapsto p(t) \end{aligned}$$

we need to invert Eq. (3.7) and insert into (3.6). In general this cannot be done analytically in closed form and one must resort to a numerical inversion. However for the usual type of Hamiltonians that are a sum of kinetic and potential energy, or more generally that can be separated as a sum of two terms $H = T(p) + V(q, t)$ the above inversion can be done explicitly:

Defining $g(p) = \partial T/\partial p$ and $f(q, t) = -\partial V(q, t)/\partial q$ we can write

$$p(t) = p(t_0) + (t - t_0)f(q(t), t_0) + \mathcal{O}((t - t_0)^2)) \tag{3.8}$$

$$q(t) = q(t_0) + (t - t_0)g(p(t_0)) + \mathcal{O}((t - t_0)^2)) \tag{3.9}$$

Remark that in (3.8) the force is evaluated at the evolved position $q(t)$, instead of at $q(t_0)$ as one would do with a simple first order Euler discretization of Hamilton's equations. This difference is crucial: the Euler algorithm is not volume preserving, while (3.8) and (3.9) is fully symplectic.

Remark. The choice of the approximating first order symplectic map is certainly not unique. In fact in the literature one finds often a similar but different algorithm, obtained by using a generating function of type $F_3(p, Q, t; t_0)$ instead of $S = F_2(q, P, t; t_0)$. Expanding it to first order we get by an analogous procedure

$$F_3(p, Q, t; t_0) = -pQ - (t - t_0)H(p, Q, t_0)$$

which can be truncated to generate a symplectic map that gives a first order approximation of the form

$$\begin{aligned} p(t) &= p(t_0) + (t - t_0)f(q(t_0), t_0) \\ q(t) &= q(t_0) + (t - t_0)g(p(t), t_0) \end{aligned} \tag{3.10}$$

In this version it is the function g that is evaluated at the evolved momentum $p(t)$ (as opposed to the force f in the previous version (3.9)).

Example: We consider the Hamiltonian of a pendulum

$$H = \frac{1}{2}p^2 + \gamma \cos 2\pi q \tag{3.11}$$

If we take $t_0 = 0$, discretize time as $t_n = n\delta t$, and define $p_n = p(t_n)$, $q_n = q(t_n)$, the symplectic map (3.10) can be written as

$$\begin{aligned} p_{n+1} &= p_n - \delta t \; 2\pi\gamma \sin 2\pi q_n \\ q_{n+1} &= q_n + \delta t \; p_{n+1} \end{aligned} \tag{3.12}$$

By a redefinition of the variable p_n this becomes identical to the so-called standard map [12], which will be an important example in coming sections: Defining

$$\hat{p}_n = \delta t \; p_n, \quad and \quad \kappa = (\delta t)^2 2\pi\gamma$$

we can write

$$\begin{aligned} \hat{p}_{n+1} &= \hat{p}_n - \kappa \sin 2\pi q_n \\ q_{n+1} &= q_n + \hat{p}_{n+1}. \end{aligned} \tag{3.13}$$

3.2. Higher Order Algorithms

A method to construct higher order integration that was proposed by Ruth [5] consists of combining several first order steps. Each first order step is allowed to depend on some arbitrary parameters. At the end, when the single steps are combined, the parameters are chosen such that all the terms that are lower than the desired order disappear. We first illustrate the idea with a second order algorithm.

An equivalent way of checking that the algorithms that we have discussed in the previous subsection are of first order in δt is by showing the new Hamiltonian obtained by canonical transformation with generating function $\tilde{F}_3 = -pQ - (t - t_0)H(p, Q, t_0)$ is of order $\mathcal{O}(t - t_0)$:

$$\begin{aligned}\hat{H} &= H(p,q,t) + \frac{\partial \tilde{F}_3}{\partial t} \\ &= H(p,q,t) - H(p,Q,t_0) \\ &= V(Q + (t - t_0)g(P), t) - V(Q, t_0) \\ &= (t - t_0)g(P)\frac{\partial V}{\partial q}(Q, t_0) + (t - t_0)\frac{\partial V}{\partial t}(Q, t_0) + \mathcal{O}((t - t_0)^2)\end{aligned}$$

One can state in general that if the new Hamiltonian $\hat{H}$ is of order $\mathcal{O}((t - t_0)^k)$ then the transformation defines an approximation of order $\mathcal{O}((t - t_0)^k)$. This follows from the fact that the right-hand side of Hamilton's equations for P,Q will also be of $\mathcal{O}((t - t_0)^k)$ and thus P and Q are of $\mathcal{O}((t - t_0)^{k+1}$. Together with the relation (3.10) this implies the claim.

To construct a second order map we compose two transformations

$$(p, q) \mapsto (P_1, Q_1) \mapsto (P_0, Q_0)$$

with generating functions

$$\begin{aligned}F_3^{(1)}(p, Q_1) &= -pQ_1 - a(t - t_0)T(p) - (t - t_0)V(Q_1, t_0 + b(t - t_0)) \\ F_3^{(2)}(P_1, Q_0) &= -P_1Q_0 - (1 - a)(t - t_0)T(P_1)\end{aligned}$$

i.e.

$$\begin{aligned}q &= Q_1 + a(t - t_0)g(p), & P_1 &= p - (t - t_0)f(Q_1, t_0 + b(t - t_0)) \\ Q_1 &= Q_0 + (1 - a)(t - t_0)g(P_1), & P_0 &= P_1\end{aligned}$$

where a and b are parameters to be determined. The new Hamiltonian obtained after the two transformations, after expansion in powers of $(t - t_0)$ is of the form

$$\hat{\hat{H}} = (t - t_0)(1 - 2a)g(P_0)f(Q_0, t_0) + (t - t_0)(1 - 2b)\frac{\partial}{\partial t}V(Q_0, t_0) + \mathcal{O}((t - t_0)^2).$$

Thus, if we choose the parameters as $a = \frac{1}{2}$ and $b = \frac{1}{2}$ the new Hamiltonian $\hat{\hat{H}}$ will be of order $\mathcal{O}((t - t_0)^2)$. After the necessary inversions and identifying $(p, q) = (p(t), q(t))$ and

$(P_0, Q_0)) = (p(t_0), q(t_0))$ the following second order symplectic algorithm that involves two steps:

1) $P_1 = p(t_0), \quad Q_1 = q(t_0) + \frac{1}{2}\delta t \; g(P_1)$

2) $p(t) = P_1 + \delta t f(Q_1, t_0 + \frac{1}{2}\delta t), \quad q(t) = Q_1 + \frac{1}{2}\delta t \; g(p(t))$.

This scheme is known as the second order Verlet or "leapfrog" algorithm. We remark that although it involves two steps, there is only one evaluation of the force f, which is often the most time consuming operation.

Forest and Ruth [6] have generalized the preceding construction (for $V(q)$ independent of t) to a procedure that can provide in principle algorithms of arbitrary order k. It consists of the composition of k transformations

$$(p(t_0), q(t_0)) \equiv (P_0, Q_0) \leftarrow (P_1, Q_1) \leftarrow \ldots \leftarrow (P_k, Q_k) \equiv (p(t), q(t))$$

defined by

$$P_{i+1} = P_i + c_i \delta t \; f(Q_i)$$
$$Q_{i+1} = Q_i + a_i \delta t \; g(P_{i+1}).$$

The condition that the transformed Hamiltonian is of order $O(\delta t^k)$ is expressed by a system of algebraic equations for the coefficients c_i, a_i. The previous example with $k = 2$ corresponds to

$$c_1 = 0, \quad a_1 = \frac{1}{2}, \quad c_2 = 1, \quad a_2 = \frac{1}{2}.$$

For $k = 3$ a solution can still be found by hand [5]

$$\begin{array}{lll} c_1 = 7/24, & c_2 = 3/4, & c_3 = -1/24 \\ a_1 = 2/3, & a_2 = -2/3, & a_3 = 1. \end{array}$$

For larger k the equations become very complicated. However, for $k = 4$ Forest and Ruth [6], and independently Candy [13], have found a solution, making extensive use of computer algebra programs. Their result is

$$\begin{array}{llll} c_1 = 0, & c_2 = 2\alpha + 1, & c_3 = -4\alpha - 1, & c_4 = 2\alpha + 1 \\ a_1 = \alpha + 1/2, & a_2 = -\alpha, & a_3 = -\alpha, & a_4 = \alpha + 1/2, \end{array}$$

where $\alpha = (2^{1/3} + 2^{-1/3} - 1)/6$.

3.3. Symplectic Algorithms with Exact Integer Arithmetics

In the algorithms we have discussed the maps are symplectic provided that they are calculated with absolute precision. However in the practical implementations using floating point arithmetics there are always rounding errors that spoil the symplectic character. These errors are small, but they can sum up and produce extraneous artifacts in the long time behavior. This effect can be prevented by modifying the maps so that they can be implemented with exact integer arithmetics. This method was pioneered by Rannou [14] and has recently found applications particularly in astronomy [15-18].

The idea is to approximate the maps we have found by other symplectic maps that have the property of mapping a lattice with integer coordinates into itself. For the first order Verlet algorithm, Eq. (3.10), and for a Hamiltonian $H = \frac{1}{2}p^2 + V(q,t)$ this can be done as follows [15]: First we choose a lattice of points for the coordinates p and q such that there are k points per unit. Then we choose the time step such that $k\delta t$ is an integer. Then the map

$$kp(t+\delta t) = kp(t) - [\![k\ \delta t\ \frac{\partial V}{\partial q}(q(t),t)]\!]$$
$$kq(t+\delta t) = kq(t) + k\ \delta t\ p(t+\delta t)$$

where the brackets indicate the nearest integer vector. This algorithm is well-defined for arbitrary initial conditions, but it has the special property that if the initial conditions are such that $kp(t_0)$ and $kq(t_0)$ have integer coordinates, then this will also be the case for all points obtained by iteration. It is clear that by taking k large enough one can approximate the original map to any desired accuracy.

Remark: If we consider situations in which the trajectories stay in a bounded region of phase space, this integer arithmetics representation has the apparently restrictive feature that there is only a finite number of points that a trajectory can visit. As a consequence, all trajectories are eventually periodic, i.e. they all end up following a cycle. One can make two comments: first, by taking k large the (fictive) periods of this orbits can be made arbitrarily large. Second, even when one uses floating point arithmetics the number of possible points is finite, since only finitely many numbers can be represented in this way. The difference is that using the integer arithmetics we have control over which points are represented and we can guarantee the symplectic character of the map restricted to the lattice. Adaptations of higher order symplectic algorithms to exact integer arithmetics have been treated in [19].

Remark: All the algorithms that we have discussed are based on approximations of the true flow by a map. The requirement that the map is symplectic guarantees that certain qualitative properties are conserved by the approximation. However, since it is only an approximation, for any chosen fixed time step δt there will be some differences between the true and the computed flow. These differences can even be of qualitative nature as illustrated by the following example. We consider the integration of the trajectories of a pendulum (3.11) using the first order Verlet algorithm (3.12). There are two easily detected qualitative differences:

1) Besides being periodic in q_n (period 1), the map (3.12) is also periodic in p_n, with period $1/\delta t$. One sees immediately that if the initial conditions are shifted as $(p_0, q_0) \mapsto (p_0 + 1/\delta t, q_0)$ then the whole trajectories are just shifted by $1/\delta t$ in the p direction. This is certainly not a property of the true flow of the pendulum. The approximation (3.12) can therefore only be a good approximation for orbits in which p stays small compared to $1/\delta t$.

2) As already mentioned, the approximating map coincides with the standard map (3.13). This map is not integrable, and as we will discuss in section 4, as soon

as $\kappa \equiv (\delta t)^2\gamma 2\pi \neq 0$ it has chaotic behavior, which is not present in the pendulum. In every neighborhood of phase space there will be spurious chaotic orbits. However if δt is small, the chaotic orbits are sparse among the regular ones and they are only observed by looking at very fine scales.

4. Structure and Chaos in Classical Phase Space

In this section we discuss some of the qualitative phenomena that one observes in the dynamics of classical Hamiltonian systems [20, 21]. We will concentrate on systems with two degrees of freedom for two reasons: because they are easy to visualize, and because some important parts, like the Aubry-Mather theory, are not developed for higher dimensions.

As illustrations we will use a system of two coupled rotators, or pendula, e.g.

$$H = \frac{1}{2}p_1^2 + \frac{1}{2}p_2^2 + c_1 \cos\varphi_1 + c_2 \cos\varphi_2 + c_{12}\cos(\varphi_1 - \varphi_2) \tag{4.1}$$

and a pendulum driven by a time periodic force

$$H = \frac{1}{2}p_1^2 + c_1 \cos\varphi_1 + c_{12}\cos(\varphi_1 - \omega_2 t) \tag{4.2}$$

This second type of problem can be transformed into a time independent one by adding a new dynamical variable $\varphi_2(t) \equiv \omega_2 t + \varphi_2(0)$ and its corresponding canonically conjugate momentum p_2. This leads to a Hamiltonian of the type

$$H = \frac{1}{2}p_1^2 + \omega_2 p_2 + V(\varphi_1, \varphi_2).$$

We will therefore consider Hamiltonians of the general form

$$H = \frac{1}{2}pMp + \omega \cdot p + V(\varphi) \tag{4.3}$$

where $\varphi = (\varphi_1, \varphi_2)$ represents two angles, p the corresponding conjugate momenta, ω is a constant vector and M a positive definite symmetric matrix. This type of models have been widely investigated [22, 23], and have natural applications in plasma physics and in particle accelerators [24, 25].

4.1. Poincaré Maps

The dynamics of systems with two degrees of freedom can be easily visualized using Poincaré maps. There is no general prescription on how to construct a Poincaré map for a general Hamiltonian, since in order to do it one needs some information on the general structure of the orbits [26]. However, some methods have been developed that apply to several classes of models. For the type of examples (4.3) it can be constructed as follows: In the four-dimensional phase space with coordinates $(p_1, p_2, \varphi_1, \varphi_2)$ we define a Poincaré section S_P as the surface defined by the conditions

$$\begin{aligned} H(p,\varphi) &= E = const \\ \varphi_2 &= const \quad (e.g. = 0). \end{aligned} \tag{4.4}$$

The points of the Poincaré section can be parametrized by the coordinates p_1, φ_1 since if they are given, Eq. (4.4) determines the value of p_2. The Poincaré map $M_P : S_P \to S_P$ is then defined as the first return map on S_P: the image of a point $(p_1, \varphi_1) \in S_P$ is the point through which the Hamiltonian flow with initial condition $(p_1, p_2, \varphi_1, \varphi_2 = 0)$ crosses again S_P for the first time (coming from the same side).

It is easy to show [27] that the Poincaré map of a Hamiltonian system is an area preserving map. Furthermore, two-dimensional area (and orientation) preserving maps are automatically symplectic. This is one of the reasons why Poincaré maps are so useful to understand the structure of the dynamics in phase space; they carry the same essential symplectic structure as the Hamiltonian flow.

4.2 Twist Maps

Although it is numerically straightforward to calculate the Poincaré map for a given Hamiltonian (4.3), it is usually not possible to obtain explicit analytical expressions. If one wants to study the general qualitative phenomena it is often convenient to study directly an area preserving map. A particularly convenient type are so-called *twist maps*, that are defined as follows: An area-preserving map

$$\begin{pmatrix} r \\ \varphi \end{pmatrix} \to \begin{pmatrix} g(r,\varphi) \\ f(r,\varphi) \end{pmatrix}$$

where φ is the angle and $r \subset [R_1, R_2]$ is called a *monotonic twist map* if

$$\frac{\partial f(r,\varphi)}{\partial r} \neq 0$$

We can choose for definiteness

$$\partial f / \partial r > 0. \tag{4.5}$$

An often used example of monotonic twist map is the standard map (3.13). The map $\begin{pmatrix} r_n \\ \varphi_n \end{pmatrix} \mapsto \begin{pmatrix} r_{n+1} \\ \varphi_{n+1} \end{pmatrix}$ can be constructed from a generating function $F(\varphi_n, \varphi_{n+1})$, e.g. depending on the old and the new angles:

$$r_{n+1} = \frac{\partial F}{\partial \varphi_{n+1}}(\varphi_n, \varphi_{n+1}) \tag{4.6}$$

$$r_n = -\frac{\partial F}{\partial \varphi_n}(\varphi_n, \varphi_{n+1}) \tag{4.7}$$

The map is recovered by first inverting Eq. (4.7) which gives $\varphi_{n+1} = f(r_n, \varphi_n)$, and then inserting this relation into (4.6). The twist condition (4.5) translates into

$$\frac{\partial^2 F}{\partial \varphi_n \partial \varphi_{n+1}} < 0.$$

There is a close relation between monotonic twist maps and mechanical systems with two degrees of freedom. It has been shown [28] that every monotonic twist map is the Poincaré map of some Hamiltonian flow. The opposite is however not true. A general Poincaré map does not necessarily satisfy the monotonicity condition (4.5) [29].

4.3. Qualitative Features of Phase Portraits

Figures 1 and 2 show typical pictures of the dynamics of Hamiltonian systems. The first one shows orbits of the Poincaré map of a periodically driven pendulum (4.2) with $c_1 = c_{12} = 0.025$, and the second one shows orbits of the standard map for $\kappa = 0.8$. Our aim is to discuss some concepts that will allow us to interpret the different features seen in these figures.

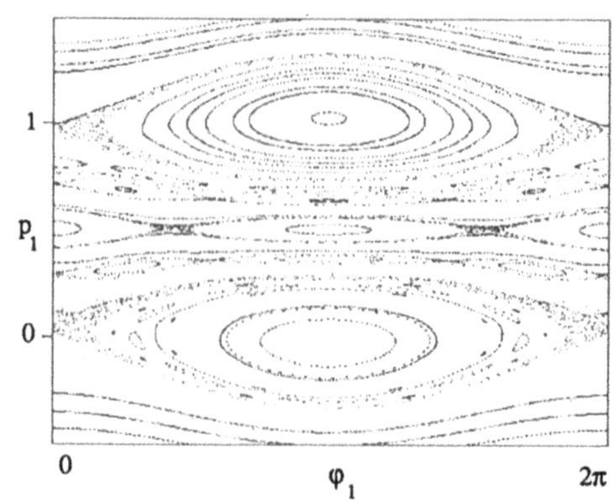

Figure 1

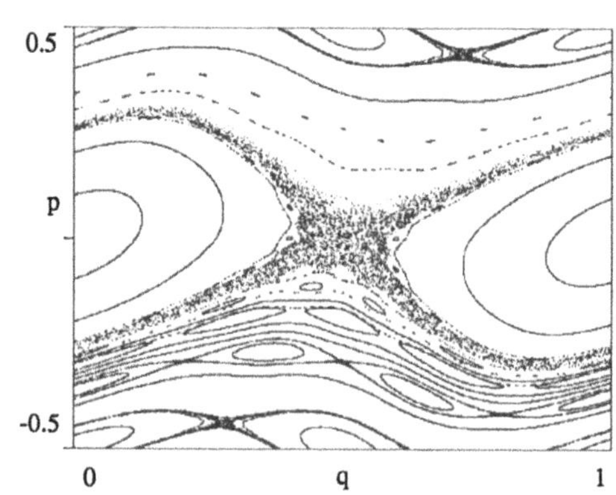

Figure 2

A Hamiltonian system is said to be integrable if it has as many independent conserved quantities I_j in involution (i.e. with vanishing Poisson brackets, $\{I_i, I_j\} = 0$) as degrees of freedom. The dynamics of such systems is simple, since one can find a canonical transformation into action-angle (A, φ) coordinates, in which the new Hamiltonian depends only on the actions [1 2]. $\hat{H} = \hat{H}(A)$. All trajectories lie on invariant tori, since they can be written in the form

$$A(t) = A(0) = const$$
$$\varphi(t) = \varphi(0) + \omega t; \qquad \omega = \omega(A) = \frac{\partial \hat{H}}{\partial A}.$$

One says therefore that the phase space of integrable systems is foliated by invariant tori. Hence, in integrable systems there can be the following kinds of orbits:

- periodic orbits (if $\alpha = \omega_1/\omega_2$ is rational),
- quasiperiodic orbits (if $\alpha = \omega_1/\omega_2$ is irrational),
- fixed points of elliptic, hyperbolic or parabolic character,
- homoclinic and heteroclinic orbits (separatrices) joining unstable fixed points.

All types evolve on invariant tori (if one considers the case of homoclinic and heteroclinic orbits as degenerated limits of tori).The quasi-periodic orbits cover the torus densely.

Integrable twist maps can be written as

$$\begin{pmatrix} r_{n+1} \\ \varphi_{n+1} \end{pmatrix} = \begin{pmatrix} r_n \\ \varphi_n + f(r_n) \end{pmatrix}. \tag{4.8}$$

To each orbit one can assign a rotation number α by

$$\alpha = \frac{1}{2\pi} \lim_{n \to \infty} \varphi_n$$

which plays the same role as the quotient ω_1/ω_2 of the continuous case.

It was shown by Poincaré that if an integrable system is perturbed generically, the system does not have a complete system of conserved quantities anymore. One calls such systems non-integrable. In non-integrable systems all the types of orbits mentioned can be present, but at least two more types of orbits appear. The first type is strongly chaotic while the second type is quasi-periodic but only in a generalized sense, since instead of generating a torus they generate a Cantor set. In the following subsections we give a more detailed characterization of the types of orbits encountered in non-integrable systems. We will discuss the results in terms of continuous flows or alternatively in terms of maps.

4.4. Quasiperiodic Orbits: K.A.M. Theory

4.4.1. Setup

We consider a particular class of integrable systems characterized by the property that the Hamiltonian expressed in action angle variables is quadratic in the actions

$$H_0 = \frac{1}{2}AMA + \omega A \tag{4.9}$$

From the equations of motion of the unperturbed problem $\dot{\varphi} = MA + \omega;\ \dot{A} = 0$ we see that all the trajectories are of the form

$$A(t) = A(0) = const.; \qquad \varphi(t) = \varphi(0) + \omega_0 t; \qquad with \quad \omega_0 = MA + \omega \tag{4.10}$$

i.e. for each choice of A the trajectory winds around a torus with a frequency vector ω_0. If the quotient of the two components of this vector $\alpha \equiv \omega_{01}/\omega_{02}$ is an irrational number, then the orbit covers the torus densely. A rotation number α is called diophantine if there exist constants c_1 and $\sigma > 1$ such that for all $n, m \in Z$

$$|\alpha n - m| > c_1/n^\sigma. \tag{4.11}$$

The K.A.M. theorem states that for those irrational α's that satisfy a diophantine condition (4.11) these tori are still present when the system is subjected to a small and sufficiently smooth perturbation. The proof of this theorem is constructive [30, 31]. We will discuss how the explicit construction of the perturbed torus can be performed for systems of the form (4.9). The algorithm is based on a proof of the K.A.M. theorem given by W. Thirring [31]. We will describe it for the case of two degrees of freedom, but it can be in principle extended to arbitrary dimension with only small modifications. The method can be formulated as follows:

One concentrates on a single torus with fixed frequency vector $\omega_0 = (\omega_{01}, \omega_{02})$ and one wants to construct a canonical transformation such that in the new coordinates there is an orbit of the form (4.10). This orbit generates the torus since it wraps densely around it. This canonical transformation cannot be defined directly. If one tries to do so, the formal expressions that appear in the classical perturbation theory do not converge due to the presence of "small denominators". The construction is done by the composition of a sequence of transformations. To be more concrete, we consider the torus with frequency vector ω_0 corresponding to the unperturbed H_0, and follow its deformation as the perturbation is turned on.

We first perform a trivial shift in the action variables: $A = A' + a$ where a is determined by the condition $Ma + \omega = \omega_0$, so that in the new variables the unperturbed Hamiltonian is $H_0 = \frac{1}{2}A'MA' + \omega_0 \cdot A'$ and the torus with frequency vector ω_0 corresponds to the value of the action $A' = 0$. (To simplify the notation we will omit the prime in A'). We consider now the perturbed Hamiltonian of the form

$$H_1 = \frac{1}{2}AMA + \omega_0 \cdot A + \varepsilon V(\varphi) \tag{4.12}$$

The algorithm proceeds by iteration. In the first step the Hamiltonian H_1 which has a perturbation of order ε is transformed (by a canonical transformation $((A,\varphi) \mapsto (A_2,\varphi_2))$ into a new Hamiltonian in which the perturbation is only of order ε^2. Iterating the procedure n times reduces the perturbation to order ε^{2^n}. When the iteration converges one can thus completely eliminate the perturbation. The composition of the transformations performed at each step yields the deformed torus with the chosen frequency. The new Hamiltonian H_2 which, by definition is H_1 but expressed as a function of the new coordinates, has in general a much more complicated form then the original one (including e.g. arbitrary powers of A). The advantage of the systems that have the special form (4.9) is that at each step the canonical transformation can be chosen such that all the Hamiltonians that appear in the iteration are of the same form, and in particular their dependence on the action is not higher than quadratic. They have the general form

$$H = \frac{1}{2}A[M + m(\varphi)]A + [\omega_0 + g(\varphi)] \cdot A + f(\varphi) \tag{4.13}$$

where $g(\varphi)$ and $m(\varphi)$ are a φ-dependent vector and a symmetric positively definite matrix respectively, with zero average $\int d^2\varphi \; g(\varphi) = 0 = \int d^2\varphi \; m(\varphi)$. The goal of the iteration is to construct a canonical transformation such that, when expressed in the final coordinates the Hamiltonian takes the form

$$H_\infty = \frac{1}{2}A_\infty[M_\infty + m_\infty(\varphi_\infty)]A_\infty + \omega_0 \cdot A_\infty \tag{4.14}$$

The corresponding equations of motion are

$$\begin{aligned} \frac{d}{dt}\varphi_\infty &= [M_\infty + m_\infty(\varphi_\infty)]A_\infty + \omega_0 \\ \frac{d}{dt}A_\infty &= -\frac{1}{2}A_\infty[\frac{\partial m_\infty(\varphi_\infty)}{\partial \varphi_\infty}]A_\infty \end{aligned} \tag{4.15}$$

For the initial condition $A_\infty = 0$ this system has an invariant torus of frequency vector ω_0 generated by the orbits

$$\varphi_\infty(t) = \varphi_\infty(0) + \omega_0 t; \qquad A_\infty(t) = 0.$$

We remark that since the torus is at $A_\infty = 0$, the φ-dependent matrix function $m(\varphi)$ does not influence it (although it certainly influences the behavior in other regions of phase space), and therefore it is not necessary to eliminate m by canonical transformation as it is done with f and g. This is what makes it possible to maintain the simple quadratic form (4.13) all along the iteration.

4.4.2. The Iteration

We start with a Hamiltonian of the form (4.13), and we treat f and g as quantities of order ε. We perform a canonical transformation

$$A = \frac{\partial S}{\partial \varphi} = A_2 + a + J^T(\varphi)A_2 + \frac{\partial \phi}{\partial \varphi} \tag{4.16}$$

$$\varphi_2 = \frac{\partial S}{\partial A_2} = \varphi + Y(\varphi) \tag{4.17}$$

defined by the generating function

$$S(A_2, \varphi) = \varphi \cdot (A_2 + a) + Y(\varphi) \cdot A_2 + \phi(\varphi) \tag{4.18}$$

where the functions $\phi(\varphi)$,$Y(\varphi)$ and the constant a are to be determined, and J^T is the transposed of the Jacobi matrix ($J_{kl} = \partial Y_k/\partial\varphi_l$). It is easy to verify that if one starts with even functions f,g and m, they will stay even along the iteration. We can use the natural Fourier representation

$$f(\varphi) = \sum_{n\in Z^2} \tilde{f}(n) \cos n \cdot \varphi, \quad g(\varphi) = \sum_{n\in Z^2} \tilde{g}(n) \cos n \cdot \varphi, \quad m(\varphi) = \sum_{n\in Z^2} \tilde{m}(n) \cos n \cdot \varphi$$

Inserting (4.16) into (4.13) we obtain

$$H = \frac{1}{2}A_2[M_2 + m_2]A_2 + [\omega_0 + g_2] \cdot A_2 + f_2 \tag{4.19}$$

with

$$\begin{aligned} M_2 + m_2 &= V^T[M + m]V \\ \omega_0 + g_2 &= V^T[\omega_0 + g + b(M + m)] \\ f_2 &= f + \frac{1}{2}b(M + m)b + [\omega_0 + g]b \end{aligned} \tag{4.20}$$

where $b \equiv a + \partial\phi/\partial\varphi$ and $V \equiv 1 + J^T$.

We now determine 1) the function ϕ such that f_2 is of order ε^2, then 2) Y such that g_2 is of order ε^2, and 3) a such that g_2 has zero average (f_2 is made of zero average by a trivial subtraction of a constant from the Hamiltonian). We recall for this purpose the order of the different quantities: M and m are of order one; g and f are of order ε; as a consequence ϕ, Y, J and a are of order ε.

1) The condition for f_2 to be of order ε^2 is thus

$$f + \omega_0 \cdot \frac{\partial \phi}{\partial \varphi} = 0 \tag{4.21}$$

or in Fourier representation $\tilde{f}(n) + \omega_0 \cdot n\tilde{\phi}(n) = 0$, which yields the solution

$$\phi(\varphi) = -\sum_{n\neq(0,0)} \tilde{f}(n)\frac{1}{n\cdot\omega_0} \sin n \cdot \varphi \tag{4.22}$$

2) The condition for g_2 to be of order ε^2 is

$$J\omega_0 + g + (M+m)b = 0 \tag{4.23}$$

Again using the Fourier representation the solution is obtained as

$$\begin{aligned} Y &= -\sum_{n\neq(0,0)} \tilde{U}(n)\frac{1}{n\cdot\omega_0}\sin n\cdot\varphi \\ J_{kl} &= -\sum_{n\neq(0,0)} \tilde{U}_k(n)\frac{n_l}{n\cdot\omega_0}\cos n\cdot\varphi \end{aligned} \tag{4.24}$$

where $\tilde{U}(n)$ are the Fourier coefficients of

$$U(\varphi) = g + (M+m)\frac{\partial\phi}{\partial\varphi} - < m\frac{\partial\phi}{\partial\varphi} > + ma \tag{4.25}$$

and the brackets indicate the average over φ.

3) We now determine a such that g_2 has zero average. We have to consider two quite different situations depending on whether the matrix M is singular or not:

(i) case 1: $\det M \neq 0$. The condition for zero average is

$$Ma + < m\frac{\partial\phi}{\partial\varphi} >= 0 \tag{4.26}$$

Since M is invertible we can write $a = -M^{-1} < m\partial\phi/\partial\varphi >$.

(ii) case 2: $\det M = 0$. Since M is not invertible Eq. (4.26) will not have a solution in general. In order to treat this case we have to modify Thirring's [31] proof. Instead of the condition (4.26) we require only that there are constants a and β such that

$$Ma + < m\frac{\partial\phi}{\partial\varphi} >= \beta\omega_0 \tag{4.27}$$

i.e. we allow g_2 to have a nonzero average, but under the restriction that it is in the direction of ω_0. This produces a change in the frequency vector ($\omega_0 \mapsto (1+\beta)\omega_0$) but not on the quotient between its two components, which is the essential condition for the method. This change of the frequency vector can be compensated by a rescaling of time, which amounts to dividing the whole Hamiltonian by the factor $1+\beta$.

We can solve (4.27) by diagonalizing $M = T\,diag(\Lambda, 0)T^{-1}$, where T is orthogonal. Eq. (4.27) becomes

$$T\,diag(\Lambda,0)T^{-1}a = - < m\frac{\partial\phi}{\partial\varphi} > + \beta\omega_0 \tag{4.28}$$

and defining $\hat{a} = T^{-1}a$, $\hat{c} = T^{-1} < m\partial\phi/\partial\varphi >$ and $\hat{\omega}_0 = T^{-1}\omega_0$, Eq. (4.27) is equivalent to

$$\begin{aligned} \Lambda\hat{a}_1 &= -\hat{c}_1 + \beta\hat{\omega}_{01} \\ 0 &= -\hat{c}_2 + \beta\hat{\omega}_{02} \end{aligned} \tag{4.29}$$

If $\hat{\omega}_{02} \neq 0$ then

$$\begin{aligned} \beta &= \hat{c}_2/\hat{\omega}_{02} \\ \hat{a}_1 &= (-\hat{c}_1 + \hat{c}_2\hat{\omega}_{01}/\hat{\omega}_{02})/\Lambda \end{aligned} \tag{4.30}$$

and $\hat{a}_2$ can be chosen arbitrarily, e.g. $\hat{a}_2 = 0$. If $\hat{\omega}_{02} = 0$ and $\hat{c}_2 = 0$, then we can choose $\beta = 0$, $\hat{a}_2 = 0$ and $\hat{a}_1 = -\hat{c}_1/\Lambda$. If $\hat{\omega}_{02} = 0$, while $\hat{c}_2 \neq 0$, then Eq. (4.27) does not have a solution. The proof of convergence for small perturbations involves showing that under suitable conditions this does not happen.

4.5. Periodic orbits: Birkhoff Fixed Point Theorem

Most of the periodic orbits that are present in the integrable case disappear when a small perturbation is turned on, due to resonance effects (see the discussion of the KAM theorem). Some of them however persist. It was proven by Birkhoff that in a monotonic twist map, for any rational number $\alpha = N/M$ there is an even number of periodic orbits with α as rotation number. Furthermore, half of them are stable (elliptic) and half of them unstable (hyperbolic). This provides an interpretation of one of the structures seen in nonintegrable phase portraits: The chains of island are built around these periodic orbits. At the center of each of the islands there is a point of an elliptic periodic orbit of Birkhoff's theorem. The islands themselves are formed by nearby trajectories that stay close by as a reflection of the stability of the periodic orbit. The points of the hyperbolic periodic orbits are located on the links between the islands.

The proof of Birkhoff's theorem for the case of small perturbations is quite simple. It can be formulated as follows [32]. We first notice that if a set of points $\{x_1, \ldots, x_n\}$ forms a periodic orbit of a map T with period N/L each of these points will be a fixed point of T^L, defined as the composition of L iterations of the map. The search for periodic orbits of T can thus be reduced to the search for fixed points of T^L. Consider first the unperturbed integrable map (4.8). There is a whole circle of fixed points of T^L that corresponds to periodic orbits of T with rotation number $\alpha = N/L$. This circle is between two circles of quasiperiodic orbits of T with rotation numbers $\alpha_- < N/L < \alpha_+$, that are also quasiperiodic under T^L with rotation numbers $\beta_- < 0 < \beta_+$, i.e. the points of these two circles rotate under T^L in opposite directions. When the perturbation is turned on, invariant curves G_-, G_+ with rotation numbers $\beta_- < 0 < \beta_+$ will still be present, as a consequence of the KAM theorem (β_-, β_+ can be chosen diophantine). Now by continuity on each radial direction there must be a point with zero rotation number under T^L. The set of those points forms a closed curve G that lies between the two invariant curves G_-, G_+. G itself is not invariant under T^L, but it will be mapped into another closed curve G'. Since the map T^L is area preserving, the two curves G and G' must intersect, and since they are closed, there will be an even number of intersections. Each intersection is a fixed point of T^L. Further, if x is a fixed point of T^L, then all the points $T^n x$, $0 < n < L$ are also fixed points of T^L. Thus the result that there always is an even number $2mL$ of periodic orbits for any rational rotation number N/L. The stability character of the orbits (elliptic, hyperbolic) follows by direct inspection of the local directions

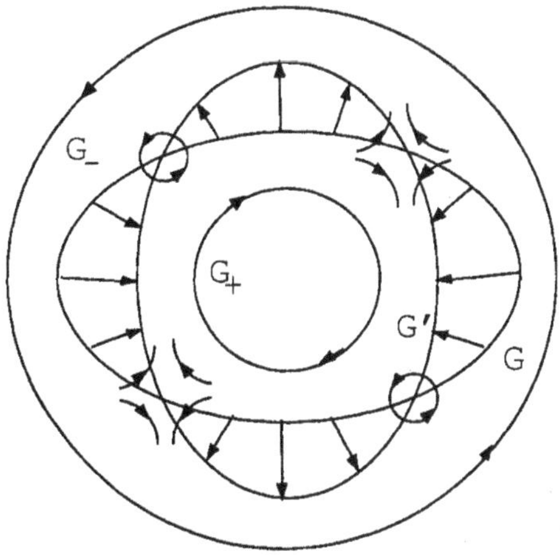

Figure 3

of the motion under T^L as indicated in Figure 3.

Remark: Notice that in this proof we have only needed the condition that the perturbation is small in order to guarantee (by the KAM theorem) the existence of the irrational invariant tori. The result is however valid for arbitrary monotonic twist maps, as can be shown by other methods.

Remark: The stability of a periodic orbit of rotation number $\alpha = N/L$ under the map T is equivalent to the stability of a fixed point $x = (r, \varphi)$ under the map T^L. It can be established by linearizing the map T^L around x:

$$x + \delta x \mapsto x + DT^L \ \delta x + \mathcal{O}(|\delta x|^2)$$

The matrix DT^L defines the tangent map , which is given by

$$DT^L = \begin{pmatrix} \frac{\partial r_{n+1}}{\partial r_n} & \frac{\partial r_{n+1}}{\partial \varphi_n} \\ \frac{\partial \varphi_{n+1}}{\partial r_n} & \frac{\partial \varphi_{n+1}}{\partial \varphi_n} \end{pmatrix}$$

evaluated at x. Its eigenvalues

$$\lambda_\pm = \frac{1}{2}(\tau \pm \sqrt{\tau^2 - 4}), \qquad \tau \equiv Trace(DT^L)$$

contain the information on the local stability of the fixed point. Notice that $det\, DT^L = 1$ since the map is area preserving. Therefore $\lambda_- = 1/\lambda_+$. One distinguishes three cases:

1) Elliptic: if $|\tau| < 2$ the two eigenvalues lie on the complex unit circle $|\lambda_\pm| = 1$. The fixed point is stable.

2) Hyperbolic: if $|\tau| > 2$ the two eigenvalues are real and $|\lambda_+| > 1$ and $|\lambda_-| < 1$. There is a contracting and an expanding direction. The fixed point is unstable.

3) Parabolic: $|\tau| = 2$ is a borderline case and the linearized map does not determine the stability, which depends on higher order terms.

4.6. Unstable Periodic Orbits, Broken Separatrices, Chaos

We have mentioned that the stable periodic orbits of Birkhoff's theorem are located at the center of the islands that appear forming chains. The fact that we see orbits that stay close and wind around the periodic points is just the expression of their stable or elliptic character. The points of the hyperbolic periodic orbits also lie on the chain, but in between the islands. Since they are unstable, the character of the orbits in their neighborhood is very different. In fact we will see that it is in their neighborhood that chaotic orbits appear.

In order to get a qualitative understanding of the phenomena in the neighborhood of an hyperbolic periodic orbit, it is useful to first consider a simple integrable model having an hyperbolic fixed point and then perturb it. We consider a pendulum perturbed by a periodic force

$$H = \frac{1}{2}p^2 + c_1 \cos\varphi + \varepsilon W(\varphi, \omega t). \tag{4.31}$$

In the unperturbed phase space diagram we see an elliptic fixed point at the center of the island and two hyperbolic fixed points (that should be identified as a single point due to the periodic condition on φ). The two special orbits that start and finish at the fixed point are the separatrices, which in general situations are called homoclinic or heteroclinic orbits depending on whether the initial and final points are the same or different. It is easier to describe our arguments if we talk about the two points as if they were different.

To an hyperbolic fixed point we can associate two special sets, W^s and W^u called the stable and unstable manifolds. They are defined by the property that all points of the stable manifold stream toward the fixed point as $t \to \infty$, while the points of the unstable manifold emanated from the fixed point at $t \to -\infty$ (or equivalently they would stream into it for $t \to \infty$ if we inverted the sense of the motion). In the example of the pendulum these manifolds coincide with the separatrices. Integrable systems have the very special property that the stable manifold of one fixed point X coincides (is identical) with the unstable manifold of another fixed point Y. This property is very unstable (it depends crucially on the existence of a complete set of conserved quantities) and we will see that it is not maintained in the perturbed system. In general the two manifolds intersect each other in a complicated way.

4.6.1. Melnikov's Criterion

Melnikov devised a method to check if the stable and unstable manifolds for a system like the pendulum intersect when a time periodic perturbation is added. His formulation also allows to see very directly that for a generic perturbation the two manifolds intersect. Melnikov's method can be described as follows: We consider a perturbed Hamiltonian of the form $H = H_0(p,\varphi) + \epsilon H_1(p,\varphi,t)$ and calculate to first order in ϵ the deformation of the stable and unstable manifolds. We first remark that the two fixed points Y, Z are still present, but possibly in slightly different location Y_ϵ, Z_ϵ.

We use the shorter notation $x = (p, \varphi)$ and write the equations of motion in the form

$$\dot{x} = J\frac{\partial H}{\partial x} = f(x) + \epsilon g(x,t).$$

We perform the following steps:

1) We expand the trajectory defining the stable and unstable manifolds up to first order in ϵ (uniformly in time)

$$x_\epsilon^{s,u}(t,t_0) = x_0(t,t_0) + \epsilon y^{s,u}(t,t_0) + \mathcal{O}(\epsilon^2).$$

We have introduced the parameter t_0 in order to be able to adjust the origin of time, so that $x_0(t_0,t_0) = x_{00}$, where x_{00} is some chosen point of the unperturbed separatrix. The analysis is done locally close to this point. By changing t_0 we can shift along the separatrix.

2) We establish the equations of motion for the first order correction $y^{s,u}$: From

$$\dot{x}_\epsilon^{s,u} \approx \dot{x}_0^{s,u} + \epsilon\dot{y}^{s,u} = f(x_0) + \epsilon A y^{s,u} + \epsilon g(x_0,t)$$

where the matrix $A_{ij} = \partial f_i/\partial x_j(x_0(t-t_0))$, which leads to

$$\dot{y}^{s,u} = A y^{s,u} + g(x_0,t) \tag{4.32}$$

together with the initial and final conditions

$$x_\epsilon^s(t\to\infty) = Z_\epsilon; \quad x_\epsilon^u(t\to-\infty) = Y_\epsilon.$$

3) With this preparation we can introduce a function that measures the distance between the stable and unstable manifolds. We first take the difference vector between points in the two manifolds

$$x_\epsilon^s(t,t_0) - x_\epsilon^u(t,t_0) = \epsilon(y^s(t,t_0) - y^u(t,t_0)) + \mathcal{O}(\epsilon^2) \equiv \epsilon d(t,t_0) + \mathcal{O}(\epsilon^2)$$

and define a distance $D(t,t_0)$ by "projecting" into the direction perpendicular to the separatrix, represented by the vector $n(t,t_0) \perp \dot{x}_0(t,t_0)$, $\quad n = (-f_2(x_0), f_1(x_0))$

$$D(t,t_0) \equiv d \cdot n \tag{4.33}$$

(this is not strictly a projection since n has not necessarily norm 1). The Melnikov distance D_M is then defined as

$$D_M(t_0) = D(t_0,t_0).$$

It is a measure of the splitting of the stable and the unstable manifold close to a point x_{00} of the unperturbed separatrix.

4) We can now use the particular Hamiltonian structure of the equations of motion to obtain a more explicit expression for the Melnikov distance. We first note that Eq (4.33) can be written as

$$D(t,t_0) = (f_1 d_1 - f_2 d_2) \equiv f \wedge d = f \wedge y^s - f \wedge y^u.$$

We can now use the equations of motion (4.32) for $y^{s,u}$ to obtain equations for D. The functions defined as

$$D^s = f \wedge y^s, \quad D^u = f \wedge y^u$$

satisfy the same equation

$$\frac{d}{dt}D^s = f \wedge g, \quad \frac{d}{dt}D^u = f \wedge g. \tag{4.34}$$

which can be obtained by direct calculation:

$$\begin{aligned}
\frac{d}{dt}D^{s,u} =& \dot{f} \wedge y^{s,u} + f \wedge \dot{y}^{s,u} \\
=& A\dot{x}_0 \wedge y^{s,u} + f \wedge (Ay^{s,u} + g) \\
=& f \wedge g + Af \wedge y^{s,u} + f \wedge Ay^{s,u} \\
=& f \wedge g + (TrA) f \wedge y^{s,u} \\
=& f \wedge g
\end{aligned}$$

where we have used the fact that since f derives from a Hamiltonian the trace of the Jacobian A is zero. Taking into account the conditions for $t \to \pm\infty$ we can integrate (4.34)

$$\begin{aligned}
D^s(t,t_0) &= -\int_t^{\infty} dt' f \wedge g \\
D^u(t,t_0) &= \int_{-\infty}^{t} dt' f \wedge g
\end{aligned}$$

and therefore

$$D_M(t_0) = -\int_{-\infty}^{\infty} dt' \; f(x_0(t'-t_0)) \wedge g(x_0(t'-t_0), t'). \tag{4.35}$$

We can write the integrand in terms of Poisson brackets:

$$f \wedge g = -\frac{\partial H_0}{\partial \varphi}\frac{\partial H_1}{\partial p} + \frac{\partial H_1}{\partial \varphi}\frac{\partial H_0}{\partial p} \equiv -\{H_0, H_1\}$$

and therefore

$$D_M(t_0) = \int_{-\infty}^{\infty} dt' \; \{H_0, H_1\}(x_0(t'-t_0)). \tag{4.36}$$

Thus, the Melnikov distance at a point x_{00} can be calculated by integrating the Poisson brackets of the unperturbed Hamiltonian and the perturbation, along the *unperturbed* separatrix. With this preparation we can formulate Melnikov's criterion [33, 27, 26].

Theorem (Melnikov): If the function $D_M(t_0)$ has simple zeros, then the perturbed system has transversally intersecting stable and unstable manifolds.

This result can be understood intuitively: If $D_M(t_0)$ has a simple zero, it means that to order ε one expects a crossing of the two manifolds with finite angle. For small ε the corrections of higher order $\mathcal{O}(\varepsilon^2)$ cannot undo this behavior. We remark that the Melnikov distance can be easily computed numerically if one knows the unperturbed separatrix. An explicit analytical calculation is also possible in some special examples [27, 26, 33, 34].

Remarks: a) Since we have a new variable $\varphi_2 = \omega t$, the situation is best pictured in a Poincaré section. There W^s and W^u appear as continuous invariant curves.

b) For Hamiltonian systems the stable and unstable manifolds always have intersections. This can be deduced from the area preserving character of the Poincaré map. Therefore the question in Melnikov's criterion is not if the two manifolds intersect, but if the intersection is transversal (i.e. at a finite angle).

c) It is intuitively understandable that generic perturbations lead to transversal crossings, since the zeros of a function that are of order higher than one are very unstable. One can expect that a small modification of the problem will produce zeros of first order.

d) We notice that the Melnikov distance is an integral of a Poisson bracket that only vanishes if the perturbation H_1 is a conserved quantity with respect to the unperturbed dynamics corresponding to H_0. As a consequence, $D_M(t_0)$ will very rarely be identically zero. If the system $H_0 + H_1$ is integrable, then $D_M(t_0) = 0$ for all t_0. Thus, Melnikov's method gives an intuitive understanding of the fact that generic Hamiltonian systems are not integrable.

4.6.2. Consequences of the Intersections of the Stable and Unstable Manifolds

If a transverse intersection of W^s and W^u has been established, it has far reaching consequences. Firstly, if there is one intersection then there are necessarily infinitely many. This follows directly from the fact that W^s and W^u are invariant manifolds (by construction, the flow maps them into themselves). Thus if a point belongs to both, then this will also be the case for all its images under the Poincaré map. Secondly, the areas of the regions defined by W^s and W^u between an arbitrary pair of successive intersections must be the same. This is a direct consequence of the area conservation by the map. But since the distance between two successive points of intersection decreases steadily, the only way to preserve the area surrounded by the curve is by making the curve longer and thus spreading further away. Since the available space is bounded this entails a very complicated pattern of foldings and tangles. This extremely complicated structure, called heteroclinic tangle, is the origin of chaotic behavior in the regions close to broken separatrices.

In fact one can show that in the neighborhood of an heteroclinic tangle there

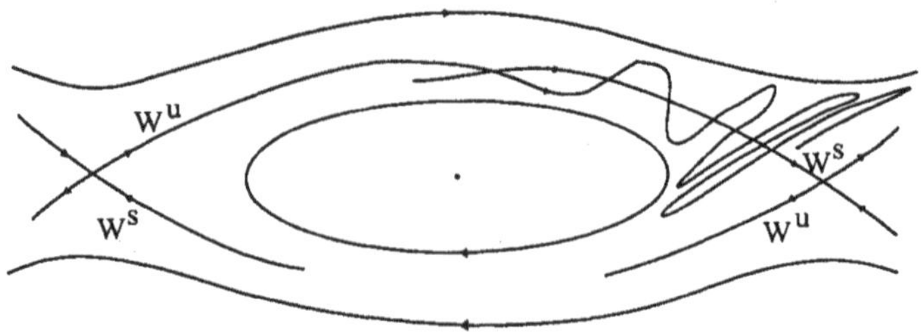

Figure 4

is a set of orbits whose evolution is isomorphic to a Bernouilli shift [35, 26]. Since a Bernouilli shift represents the randomness of a fair coin toss, this means that close to a broken separatrix, or close to a hyperbolic fixed point, there is complete unpredictability of the evolution, i.e. chaos of the strongest type.

4.7. Aubry-Mather Theory

4.7.1. Twist Maps

For an integrable monotonic twist map Φ_0 the plane is foliated by invariant curves, which are circles in the appropriate coordinates (c.f. Eq. (4.8)). To each invariant curve corresponds a rotation number defined by $\alpha = (1/2\pi)\ lim_{n\to\infty}\varphi_n/n$. The KAM theorem in this context is formulated as follows: Consider the perturbed map $\Phi = \Phi_0 + \varepsilon\Phi_1$, and assume that it is sufficiently differentiable (more than C^3). A rotation number α is called diophantine if there exist constants c_1 and $\sigma > 1$ such that for all n,m $|\alpha n - m| > c_1/n^{\sigma}$. Then for any diophantine number α, if the perturbation is small enough there will be an invariant curve with α as rotation number (a circle in appropriately chosen coordinates).

In general one finds that as the strength of the perturbation ε is increased, more and more of these invariant curves are destroyed. The theory developed by S. Aubry and J. Mather allows to establish the existence of a more general kind of invariant set for arbitrary perturbation strength ε and arbitrary real rotation number α. These invariant sets (Aubry-Mather sets) are in general not full curves, but Cantor sets, which can be thought of as constructed by starting with a curve and eliminating a sequence of subsets.

This theory provides a framework in which one can give an intuitive picture of what happens when invariant curves are destroyed by a perturbation. The basic mechanism is that at a critical value of the perturbation the curve becomes porous, i.e. it develops a complicated structure of holes, whose size grows if one increases further the perturbation.

The Aubry-Mather sets can be described more precisely as follows [28]. First we describe the usual invariant curves Γ in a form that lends itself easily to generalization: Instead of the usual form $r = w(\varphi)$ we write the curve in parametric representation

$$\begin{aligned} r &= v(\theta) \\ \varphi &= u(\theta) \end{aligned} \tag{4.37}$$

where $u(\theta)$ is monotonic and the functions $u(\theta) - \theta$ and $v(\theta)$ are continuous and of period 2π. We require that the parametrization is constructed such that the twist map Φ acts on the points of the invariant curve with rotation number α as follows: if

$$\begin{aligned} r_n &= v(\theta) \\ \varphi_n &= u(\theta) \end{aligned}$$

then

$$\begin{aligned} r_{n+1} &= v(\theta + \hat{\alpha}) \\ \varphi_{n+1} &= u(\theta + \hat{\alpha}) \end{aligned}$$

where we have introduced the notation $\hat{\alpha} \equiv 2\pi\alpha$. This condition implies that the functions u and v satisfy the equations

$$v(\theta + \hat{\alpha}) = g(u(\theta), v(\theta)) \tag{4.38}$$

$$u(\theta + \hat{\alpha}) = f(u(\theta), v(\theta)) \tag{4.39}$$

We remark that if u is known, then v is determined from it by Eq. (4.39) (since by the twist condition $\partial f / \partial v \neq 0$, we can apply the implicit function theorem and express $v(\theta)$ as a function of $u(\theta)$ and of $u(\theta + \hat{\alpha})$). In fact, using the generating function $F(\varphi_n, \varphi_{n+1})$, we can write

$$\begin{aligned} r_n \equiv v(\theta) &= -\frac{\partial F(u(\theta), u(\theta + \hat{\alpha}))}{\partial \varphi_n} \\ r_{n+1} \equiv v(\theta + \hat{\alpha}) &= \frac{\partial F(u(\theta), u(\theta + \hat{\alpha}))}{\partial \varphi_{n+1}} . \end{aligned}$$

By renaming in the second equation $\theta + \hat{\alpha} \to \theta$ and adding the two equations we obtain

$$\frac{\partial F(u(\theta), u(\theta + \hat{\alpha}))}{\partial \varphi_n} + \frac{\partial F(u(\theta - \hat{\alpha}), u(\theta))}{\partial \varphi_{n+1}} = 0. \tag{4.40}$$

The solution of this equation with the conditions that $u(\theta)$ is monotonically increasing and that $u(\theta) - \theta$ is of period 2π gives us the required parametric representation.

The continuity properties of this solution are the essential feature that bring the distinction between usual invariant curves and Aubry-Mather sets: If $u(\theta)$ is a continuous solution of (4.40), then $v(\theta)$ is also continuous (by Eq. (4.39)) and the parametrization (4.37) represents an invariant curve. On the other hand, if $u(\theta)$, and therefore also $v(\theta)$, is discontinuous, then Eq. (4.37) does not represent a curve anymore. It does however represent a subset of the plane that can be embedded in a curve. We remark that since we have required that $u(\theta)$ is monotonic, it can have at most countably many discontinuities [36]. This leads us to the following:

Definition: An *Aubry-Mather set of rotation number* α is the closed invariant set described by the parametrization $\varphi = u(\theta)$, $r = v(\theta)$ where v and u satisfy Eqs. (4.38) and (4.39); u is monotonic and $u(\theta) - \theta$ and $v(\theta)$ are of period 2π, but not necessarily continuous.

The main result of S. Aubry and J. Mather states the existence of such invariant sets for arbitrary monotonic twist maps, and any rotation numbers.

Theorem (S. Aubry, P.Y. Le Daeron [37], J. Mather [38]): Let Φ be an area-preserving monotone twist map of an annulus into itself, α any real number in the interval (α_1, α_2), where α_1 and α_2 are the rotation numbers of the inner and outer circles of the annulus. Then there is an invariant Aubry-Mather set Γ_α with rotation number α. The set Γ_α is a subset of a closed Lipschitz continuous curve $r = W(\varphi)$.

Remarks: For rational α this theorem gives periodic orbits, whose existence had already been established by Birkhoff's fixed point theorem. If they are isolated, then $u(\theta)$ and $v(\theta)$ are discontinuous piecewise constant, taking only finitely many values. If α is irrational, one has to distinguish two cases: (i) if u and v are continuous, then there is an invariant curve, and each of its orbits cover it densely; (ii) if u and v can have countably many discontinuities, then Γ_α is a Cantor set (which can however be embedded in a continuous curve), where the gaps correspond to the discontinuities.

The proof of this result was obtained from variational principles. In Mather's proof the solution $u(\theta)$ is characterized as the minimizer of the functional

$$\int_0^1 F(u(\theta), u(\theta + \hat{\alpha}))d\theta.$$

In Aubry's approach the orbits that generate the Aubry-Mather set are constructed as those sequences $\{\varphi_n\}$ that minimize the "energy functional"

$$\sum_{n=-\infty}^{\infty} F(\varphi_n, \varphi_{n+1}). \tag{4.41}$$

The motivation in Aubry's approach came from solid-state physics. The minimizing sequences correspond essentially to ground states of models for incommensurate crystals, like e.g. the Frenkel-Kontorova model [37, 39].

4.7.2. Continuous Time Hamiltonian Systems

The Aubry-Mather sets that we have described for monotonic twist maps can be expected to have their counterparts in Poincaré maps of Hamiltonian systems with two degrees of freedom. The connection is however not completely straightforward, the problem being that the Poincaré maps do not necessarily satisfy the monotonicity condition (4.5). The theory of Aubry and Mather can be adapted to this case by formulating it in terms of a variational problem similar to (4.41) but for continuous time trajectories [28 29]. Consider as an example a system described by a Lagrangean

$$L(\varphi,\dot{\varphi},t)=\frac{1}{2}m\dot{\varphi}^2-V(\varphi,t) \tag{4.42}$$

where V is bounded and periodic in φ and in t. We consider first the trivial integrable case in which V is independent of t:

$$L_0(\varphi,\dot{\varphi})=\frac{1}{2}m\dot{\varphi}^2-V_0(\varphi).$$

(e.g. $V_0(\varphi)=1-\cos\varphi$). A trajectory with energy $E>max_\varphi V_0(\varphi)$ can be calculated from

$$t-t_0=\int_0^\varphi d\varphi'\frac{1}{\sqrt{2(E-V_0(\varphi'))}}=G(\varphi)$$

by inverting $\varphi(t)=G^{-1}(t-t_0)$. The function $G(\varphi)$ has the following properties:

(i) Since it is the integral of a periodic function, it can be decomposed as

$$G(\varphi)=\gamma\varphi+g(\varphi)$$

where $g(\varphi+2\pi)=g(\varphi)$ and γ is the zeroth Fourier coefficient of the integrand:

$$\begin{aligned}G(\varphi)&=\int_0^\varphi d\varphi'\sum_{n=-\infty}^{\infty}\tilde{c}_ne^{in\varphi'}=\\&=\tilde{c}_0\varphi+\sum_{n\neq 0}\tilde{c}_n\frac{1}{in}(e^{in\varphi}-1)\equiv\gamma\varphi+g(\varphi).\end{aligned}$$

(ii) $G(\varphi)$ is monotonic, since

$$\frac{dG}{d\varphi}=\frac{1}{\sqrt{2(E-V_0(\varphi))}}>0.$$

Therefore its inverse G^{-1} is also monotonic.

(iii) The inverse of the relation

$$(t-t_0)=\gamma\varphi+g(\varphi) \tag{4.43}$$

can be written as

$$\varphi=\theta+f(\theta)\quad with\ \theta=\frac{t-t_0}{\gamma}$$

with $f(\theta + 2\pi) = f(\theta)$ periodic. This follows from (4.43) and the application of the usual iteration that appears in the proof of the inverse function theorem: Rewriting (4.43) as $\theta = \varphi + \hat{g}(\varphi)$ we write the iteration as

$$\begin{aligned}\varphi &= \theta - \hat{g}(\theta - \hat{g}(\varphi)) \\ &= \theta - \hat{g}(\theta - (\theta - \hat{g}(\varphi))) \\ &= \ldots = \theta + f(\theta) \equiv U_0(\theta)\end{aligned}$$

We can therefore write the trajectories as

$$\varphi(t) = U_0(\alpha t + \beta) \equiv \alpha t + \beta + f(\alpha t + \beta)$$

where $\beta = -t_0/\gamma$ and $\alpha = 1/\gamma$ is equal to the rotation number

$$lim \frac{1}{\tau}\varphi(\tau) = \alpha \equiv [\int_0^{2\pi} \frac{d\varphi'}{2\pi} \frac{1}{\sqrt{2(E - V_0(\varphi'))}}]^{-1}.$$

The function $U_0(\theta)$ is monotonic, since it is the inverse of the monotonic function G. Using the methods of the classical calculus of variations [40 - 43] one sees immediately that all these solutions are minimals of the action functional

$$\int_{-\infty}^{\infty} dt L_0(\varphi, \dot{\varphi}).$$

If we extend the phase space trivially by defining a new angle $\varphi_2(t) = t + \varphi_2(0)$, the equations

$$\begin{aligned}\varphi(t) &= U_0(\alpha t + \beta) \\ \varphi_2(t) &= \varphi_2(0) + t\end{aligned}$$

generate the Aubry-Mather sets, which since this case is integrable are invariant tori (for irrational α).

The natural generalization to the time-dependent $V(\varphi, t)$ (non-integrable) case is to look for trajectories of the form

$$\varphi(t) = U(t, \alpha t + \beta).$$

The Aubry-Mather theorem adapted to this situation can be formulated as [28]:

Theorem: Consider the system described by a Lagrangean (4.42). For any $\alpha \in R$ there exists a function $U(t, \theta)$, strictly monotonically increasing in θ, with $U(t, \theta) - \theta$ periodic in t and θ such that for any β

$$\varphi(t) = U(t, \alpha t + \beta)$$

is a solution of the Euler equation

$$m\frac{d^2}{dt^2}\varphi = -\frac{\partial}{\partial\varphi}V(\varphi, t).$$

As a function of θ, $U(t,\theta)$ is monotonic but not necessarily continuous. In the case that U is continuous we recover, for α irrational, invariant tori (that are densely covered by each orbit).

If U is discontinuous, the Aubry-Mather sets generated by these orbits can be thought of as Cantor sets that can be embedded in tori, and that have a countable number of gaps.

We remark that the above theorem can be interpreted as stating the existence of generalized quasi-periodic solutions with any prescribed ratio between the frequencies.

$$\varphi(t) = \alpha t + \beta + \sum_{m,n} \tilde{c}_{m,n} e^{i(m+n\alpha)t}$$

This theorem is again proven by methods of the calculus of variations [28, 29]. The orbits generating the Aubry-Mather sets can be characterized as minimizers of the action

$$\int_{-\infty}^{\infty} dt L(\varphi, \dot{\varphi}, t).$$

We finally remark that the Aubry-Mather theory is limited to Hamiltonian systems with two degrees of freedom. An extension to higher dimensional systems is not obvious.

QUANTUM SYSTEMS

5. Time Dependent Hamiltonians

It is a very natural question to ask whether the richness of phenomena that has been discovered in classical Hamiltonian dynamics, involving both highly structured and strongly chaotic aspects, has any counterpart in the quantum mechanical treatment of the same Hamiltonian systems. We will formulate some aspects of this question more specifically.

1) Integrable and non-integrable classical systems show an intrinsically different dynamical behavior. Is there any well-defined qualitative difference in the quantized version of the two kinds of models?

2) In classical systems the existence of a complete set of conserved quantities defines its integrability, guarantees the global regularity of the dynamics in phase space, and strongly limits the possible qualitative types of trajectories. Furthermore it allows in principle the exact solution of the model, i.e. the determination of all the possible trajectories by quadrature, i.e by performing integrals, and inverting functions. One can ask if there is an analogous role of conserved quantities in quantum mechanics [44]. It is well-known that conserved quantities are related to symmetries by Noether's theorem. These symmetries and their group theoretical formulation have been used since the beginnings of quantum mechanics to classify eigenvalues (to find "good quantum numbers") and to simplify and improve the

methods to determine them. But it is not clear if or under which conditions there is an analogue to being able to solve the problem exactly using only "quadratures".

3) Since in orthodox quantum mechanics the trajectories of particles are not part of the formalism, the question of regularity or chaos of trajectories cannot be formulated in this context. One would like however to find some other properties whose dynamics could be attributed varying degrees of regularity or of randomness. For a recent discussion on the interpretation of particle trajectories in quantum mechanics see Dürr, Goldstein and Zanghi [45].

4) One can think also that in quantum mechanics one can find new types of instabilities or other static or dynamical structures, that depend on some integrability character, but are not present in classical mechanics. This can be the case, among others, in quantum spin 1/2 systems that do not have a classical correspondence.

This field is one of intense active research [46–51]. Many people are working on clarifying some the questions we have mentioned, among many others. At this point the subject is in a stage of exploration, of gathering of the phenomena, and probing out concepts and tools that might lead to a more general unified understanding of the basic questions.

In the following sections we will only discuss one particular type of quantum problem for which some progress has been made in the characterization of the regularity or instability of the dynamics. It concerns the quantized versions of systems of the type of one of the examples we have discussed in classical mechanics, namely the periodically driven pendulum. We consider quantum systems subjected to time dependent external forces [52], and try to get information about the regularity of their time evolution. Since we do not have access to particle trajectories we analyze the dynamics of other objects. Some of them are observable, like expectation values and correlation functions and others, like the evolution of the wave functions in Hilbert space, are of mathematical nature. In particular we will discuss spectral methods as tools to study stability.

The reason to study quantum systems with time dependent forces is twofold. Firstly, in the last couple of years a lot of new experimental results have been obtained involving driven quantum systems. The other related motivation is that their behavior can be radically different from autonomous quantum systems.

It is well known that the energy levels of isolated quantum system in a bounded spatial domain, or confined by a potential going rapidly to infinity, are discrete, the evolution of the wave function and the density matrix will necessarily be almost periodic in time [53]. This rules out any good ergodic properties or other forms of classical chaoticity, which require decay of correlation functions in suitable stationary ensembles. Of course as the size of the system becomes large and the spacing between levels becomes very small, the ensemble averages of some classes of quantum variables may exhibit good decay properties over long time periods. These can become, in suitable limits, indistinguishable from those given by chaotic dynamics.

The situation is quite different for quantum systems subjected to time dependent perturbations. In this case the relevant spectrum for the dynamics is the one of the quasienergy operator [54, 55], which, as we will see, can be continuous and thus lead to decaying correlations and even strong ergodic properties of the quantum evolution.

In our discussion of classical mechanics we have seen that the models of two coupled pendula and of a periodically driven pendulum belong to the same class. They can be treated in a unified way with the same tools and their general qualitative behavior is of the same type. The preceding comment means that this is not the case in the quantum treatment. We have to develop new tools for the time dependent case, and its properties can be qualitatively different.

The interest in driven quantum systems has been strongly motivated by the multi-photon ionization experiments on Rydberg atoms in a microwave field, first carried out by Bayfield and Koch [56]. These were first analyzed in terms of classical chaotic dynamics, and then within quantum mechanics [57–63]. Other examples include ions localized in a Paul or in a Penning trap [64], electrons attracted to the surface of liquid Helium [65, 66], and charge transfer in atomic collisions [67].

A natural framework for the study of typical long time behavior of systems subjected to general time dependent perturbations, which includes both periodic and random potentials as special cases, is to write the Hamiltonian in the form

$$H(t) = H_0(x) + V(x; \underline{\theta}(t)) \tag{5.1}$$

Here x stands for the internal dynamical variables of the system and $\underline{\theta}(t) \in \Omega$ is a time dependent "external" classical variable, corresponding to the trajectory of a classical dynamical system on a domain or a manifold Ω, having an invariant ergodic measure μ. One considers then typical or averaged behavior with respect to μ. This situation can be thought of as a limiting case of the system being in contact with an external bath whose relevant time dependence, described by $\underline{\theta}(t)$, is independent of the state of the system evolving according to the Hamiltonian (5.1).

We are interested primarily in the problem of stability which can be formulated in terms of the following two questions:

(i) Assuming that the unperturbed system has a discrete spectrum, i.e. an almost-periodic time evolution, does the perturbed system also have an almost-periodic dynamics or is there a transition to a more complicated chaotic behavior?

(ii) Given a system initially in a state which is localized in "phase space", does the time evolution under (5.1) lead to delocalization? This would correspond in some cases, like the kicked rotator, to an unbounded growth of the energy, and in others, like Rydberg atoms, to ionization. The two questions are related but they are not equivalent except in the case where the time dependent perturbation is periodic.

In classical mechanics we have seen that time dependent systems can be transformed in autonomous systems by substituting time by a new dynamical variable. In the example

$$H(t) = \frac{1}{2}p_1^2 + \gamma \cos\varphi_1 \cos\omega t$$

one defines $\varphi_2 \equiv \omega t$, and its canonically conjugate momentum p_2. The dynamics is then equivalently described by the time independent Hamiltonian

$$H' = \frac{1}{2}p_1^2 + \omega p_2 + \gamma \cos\varphi_1 \cos\varphi_2$$

One can try to naively quantize the new Hamiltonian H' by the correspondence

$$p_1 \to -i\hbar\frac{\partial}{\partial\varphi_1}, \qquad p_2 \to -i\hbar\frac{\partial}{\partial\varphi_2}$$

which would lead to a Hamiltonian of the form

$$H' = -\frac{\hbar^2}{2}\frac{\partial^2}{\partial\varphi_1^2} - i\hbar\frac{\partial}{\partial\varphi_2} + \gamma \cos\varphi_1 \cos\varphi_2 \tag{5.2}$$

This is a formal manipulation that is a priori not justified. The problem is that it effectively quantizes the time variable, while it is well established that time is a classical superselection rule [68], and it cannot be assigned a time operator. The formalism that we are going to describe works around this difficulty. It is based on the analysis of the spectrum of operators like (5.2). The goal is to extract from this spectrum information on the dynamics of the system, without needing to resort to a direct physical interpretation of the variable φ_2. In what follows we will set $\hbar = 1$.

5.1. Floquet Formalism, Quasienergy Operator

We will start setting up the formalism [69] for quite general time dependent driving forces. We then will consider particular properties for the practically important cases of periodic forces and of quasiperiodic forces, i.e forces with two or more incommensurate frequencies, like for example

$$H(t) = H_0 + \gamma_1 \cos(\varphi - \omega_1 t) + \gamma_2 \cos(\varphi - \omega_2 t)$$

with $\alpha \equiv \omega_1/\omega_2$ irrational.

We consider the time dependent Hamiltonian (5.1) acting on a Hilbert space $\mathcal{H}$ with $\theta(t) = g_t\theta \in \Omega$, where g_t is an invertible flow, i.e. one considers in fact a family of Hamiltonians depending parametrically on the initial point $\theta \in \Omega$.

Let $U(t,s;\theta)$ be the corresponding propagator, which we assume to be unitary and strongly continuous in t and s. It satisfies

$$U(t+a, s+a;\ \theta) = U(t,s;\ g_a\theta); \qquad a \in R \tag{5.3}$$

One considers then, in analogy with the usual construction of classical mechanics, the enlarged Hilbert space

$$\mathcal{K} = \mathcal{H} \otimes L_2(\Omega, d\mu) \tag{5.4}$$

where $L_2(\Omega, d\mu)$ is the space of functions on Ω that are square integrable with respect to the measure μ. We then consider the one-parameter family of operators $W(t)$ acting on $\Psi \in \mathcal{K}$ by

$$[W(t)\Psi](\theta) = U(0,-t;\ \theta)\mathcal{T}_{-t}\Psi(\theta) \equiv \mathcal{T}_{-t}U(t,0;\ \theta)\Psi(\theta). \tag{5.5}$$

with

$$[\mathcal{T}_{-t}\Psi](\theta) = \Psi(g_{-t}\theta). \tag{5.6}$$

Under suitable conditions on the Hamiltonian, $W(t)$ is a strongly continuous family of unitary operators [70, 69]. By Stone's theorem it can therefore be represented as

$$W(t) = e^{-iKt} \tag{5.7}$$

The selfadjoint generator K is by definition the quasi-energy operator. It acts as

$$[K\Psi](\theta) = -i\frac{d\Psi(g_t\theta)}{dt}\bigg|_{t=0} + H(\theta)\Psi \tag{5.8}$$

If the flow g_t originates from a differential equation, i.e. $\theta(t) = g_t\theta$ is the solution of $d\theta(t)/dt = G(\theta(t))$ then (5.8) can be written as

$$K = -iG(\theta)\cdot\frac{\partial}{\partial\theta} + H(\theta).$$

The problem is then to determine the relation between spectral characteristics of K and stability properties of the time evolution of the system. Since K is selfadjoint, we know from the spectral theorem that essentially three types of spectrum can appear: point spectrum (i.e. eigenvalues), absolutely continuous spectrum, and singularly continuous spectrum.

Examples:

1) Periodic force: $H = H_0 + f(x)\cos(\omega t + \theta)$. Here $\Omega = S^1$ is the unit circle, $\theta = \theta + \omega t$, $d\mu = d\theta$, and

$$K = -i\omega\frac{\partial}{\partial\theta} + H(\theta). \tag{5.9}$$

with $H(\theta) = H_0 + f(x)\cos(\theta)$.

2) Quasi-periodic force with two frequencies ω_1, ω_2: $H = H_0 + f(x)(\cos(\omega_1 t + \theta_1) + \cos(\omega_2 t + \theta_2))$. Then Ω is a torus $S^1 \times S^1$, $T_j = 2\pi/\omega_j$, the flow $g_t\ (\theta_1, \theta_2) = (\theta_1 + \omega_1 t, \theta_2 + \omega_2 t)$, $d\mu = d\theta_1 d\theta_2$, and

$$K = -i\omega_1\frac{\partial}{\partial\theta_1} - i\omega_2\frac{\partial}{\partial\theta_2} + H(\theta_1, \theta_2).$$

with $H(\theta) = H_0 + f(x)(\cos(\theta_1) + \cos(\theta_2))$

There are some immediate consequences that can be deduced from the nature of the spectrum.

5.2. Point Spectrum and Stability

If the spectrum is pure point, then since K is selfadjoint the eigenfunctions $\{\psi_m\}$ form a basis of the enlarged Hilbert space $\mathcal{K}$. We want to use this set to develop functions of the "small" Hilbert space $\mathcal{H}$, and then represent their time evolution in a simple way. This is in close analogy with the usual expansion in eigenfunctions of time independent Hamiltonians H_0 with discrete spectrum, which we first recall for comparison: If $H_0\varphi_n = E_n\varphi_n$ then the evolution given by the Schrödinger equation $i\partial\varphi/\partial t = H_0\varphi$ of an arbitrary state $\varphi \in \mathcal{H}$ can be represented as

$$\varphi(t) = \sum_n c_n e^{-iE_n t}\varphi_n$$

where the coefficients c_n are determined by the initial condition

$$c_n =< \varphi_n, \varphi(0) >_{\mathcal{H}} .$$

In the time dependent case we have for the evolution of an arbitrary state $\varphi(t) = U(t,0\ ;\ \theta)\varphi(0)$ the following expansion

$$\varphi(x,\theta,t) = T_t \sum_m c_m\ e^{-i\lambda_m t}\psi_m(x,\theta) = \sum_m c_m\ e^{-i\lambda_m t}\psi_m(x, g_t\theta) \tag{5.10}$$

where the coefficients c_m are determined by the initial condition. Since the eigenfunctions $\psi_m \in \mathcal{K}$ are in a larger space than the state $\varphi(t)$, we first need to embed φ in $\mathcal{K} = \mathcal{H} \otimes L_2(\Omega, d\mu)$. We can embed it by tensoring it with the function of $L_2(\Omega, d\mu)$ which assigns the trivial value 1 to all θ's: $\varphi \otimes 1 \in \mathcal{K}$. We can then expand

$$\varphi(0) \otimes 1 = \sum_m c_m\psi_m, \qquad with \qquad c_m =< \psi_m\ ,\ \varphi(0) \otimes 1 >_{\mathcal{K}} . \tag{5.11}$$

i.e. c_m is determined by a scalar product in the enlarged space. Notice that since

$$\begin{aligned} c_m &= < \psi_m\ ,\ \varphi(0) \otimes 1 >_{\mathcal{K}} \equiv \int d\mu(\theta)\ \ < \psi_m\ ,\ \varphi(0) >_{\mathcal{H}} \\ &= < \left(\int d\mu(\theta)\ \psi_m\right)\ ,\ \varphi(0) >_{\mathcal{H}} \end{aligned}$$

we can say informally that the coefficients c_m are determined by a scalar product, in the "small" Hilbert space, of the initial condition $\varphi(0)$ with the "average" over the phases θ of the eigenfunctions ψ_m.

The expansion (5.10) can be deduced directly from the defining relation (5.5)(5.7) between the propagator and the quasienergy operator. We can write

$$\begin{aligned} \varphi(t) &= U(t,0;\theta_1,\theta_2)(\varphi(0) \otimes 1) = T_t e^{-iKt}\sum_m c_m\psi_m \\ &= T_t \sum_m c_m e^{-i\lambda_m t}\psi_m = \sum_m c_m e^{-i\lambda_m t} T_t\psi_m \end{aligned} \tag{5.12}$$

It is important that the convergence of this series is uniform in t: $\forall\varepsilon > 0$ there is an M independent of t such that

$$\begin{aligned}||\varphi(t) - \sum_{m<M} c_m e^{-i\lambda_m t}\mathcal{T}_t\psi_m||^2_{\mathcal{K}} &= ||\sum_{m\geq M} c_m e^{-i\lambda_m t}\mathcal{T}_t\psi_m||^2_{\mathcal{K}} \\ &\leq \sum_{m\geq M} |c_m|^2 < \varepsilon^2.\end{aligned} \tag{5.13}$$

This uniformity in time means that if we approximate the initial state by taking only a finite number of terms in (5.11), such that the error is ε, then the size of the error will have the same bound for all time if we approximate the time evolution by (5.10) with the same finite set of terms. This means that the time evolution happens essentially in a subspace of finite dimension. This is a statement of the stability of the system. This fact can be expressed by saying that the trajectories in Hilbert space are precompact [71]. Notice that the only difference with the time independent case is the presence of the translation operator $\mathcal{T}_t$, which acts exclusively on the θ variables.

5.3. Correlation Functions

Another property that directly reflects the nature of the quasienergy spectrum is the following correlation function introduced by Birkhoff and von Neumann [69]:

$$S(t) = \lim_{T\to\infty} \frac{1}{T}\int_0^T ds < \varphi,\ U(t+s,s;\ \theta)\ \varphi >_{\mathcal{H}}. \tag{5.14}$$

The subscript on the scalar product bracket indicates the space in which it acts. Let $\mathcal{K}_c, \mathcal{K}_{ac}, \mathcal{K}_{pp}$ denote respectively the continuous, absolutely continuous and pure point subspaces of $\mathcal{K}$. Then,

(i) If $\varphi\otimes 1 \in \mathcal{K}_{ac}$ then $S(t) \underset{t\to\infty}{\longrightarrow} 0$, for almost all θ.

(ii) If $\varphi\otimes 1 \in \mathcal{K}_c$ then $S(t) \underset{t\to\infty}{\longrightarrow} 0$ in Cesaro mean, for almost all θ.

(iii) If $\varphi\otimes 1 \in \mathcal{K}_{pp}$ then $S(t)$ is almost periodic for almost all θ.

This statements follow immediately from standard results of ergodic theory and the relation, valid for almost all θ:

$$\begin{aligned}S(t) = \lim_{T\to\infty} \frac{1}{T}\int_0^T ds < \varphi,\ U(t+s,s;\ \theta)\ \varphi >_{\mathcal{H}} &= \int_\Omega \mu(\theta) < \varphi,\ U(t,0;\ \theta)\ \varphi >_{\mathcal{H}} \\ &= < \varphi\otimes 1,\ e^{iKt}\varphi\otimes 1 >_{\mathcal{K}}.\end{aligned} \tag{5.15}$$

Remark: Recently some new connections have been reported between finer properties of the spectrum, like its fractal dimension, and the long time dynamics of the system [72, 73].

6. The Time-periodic Case

The case of time-periodic external force has been studied most extensively both for classical and quantum systems. The stability problem can then be expressed in terms of the properties of the Floquet operator, defined as

$$U_F = U(T, 0;\ 0)$$

which gives the evolution of the system over one period $T = 2\pi/\omega$. Classically U is a canonical volume preserving Poincaré map in phase space, while quantum mechanically it is a unitary operator on the Hilbert space $\mathcal{H}$ of the unperturbed system. The long time behavior of the system and its stability is determined by iteration of this map. Classically the system can be studied by standard methods, including direct numerical simulation. It can exhibit varieties of behavior, ranging from integrable to chaotic. The quantum problem is considerably more difficult to deal with numerically than the classical one, and one of the outstanding questions in the field is the existence of qualitatively different behavior of quantum and classical systems. There are models, such as the kicked rotator, which classically show a linear growth of the energy, while the quantum analogues have been found to saturate (quantum limitation of diffusion [57, 74]). The interpretation of this discrepancy is however a quite subtle problem. It has been proven recently that if instead of kicks ($\delta(t - nT)$) one considers smooth periodic forces, then in the classical case the diffusion is also limited [75, 76]. This is a consequence of the existence – at sufficiently high energies – of KAM tori that block the passage in phase space. This suggests that the qualitatively different behavior in the quantum and classical kicked rotator comes from the way that a δ function affects the dynamics. For an example of unbounded energy growth in a classical nonlinear model see M.Levi [77].

The question of quantum mechanical stability for periodic forces can be expressed in terms of the spectral properties of the Floquet operator [55, 78, 70, 71]. Its spectrum in the Hilbert space $\mathcal{H}$ of H_0 is directly related to the spectrum of the quasienergy operator in the space $\mathcal{K} \equiv \mathcal{H} \otimes L_2(S^1, d\theta)$. The eigenvalues and eigenvectors of the two operators are related by

$$\begin{aligned} K\psi &= \lambda\psi \\ U_F\phi &= e^{-i\lambda T}\phi \\ \psi(x,\theta) &= e^{i\lambda\theta/\omega}U(\theta/\omega, 0;\ 0)\phi(x) = e^{i\lambda\theta/\omega}U(0, -\theta/\omega;\ \theta)\phi(x). \end{aligned} \tag{6.1}$$

As we will see, the energy remains bounded if the spectrum is pure point, and it grows unbounded if the spectrum is continuous.

To make this statement more precise we introduce the following subspaces: Let $\mathcal{H}_{pp}$ and $\mathcal{H}_c$ be the pure point and continuous spectrum subspaces respectively corresponding to the Floquet operator $U(T,0)$. The subspace $\mathcal{H}_{be}$ of state trajectories with bounded energy is defined as

$$\mathcal{H}_{be} = \{\psi \in \mathcal{H} \mid \lim_{E\to\infty} \sup_{t\geq 0} \|F(H_0 > E)U(t, 0;\ \theta)\psi\| = 0\}$$

and the subspace of states with precompact trajectories by

$$\mathcal{H}_{pc} = \{\psi \in \mathcal{H} \mid \textit{closure of } \{U(t,0;\ \theta)\psi\ ,\ t \geq 0\} \textit{ is compact}\}$$

i.e. the trajectories evolve except for a small correction in a space of finite dimension. More precisely, given $\psi \in \mathcal{H}_{pc}$, for any $\varepsilon > 0$ there is a decomposition

$$\psi(t) = \psi_f(t) + \vartheta(t) \tag{6.2}$$

such that for all $t \geq 0$, $\psi_f(t)$ is in a finite dimensional subspace and $||\vartheta(t)|| < \varepsilon$.

Theorem 6.1: Consider the time dependent Hamiltonian

$$H(t) = H_0 + V(x,t); \qquad V(x,t+T) = V(x,t)$$

such that the spectrum of the unperturbed H_0 is discrete and bounded from below (e.g. having a confining potential, or defined on a compact manifold) and such that the propagator $U(t,s)$ exists as a strongly continuous family of unitary operators. Then

$$\mathcal{H}_{pp} = \mathcal{H}_{be}; \qquad \mathcal{H}_c = \mathcal{H}_{be}^{\perp}$$

Remark: Point spectrum does not guarantee that the expectation value of the energy $< \varphi(t),\ H_0\varphi(t) >_{\mathcal{H}}$ is bounded. It is in principle possible that a state is well localized in energy space (i.e. $\in \mathcal{H}_{be}$), but has a tail that decays too slowly and therefore has an infinite expectation of the energy. An equivalent statement is that not all of the eigenfunctions of K are necessarily in the domain of the energy operator. This phenomenon can become manifest already in finite time or only in the limit $t \to \infty$.

For the proof [79] one uses the following two theorems due to V. Enss and K. Veselić:

Theorem 6.2 [71]: Let C be any compact operator and $\psi \in \mathcal{H}_c$, then there is a non-negative function $f(\tau)$ such that

$$\frac{1}{\tau}\int_0^{\tau} dt\ ||C\ U(t,0)\psi|| \leq f(\tau)\ ||\psi||$$

with

$$\lim_{|\tau|\to\infty} f(\tau) = 0$$

This is a formulation of a RAGE theorem [80] for time dependent Hamiltonians. [78, 71]

Theorem 6.3 [71]: $\qquad \mathcal{H}_{pp} = \mathcal{H}_{pc}$.

Remark: There is a recent new application of the Floquet formalism as a method to calculate non-adiabatic Berry phases [81].

There has been much progress in solving the time periodic problem in some special cases. For Hamiltonians that are quadratic in p and q with time-periodic coefficients, the possible asymptotic behaviors have been completely classified [82], and the spectrum was related to properties of the corresponding classical model. In particular, models describing ions in Paul or in Penning traps (involving parametrically driven quadratic potentials and $1/|x|^2$ interactions [83, 64]) have been studied in detail. In addition the spectrum of the Floquet operator has been proven to be pure point for the smoothly kicked rotator with small coupling [84]. Some general criteria for stability of the point spectrum have been developed in Refs. 85, 86.

For a harmonic oscillator driven by a periodic force $F(t+T) = F(t)$,

$$H(t) = \frac{1}{2}p^2 + \frac{1}{2}\omega_0^2 x^2 - xF(t) \equiv H_0 - xF(t)$$

the spectral properties of the Floquet operator can be calculated explicitly [79]. The result is that in case of resonance ($\omega_0 = k\omega,\ k \in Z,\ \omega \equiv 2\pi/T$) the spectrum is absolutely continuous, whereas otherwise it is pure point. This result is obtained as follows:

The propagator can be expressed as [71]

$$U(t,0) = e^{-i\delta(t)} e^{i\sigma_2(t)x} e^{-i\sigma(t)\hat{p}} e^{-iH_0 t}$$

where $\hat{p} = -i\partial/\partial x$ and

$$\sigma_1(t) = \frac{1}{\omega_0}\int_0^t ds\ F(s)\sin\omega_0(t-s)$$

$$\sigma_2(t) = -\int_0^t ds\ F(s)\cos\omega_0(t-s)$$

$$\delta(t) = -\frac{1}{2}\int_0^t ds\ \left[\sigma_2^2(s) - \omega_0\sigma_1^2\right]$$

We distinguish two cases:

I) Resonance $\omega_0 = k\omega, k \in Z$: The operator e^{-iH_0T} reduces to multiplication by a constant. This is easily seen by applying it to a basis of eigenfunction $\{\varphi_n\}$ of H_0:

$$e^{-iH_0T}\varphi_n = e^{-i(n+1/2)\omega_0 T}\varphi_n = e^{-i\pi k}\varphi_n = \pm\varphi_n$$

The spectrum is therefore transient absolutely continuous [82].

II) Non-resonant case: We use the following identity, which is valid when $\omega_0 t \neq 2\pi k, k \in Z$:

$$e^{i\beta x} e^{-i\alpha\hat{p}} e^{-iH_0 t} = e^{i\theta} e^{i(\mu x - \gamma\hat{p})} e^{-iH_0 t} e^{-i(\mu x - \gamma\hat{p})}$$

with

$$\begin{pmatrix} \gamma \\ \mu/\omega_0 \end{pmatrix} = (1-R)^{-1}\begin{pmatrix} \alpha \\ \beta/\omega_0 \end{pmatrix}$$

where R_t is the rotation matrix by an angle $\omega_0 t$ and θ is a constant. (This is easily verified by application on a basis of coherent states). As a consequence the Floquet operator is unitarily equivalent (up to a constant phase ρ) to e^{-iH_0T} and has pure point spectrum with eigenvalues

$$e^{-i\lambda_n T} = e^{-i\omega_0(n+1/2)T - i\rho T}, \qquad n = 1, 2, , \ldots$$

The eigenfunctions ϕ_n can be expressed in terms of the eigenfunctions φ_n of H_0:

$$\phi_n(x) = e^{i(\mu x - \gamma \hat{p})}\varphi_n(x) = ce^{i\mu x}\varphi_n(x-\gamma)$$

where

$$\begin{pmatrix} \mu \\ \gamma/w_0 \end{pmatrix} = (1-R_T)^{-1}\begin{pmatrix} \sigma_1(T) \\ \sigma_2(T)/\omega_0 \end{pmatrix}.$$

Using the relation (6.1) between Floquet and quasienergy properties, we obtain that in the case of the non-resonant driven harmonic oscillator, the complete basis of eigenfunctions and the corresponding eigenvalues of the quasi-energy operator K are given by

$$\lambda_{n,m} = \omega m + \omega_0(n+\frac{1}{2}) + \rho; \qquad n = 1,2,\ldots; m \in Z$$
$$\psi_{n,m}(x,\theta) = e^{im\theta}\psi_n(x,\theta)$$

where

$$\psi_n(x,\theta) = e^{i\lambda_n\theta/\omega}U(\theta/\omega, 0;\ 0)\phi_n(x)$$

This example illustrates a property that is expected to be quite general: If ω and ω_0 are not commensurate, the point spectrum is non-degenerate and dense. This is the origin of small denominator problems in perturbation analysis of such operators. If ω and ω_0 are commensurate (but non-resonant) then there is a finite number of eigenvalues λ and they have infinite degeneracy.

This example also illustrates that in general the usual methods of perturbation theory will not be applicable to the quasienergy spectrum. This can be seen already from a low order perturbation expansion. Consider a perturbation of the harmonic oscillator leading to a quasienergy operator of the form

$$K_\epsilon = K + \varepsilon Y$$

The lower order terms for the formal perturbation expansion for the eigenvalues are

$$\lambda^\epsilon_{n,m} = \lambda_{n,m} + \varepsilon Y_{n,m;n,m} + \varepsilon^2 \sum_{(n',m')\neq(n,m)} \frac{|Y_{n',m';n,m}|^2}{\lambda_{n,m} - \lambda_{n',m'}}$$

and for the eigenfunctions

$$\psi^\epsilon_{n,m} = \psi_{n,m} + \varepsilon \sum_{(n',m')\neq(n,m)} \frac{Y_{n',m';n,m}}{\lambda_{n,m} - \lambda_{n',m'}}\psi_{n',m'}$$

As a consequence of the fact that the unperturbed eigenvalues are dense, the convergence of the sums appearing above is not clear. We encounter here a problem with "small denominators" that is mathematically similar to the one encountered in perturbation expansions in classical mechanics. There one manages to get control of the perturbation by means of the K.A.M. iteration. A similar type of K.A.M. technique has been developed for the present kind of quantum perturbation theory [84, 85].

6.1. Howland's Criterion

It is not easy in general to determine the spectrum of the quasienergy operator. For periodic forces some tools have been developed that give concrete information on the type of spectrum that one finds for some classes of models. One of the more powerful tools is the following criterion devised by J. Howland [86]. One considers Hamiltonians of the form

$$H = H_0 + V(x, \omega t)$$

where H_0 is time independent with a purely discrete nondegenerate spectrum $0 < E_0 < E_1 < \ldots$, satisfying the growth property

$$E_n - E_{n-1} \geq cn^{\alpha} \tag{6.3}$$

where c and $\alpha > 0$ are constants. Further, we assume that the perturbation V is bounded, time periodic with period T, and with zero average

$$\int_0^T dt\ V(x, \omega t) = 0$$

Theorem 6.4 (Howland): If the difference between successive eigenvalues of H_0 grows as cn^{α} with some $\alpha > 0$ and $V(t)$ is r times continuously differentiable (C^r), such that $r \geq [1/\alpha] + 1$ (where the bracket denotes the integer part), then the quasienergy operator

$$K = -i\omega\frac{\partial}{\partial\theta} + H_0 + V(x, \theta)$$

has no absolutely continuous spectrum.

Remarks:

1) This result in not of perturbative nature; it applies to bounded but arbitrarily large V's.

2) This result is formulated apparently for Hamiltonians H_0 of an arbitrary number of degrees of freedom. However, the condition (6.3) on the growth of the eigenvalues is not satisfied by most systems of more than one degree of freedom. The result is thus useful only for one dimensional systems.

3) The theorem requires that the unperturbed H_0 has non-degenerate eigenvalues. This excludes e.g. examples like the periodically driven rotator, which has doubly degenerate eigenvalues. This requirement is not essential for the method, and an improved version has been announced that doesn't require it [87].

4) The growth requirement (6.3) excludes e.g. the harmonic oscillator, since there $\alpha = 0$.

5) The theorem does not exclude the presence of singular continuous spectrum.

6) In spite of the mentioned limitations the result is quite strong. Together with other less rigorous arguments and with numerical simulations the general

emerging picture is that there will be few rather exceptional cases of periodically driven systems of one degree of freedom with unstable behavior. Numerical simulations indicate [88] that higher dimensional systems are more prone to instabilities. Also if the system is driven with two or more incommensurate frequencies new instabilities have been observed [89]. We will discuss this case in Section 7.

6.2. Application: Quantized Fermi Acceleration

Fermi devised a simple model meant to describe the acceleration of cosmic rays by collisions. The model involves a particle bouncing on a wall that moves periodically. The model has been extensively studied in the framework of classical mechanics. It shows strong chaotic behavior when the amplitude of the motion of the wall is sufficiently large [27]. The unbounded acceleration which was originally expected does not take place, but considerable energies can be reached. Different properties of the quantum version of the model have been studied [90, 91]. Recently P. Seba [92] showed using Howland's criterion that if the motion of the wall is sufficiently smooth, the corresponding quasienergy spectrum has no absolutely continuous part. From the results we have discussed at the beginning of this section one concludes that the energy growth is also bounded in the quantum problem (if we disregard the possibility of singular continuous spectrum, which in any case would correspond to a very slow process). Further he showed that the condition of differentiability is essential. He constructed a simple model in which the motion of the wall has discontinuities, which has absolutely continuous spectrum and an unbounded growth of energy. We summarize here some of the main steps of this analysis. The quantized Fermi accelerator is described by a free particle with Hamiltonian $H(t) = -\partial^2/\partial x^2$ inside of a box with a fixed wall at $x = 0$ and another one at $x = L(t)$ that moves periodically: $L(t+T) = L(t)$. We assume that $L(0) = 1$ and $\dot{L}(0) = 0$. The propagator U satisfies $i\partial U/\partial t = H(t)U$. The problem can be transformed into one with fixed boundary conditions on the interval $[0,1]$ by the time dependent unitary transformation $W(t) : L_2([0,1], dx) \to L_2([0, L(t)], dx)$ defined by

$$[W^{-1}\varphi](x) = L(t)^{1/2}\, exp(-iL(t)\dot{L}(t)x^2/4)\varphi(L(t)x)$$

We then define a unitary operator $\tilde{U}$ on $L_2([0,1], dx)$ by

$$\tilde{U}(t) = W^{-1}(t)U(t)$$

which satisfies

$$\begin{aligned} i\frac{\partial \tilde{U}(t)}{\partial t} &= [W^{-1}(t)H(t)W(t) - iW^{-1}(t)\frac{\partial W(t)}{\partial t}]\tilde{U}(t) \\ &= \frac{1}{L(t)^2}[-\frac{\partial^2}{\partial x^2} + L(t)^3\ddot{L}(t)\ x^2/4]\tilde{U}(t) \end{aligned}$$

Finally we rescale time by introducing

$$\tau = \zeta(t) \equiv \int_0^t ds\ \frac{1}{L(s)^2} \qquad \textit{and} \qquad V(\tau) = \tilde{U}(\zeta^{-1}(\tau))$$

and obtain the new equation

$$i\frac{\partial}{\partial\tau}V(\tau)=\tilde{H}(\tau)V(\tau)$$

with

$$\tilde{H}(\tau)=-\frac{\partial^2}{\partial x^2}+g(\tau)x^2,\qquad g(\tau)=[L(\zeta^{-1}(\tau))]^3\ddot{L}(\zeta^{-1}(\tau))/4 \tag{6.4}$$

which is a Schrödinger equation with periodic time dependent Hamiltonian but now with fixed Dirichlet boundary conditions $\varphi(x=0,t)=0=\varphi(x=1,t)$. The function g is periodic with a rescaled period $\tilde{T}=\zeta(T)$. The Floquet operator associated to the new problem $V_F=V(\tilde{T})$ turns out to be identical with the Floquet operator $U_F=U(T)$ of the original problem:

$$V_F=V(\tilde{T})=V(\zeta(T))=\tilde{U}(T)=W^{-1}(T)U(T)=U(T)=U_F.$$

Here we have used $W(T)=1$. The boundary conditions are also the same, since $L(T)=L(0)$ and thus in the original problem they always have a constant value at times that are integer multiples of the period T. Thus we can obtain information about the stability of the original problem by studying the spectrum of V_F.

We can now apply Howland's criterion, since the eigenvalues of the unperturbed operator $\tilde{H}_0=-\partial^2/\partial x^2$ in $L_2([0,1],dx)$ are $E_n=\pi^2n^2$, and thus

$$E_n-E_{n-1}\sim n^2-(n-1)^2\sim n$$

i.e. $\alpha=1$. We remark also that, if $g(\tau)$ is e.g. differentiable, the perturbation $g(\tau)x^2$ is bounded since x^2 acts only on the finite interval $[0,1]$. Howland's criterion implies therefore that if $g(\tau)$ is twice continuously differentiable (C^2), which in terms of the original problem means $L(t)\in C^4$, then the Floquet operator does not have any absolutely continuous spectrum.

In order to show that smoothness plays an essential role, Seba [92] constructed the following counterexample [92]: Consider a wall oscillating according to the function

$$L(t)=\begin{cases} b+(t-2nT)(a-b)/T, & \text{if } t\in[2nT,(2n+1)T],\quad n\in Z\\ a-((t-(2n+1)T)(a-b)/T, & \text{if } t\in[(2n+1)T,(2n+2)T],\quad n\in Z\end{cases}$$

where a,b are constants and T is the period. Applying the above transformation to this example one obtains a Hamiltonian (6.4) with

$$g(\tau)=c_1\sum_{n\in Z}\delta(\tau-2nT/ab)-c_2\sum_{n\in Z}\delta(\tau-(2n+1)T/ab)$$

where $c_1=b(a-b)/2T$, $c_2=a(a-b)/2T$, and δ is the Dirac δ-function. The Floquet operator can then be written as

$$V_F=e^{i(Tc_2/ab)x^2}e^{-i(T/ab)\tilde{H}_0}\,e^{-i(Tc_1/ab)x^2}e^{-i(T/ab)\tilde{H}_0}$$

If we now make the special choice $T = T_0 = 2ab/\pi$ then the exponential containing the Laplacean becomes trivial: The spectral decomposition in terms of the eigenfunctions φ_n of $\tilde{H}_0$, and eigenvalues $E_n = \pi^2 n^2$ is

$$e^{-i(T_0/ab)\tilde{H}_0} = \sum_n e^{-i(T_0/ab)\pi^2 n^2} |\varphi_n >< \varphi_n| = \sum_n e^{-i2\pi n^2} |\varphi_n >< \varphi_n|$$
$$= \sum_n |\varphi_n >< \varphi_n| = 1.$$

Thus the Floquet operator is just a multiplication operator

$$V_F = e^{i\left((a-b)^2/2ab\right)x^2}$$

and therefore it has absolutely continuous spectrum. One can also easily show that the expectation value of the kinetic energy grows quadratically with the number of oscillations.

7. Quasiperiodic Driving Forces

Existing studies indicate that the quantum limitation of diffusion is weaker or non-present for nonperiodic perturbations. This has been observed numerically for the quasi-periodically kicked rotator, with two or three incommensurate frequencies [89]. Luck, Orland and Smilansky showed that for a class of two-level systems with an almost-periodic force generated by the Fibonacci sequence the response generically is not almost-periodic [93]. Other examples of two-level systems were studied numerically [94-96]. Kicked systems modulated by substitution sequences (e.g. Morse-Thue) of different degrees of randomness were studied by M. Combescure [97, 98]. In these examples the correlation functions have singularly continuous spectrum. For the randomly kicked rotator, it has been shown that the energy grows unbounded [99]. This is in fact the case for a general class of quantum systems, in which the time-dependence is given by a Markov process [100, 101]. It appears from that analysis that quantum systems may be even more unstable than classical systems under such random perturbations.

The problem of stability was investigated [79] in the context of a harmonic oscillator subjected to a general time dependent force. The simplicity of the model, in which the classical and quantum behavior are similar, made it possible to establish some general relations between the ergodic properties of the dynamical system $\theta(t)$ and the growth of the energy of the oscillator. It was found that in general when the autocorrelation of the force decays fast enough the system behaves chaotically and the energy grows linearly. In fact, it was shown there for an explicit example with good ergodic properties, that the energy behaves like the square of a Gaussian random variable with variance proportional to time. When the correlations do not decay rapidly the situation is more complicated. The study of some quasiperiodic examples in Ref. 79 indicate that the asymptotic behavior depends on the fine details of the particular model. There are examples both of bounded and of unbounded energy.

Here we will discuss the case of quasiperiodic perturbations. We will see that there are important qualitative differences between the periodic and quasi-periodic case. This manifests itself particularly clearly when the Hilbert space $\mathcal{H}$ is finite dimensional, like e.g. for a spin 1/2 in a variable magnetic field, or for an N-level system in a radiation field.

In the periodic case the Floquet operator is a finite dimensional unitary matrix, and thus the spectrum of the quasi-energy operator is always pure point. This need not be so for quasi-periodic fields. We will discuss two different types of mechanism that lead to continuous spectrum: the first one involves phenomena that are close to resonances, i.e. they appear when the ratio of the frequencies are well approximated by rationals. The second mechanism is of topological nature and can appear for arbitrary frequencies. The close to resonant mechanism appears already in the simplest case in which $\mathcal{H}$ is one-dimensional. By diagonal composition one obtains also examples for the N-level models. For the second kind one needs at least two degrees of freedom.

We first discuss some general properties of quasiperiodic systems. It is useful to introduce a generalized Floquet operator [102], whose spectral properties are equivalent to the ones of the quasi-energy operator. It acts on $\mathcal{K}_1 \equiv \mathcal{H} \otimes L_2(S^1, d\theta_1)$ and is defined by

$$U_F = T^1_{-T_2} u_1(\theta_1) \tag{7.1}$$

where $u_1(\theta_1) \equiv U(T_2, 0; \theta_1, 0)$ ($\equiv$ monodromy operator) and

$$[T^1_{-T_2}\phi](\theta_1) = \phi(\theta_1 - \omega_1 T_2).$$

The relation (6.1) can be generalized to this case [102]:

(i) If $\phi \in \mathcal{K}_1$ is an eigenfunction of

$$U_F \phi = e^{-i\lambda T_2}\phi \tag{7.2}$$

then the function ψ defined by

$$\psi(\theta_1, \theta_2) = e^{i\lambda\theta_2/\omega_2} U(0, -\theta_2/\omega_2;\ \theta_1, \theta_2)\phi(\theta_1 - \theta_2\ \omega_1/\omega_2) \tag{7.3}$$

belongs to $\mathcal{K}$ and is an eigenfunction of the quasienergy operator, with eigenvalue λ:

$$K\psi = \lambda\psi. \tag{7.4}$$

(ii) Conversely, if $\psi \in \mathcal{K}$ is an eigenfunction of the quasienergy operator, then there is a function $\phi \in \mathcal{K}_1$ such that ψ can be represented by (7.3), and ϕ is an eigenfunction of the generalized Floquet operator.

7.1. Point Spectrum and Stability

Theorem 7.1: For initial states $\varphi \in \mathcal{H}$ such that $\varphi \otimes 1$ is in the pure point subspace $\mathcal{K}_{pp} \subset \mathcal{K}$ associated with the quasienergy operator the time evolution $\varphi(t)$ is almost-periodic (for almost all θ) [103].

This is a direct consequence of the general property (5.10), the particular form of the translation $\mathcal{T}_t$, and the uniform convergence of the expansion: A uniformly convergent series of quasiperiodic functions is almost periodic [103].

Corollary 7.2: If $\varphi \otimes 1 \in \mathcal{K}_{pp}$ then φ has bounded energy (for almost all θ). More precisely,

$$\varphi \in \mathcal{H}'_{be} = \{\varphi \in \mathcal{H} \mid \lim_{E\to\infty} \sup_{t\geq 0} \|F'(H_0 > E)U(t,0;\theta_1,\theta_2)\varphi\|_{\mathcal{H}} = 0\} \tag{7.5}$$

where $F(H_0 > E) : \mathcal{H} \to \mathcal{H}$ is the projector into the subspace generated by eigenfunctions of H_0 with energy larger than E.

Remark: As is the case of periodic force, one cannot conclude that the expectation value of the energy is finite, since the decay of the tail might be too slow.

7.2. Continuous Spectrum, RAGE Theorem

As opposed to the periodic case [79, 71], continuous spectrum of the quasienergy operator does not necessarily imply unbounded growth of the energy of the system. As we will see there are counterexamples, e.g. for a spin 1/2 in a quasi-periodic magnetic field, for which the energy is trivially bounded but the spectrum of the quasienergy operator is continuous [104]. Here we consider some energy-like operators that play a role analogous to that of the unperturbed energy operator H_0 in the periodic case.

Let $\mathcal{K}_{1,pp}$ and $\mathcal{K}_{1,c}$ be the subspaces of $\mathcal{K}_1$ corresponding respectively to the point and continuous spectrum of the generalized Floquet operator U_F.

(i) Let $\Pi_0(E_0) : \mathcal{H} \to \mathcal{H}$ be the projector into the subspace of $\mathcal{H}$ generated by the eigenfunctions of $H_0\varphi^{(0)} = E\varphi^{(0)}$ with $E < E_0$.

(ii) Let $\Pi_1(E_1)$ be the projector of $L_2(S^1_{T_1}, d\theta_1)$ into the subspace generated by the eigenfunctions of $H_1 \equiv -\partial^2/\partial\theta_1^2$ with energy smaller than E_1.

(iii) $\Pi(E_0, E_1) = \Pi_0(E_0) \otimes \Pi_1(E_1)$ is then a projector in $\mathcal{K}_{1,c}$ into the subspace in which both energies are bounded $H_0 < E_0$, $H_1 < E_1$. Continuous spectrum implies that one or both of the energies H_0, H_1 grows unbounded:

Theorem 7.3: Let $\varphi \in \mathcal{H}$ be such that $\varphi \otimes 1 \in \mathcal{K}_{1,c}$. Then for any fixed energies E_0, E_1, $\quad sup_{t\geq 0} \left\|[1_1 - \Pi(E_0, E_1)]\ U(t,0;\ \theta_1, 0)\varphi\right\|^2_{\mathcal{K}_1} = 1$.

Corollary 7.4: Let $\varphi(t) = U(t,0;\theta_1,0)\varphi$. The expectation value $< H_0 + H_1 > (t) \equiv < \varphi(t),\ [H_0 + H_1]\varphi(t) >_{\mathcal{K}_1}$ is unbounded. More precisely, $\forall E > 0$ there is a time t_1 such that either $< H_0 + H_1 > (t_1) > E$ or $\varphi(t_1)$ is not in the domain of $H_0 + H_1$ (i.e. $< H_0 + H_1 > (t_1)$ is infinite).

7.3. Scalar Examples with Continuous Spectrum

We consider two-level models whose general Hamiltonian acting on $\mathcal{H} = C^2$ is of the form

$$H = h_0(t)1 + \sum_{j=1}^{3} h_j(t)\sigma_j \tag{7.6}$$

where σ_j are the Pauli matrices, and $h_j(t)$ are real quasiperiodic functions, i.e. $h_j(t) = \bar{h}_j(\omega_1 t + \theta_1, \omega_2 t + \theta_2)$, where $\bar{h}_j(\theta_1, \theta_2)$ are continuous and 2π-periodic in the two arguments $\theta_1, \theta_2 \in S^1$. Without loss of generality we can take $h_0 = 0$ since its effect is only a global time dependent phase that can be studied with the methods of the previous section. The Hamiltonian is then an hermitean 2×2 matrix with trace zero, and thus the propagator $U(t,s;\theta)$ is unitary with determinant one (i.e. $\in SU(2)$).

The special case

$$H = h_3(t)\sigma_3 = h_3(t)\begin{pmatrix} 1 & 0 \\ 0 & -1 \end{pmatrix} \tag{7.7}$$

can be reduced to the solution of a scalar problem. We will show that already in this simplest case it is possible to have continuous quasienergy spectrum. In the scalar case the Hamiltonian is a function

$$H = -f(\omega_1 t + \theta_1, \omega_2 t + \theta_2) \tag{7.8}$$

that acts mutiplicatively on $\mathcal{H} = C$. The propagator can be written explicitly as

$$U(t,s\ ;\ \theta) = exp(+i\int_s^t dt'\ f(\omega_1 t' + \theta_1, \omega_2 t' + \theta_2)) \tag{7.9}$$

The question of stability reduces thus to the question if the integral of a quasi-periodic function is almost-periodic [103, 105]. The Floquet operator is

$$U_F = \mathcal{T}^1_{-T_2}\, exp(-iv(\theta_1)) \tag{7.10}$$

with

$$v(\theta_1) = +\int_0^{T_2} dt'\ f(\omega_1 t' + \theta_1, \omega_2 t'). \tag{7.11}$$

We first formulate a sufficient condition for point spectrum:

Theorem 7.5: If $\alpha = \omega_1/\omega_2$ satisfies a diophantine condition with $\gamma > 0$ and $\sigma > 1$ such that for all $n \in Z^2$, $(n_1 \neq 0)$

$$|\alpha n_1 + n_2| > \frac{\gamma}{|n_1|^\sigma} \tag{7.12}$$

and if f has a Fourier representation

$$f(\theta_1, \theta_2) = \sum_{n \in Z^2} \tilde{f}(n) e^{in\cdot\theta} \tag{7.13}$$

such that

$$\sum_{n\in Z^2} |\tilde{f}(n)||n_1|^{\sigma} < \infty \tag{7.14}$$

then the spectrum of the Floquet operator is pure point.

Remark: The complement of the set of α's satisfying (7.12) has zero measure. We give the proof of this result since it illustrates some of the general mechanisms that one encounters in this kind of problems.

Proof: We will use the fact that the eigenfunctions of U_F have constant absolute value [102]. We can therefore represent them as

$$\phi(\theta_1) = e^{i\varphi(\theta_1)} \tag{7.15}$$

The eigenvalue equation can then be expressed as

$$v(\theta_1) + \varphi(\theta_1) = \mu + \varphi(\theta_1 + 2\pi\alpha), \qquad (mod\ 2\pi) \tag{7.16}$$

with $\mu = -\lambda T_2$. The condition that $\varphi \in L_2(S^1, d\theta_1)$ is equivalent to the requirement that ϕ is a measurable function. We will try to find eigenfunctions representing them by Fourier series:

$$\varphi(\theta_1) = \sum_{n_1\in Z} \tilde{\varphi}(n_1)e^{in_1\theta_1}, \qquad v(\theta_1) = \sum_{n_1\in Z} \tilde{v}(n_1)e^{in_1\theta_1}. \tag{7.17}$$

Inserting into (7.16) with *mod* 2π omitted we get

$$\tilde{v}(n_1) + \tilde{\varphi}(n_1) = \mu\delta_{n_1,0} + \tilde{\varphi}(n_1)e^{2\pi i n_1\alpha}. \tag{7.18}$$

The equation for $n_1 = 0$ determines the eigenvalue: $\mu = \tilde{v}(0)$. We set $\tilde{\varphi}(0) = 0$, which is only a trivial phase. The other n_1 yield

$$\tilde{\varphi}(n_1) = \frac{\tilde{v}(n_1)}{e^{2\pi i n_1\alpha} - 1} \qquad n_1 \neq 0 \tag{7.19}$$

To show the convergence of the Fourier series we first write

$$v(\theta_1) = \sum_n \tilde{f}_n e^{in_1\theta_1} \int_0^{T_2} dt'\ e^{in\cdot\omega t'} \equiv \sum_n \tilde{f}_n\Gamma_n e^{in_1\theta_1} \tag{7.20}$$

and therefore

$$\tilde{v}(n_1) = \sum_{n_2} \tilde{f}_n\Gamma_n \tag{7.21}$$

Since $|\Gamma_n| \leq |T_2|$ we have

$$|\tilde{v}(n_1)| \leq |T_2| \sum_{n_2} |\tilde{f}_n| \tag{7.22}$$

The denominator in (7.19) can be estimated by

$$\pi(\alpha n_1)_{mod\ 1} \leq sin(2\pi\alpha n_1) \leq |e^{2\pi i\alpha n_1} - 1| \leq 2\pi(\alpha n_1)_{mod\ 1} \tag{7.23}$$

which combined with the diophantine condition (7.12) gives

$$|\tilde{\varphi}(n_1)| < \frac{|\tilde{v}(n_1)|}{\pi(\alpha n_1)_{mod\ 1}} < c_1 |n_1|^{\sigma} \sum_{n_2} |\tilde{f}_n|. \tag{7.24}$$

Therefore

$$\sum_{n_1} |\tilde{\varphi}(n_1)| < c_1 \sum_{n} |\tilde{f}_n| |n_1|^{\sigma} < \infty. \tag{7.25}$$

and thus the Fourier series converges.

Remarks:

1) It is clear from (7.19) and (7.21) that if $f(\theta)$ is a trigonometric polynomial the spectrum is pure point for any irrational α.

2) From (7.23) (7.19) we can deduce an inequality in the opposite direction:

$$|\tilde{\varphi}(n_1)| > \frac{|\tilde{v}(n_1)|}{2\pi(\alpha n_1)_{mod\ 1}} \tag{7.26}$$

By chosing an α that is well approximable by rationals (i.e. such that the terms in its continuous fraction representation grow fast enough), always is an infinite subsequence of n_1 such that for v arbitrarily smooth but not a trigonometric polynomial

$$(\alpha n_1)_{mod\ 1} < \frac{|\tilde{v}(n_1)|}{2\pi} \tag{7.27}$$

Then the Fourier series is not convergent. Although we cannot conclude from this that in that case there are no eigenfunctions (since the only requirement for $\varphi(\theta_1)$ is that it is a measurable function, which does not entail the existence of a Fourier representation), it is a good indication. The proof can in fact be done using other methods [104].

7.4. Two-level Systems with Continuous Quasienergy Spectrum

In this section we will discuss an example proposed by Rychlik [106], that has continuous spectrum for any frequency ratio α. We start with some general remarks. The Floquet operator is $U_F = T_{-T_2} u_1$ with the monodromy operator $u_1 \equiv U(T_2, 0\ ;\ \theta_1, 0) \in SU(2)$, i.e. it can be represented in general as

$$u_1(\theta_1) = \begin{pmatrix} a & b^* \\ -b & a^* \end{pmatrix}; \qquad |a|^2 + |b|^2 = 1 \tag{7.28}$$

We first remark that for any choice of functions $a(\theta), b(\theta)$ with $|a|^2 + |b|^2 = 1$ there is some quasiperiodic H of the form (7.6) such that (7.28) is the corresponding monodromy operator. In fact there is large family of them, and they can be chosen to be as smooth as a and b. In particular there are examples that are C^{∞} [106, 104].

The construction is based on the fact that $SU(2)$ is simply connected. It cannot be done in the scalar case because there we have $U(1)$ instead, which is not simply connected.

We consider the following example [106]:

$$u_1(\theta_1) = \begin{pmatrix} e^{i\theta_1} & 0 \\ 0 & e^{-i\theta_1} \end{pmatrix} \tag{7.29}$$

The corresponding Floquet operator $U_F = T^1_{-T_2} u_1$ has absolutely continuous spectrum for any irrational α.

Proof: We must show that there are no eigenvalues: Since the matrix u_1 is diagonal we can write any candidate for eigenvector of U_F as $(y(\theta_1), 0)$ or $(0, z(\theta_1))$. For the first type, the eigenvalue equation is equivalent to

$$e^{i\theta_1} y(\theta_1) = e^{-i\mu} y(\theta_1 + 2\pi\alpha) \tag{7.30}$$

If there were an eigenfunction $y \in L_2(S^1, d\theta_1)$, it would necessarily have a converging Fourier series, with coefficients $\tilde{y}_n$. Insertion into (7.30) yields

$$\tilde{y}_{n+1} = e^{-i\mu} e^{-2\pi i\alpha n} \tilde{y}_n, \tag{7.31}$$

which implies $|y_{n+1}| = |y_n|$ and therefore the Fourier series would not be convergent. As a consequence there are no eigenvalues.

Remark: We say that the continuous spectrum of this example has a topological origin because of the following property: We have seen that $y(\theta_1)$ and $z(\theta_1)$ have constant absolute value for almost all θ_1, and therefore we can write e.g. $y(\theta_1) = exp(i\varphi(\theta_1))$. The index of a function from S^1 to S^1 is defined as the number of times that the image wraps around the circle when θ_1 goes from 0 to 2π. In equation (7.30) the index of the left hand side is larger by one than that of the right hand side, and therefore the equation cannot have a solution.

8. Stability of the Point Spectrum: KAM Techniques in Quantum Mechanics

As already mentioned, the ordinary methods of perturbation theory cannot be applied to the quasienergy spectrum, because it is usually dense. In this section we illustrate a perturbation theory that was developed in order to control the small denominator problems that appear when the unperturbed eigenvalues form a dense set.

8.1. Properties of the Eigenfunctions of the Quasienergy Operator

We first summarize some general properties of the eigenfunctions of the Floquet operator of two-level systems, that are mostly a consequence of the ergodicity of the dynamical system on the torus Ω:

1) If ϕ_1, ϕ_2 are eigenfunctions of the Floquet operator U_F, then the absolute value of the $\mathcal{H}$-scalar product $|< \phi_1(\theta_1)\ ,\ \phi_2(\theta_1) >_{\mathcal{H}}| \equiv |\sum_{j=1}^N \phi_1^{(j)*}(\theta_1)\phi_2^{(j)}(\theta_1)|$ is constant for almost all $\theta_1 \in S^1$.

2) The eigenfunctions of U_F have $\mathcal{H}$-norm $||\phi(\theta_1)||_{\mathcal{H}}$ that is constant for almost all $\theta_1 \in S^1$.

3) For the case $\mathcal{H} = C^N$ with $N = 1, 2$, if U_F has one eigenfunction, then the whole spectrum is pure point.

Proof:

(i) N=1: If ϕ is an eigenfunction with eigenvalue Λ then $\phi_m = e^{im\theta_1}\phi, \quad m \in Z$ are also eigenfunctions with eigenvalue $\Lambda e^{-im2\pi\alpha}$. We will show that the $\{\phi_m\}$ form a complete basis of $\mathcal{K}_1 = L_2(S^1, d\theta_1)$: Since $\{e^{im\theta_1}\}$ is a basis and ϕ can be chosen such that $|\phi(\theta_1)| = 1$ for all θ_1, and thus $1/\phi$ is well defined. Therefore any function $f \in \mathcal{K}_1$ can be expanded as

$$f(\theta_1) = \sum_m a_m \phi_m(\theta_1), \tag{8.1}$$

where a_m is the Fourier coefficient of f/ϕ:

$$a_m = \frac{1}{2\pi}\int_{S^1} d\theta_1 \frac{f(\theta_1)}{\phi(\theta_1)} e^{-im\theta_1}. \tag{8.2}$$

(ii) N=2: We generalize the preceding argument as follows: First remark that if $\phi_1 = (y(\theta_1), z(\theta_1))$ is an eigenfunction with eigenvalue Λ, then we get a second one with eigenvalue Λ^* of the form $\phi_2 = (-z(\theta_1), y(\theta_1))$. This can be verified by direct insertion into the eigenvalue equation. Since the $\mathcal{H}$-norm of the eigenfunctions is constant almost everywhere, we can choose them such that $|y(\theta_1)|^2 + |z(\theta_1)|^2 = 1$ for all θ_1. We will prove that the set $\{\phi_1 e^{in\theta_1}, \phi_2 e^{in\theta_1}\}_{n\in Z}$ is a basis of $\mathcal{K}_1$, i.e. that any function $(\mu(\theta_1)\ ,\ \nu(\theta_1)) \in \mathcal{K}_1$ can be represented as a linear combination

$$\begin{pmatrix}\mu(\theta_1)\\ \nu(\theta_1)\end{pmatrix} = \sum_{n\in Z} a_n\phi_1 e^{in\theta_1} + b_n\phi_2 e^{in\theta_1}; \qquad a_n, b_n \in C \tag{8.3}$$

We construct a 2×2 matrix R that has the eigenfunctions ϕ_1 and ϕ_2 in the columns, and $R^\dagger$ is its conjugate

$$R = \begin{pmatrix} y & -z^* \\ z & y^* \end{pmatrix}, \qquad R^\dagger = \begin{pmatrix} y* & z^* \\ -z & y \end{pmatrix} \tag{8.4}$$

Viewed as operators in $\mathcal{K}_1$, they satisfy $RR^\dagger = 1 = R^\dagger R$. Equation (8.3) can be equivalently written as

$$\begin{pmatrix}\mu(\theta_1)\\ \nu(\theta_1)\end{pmatrix} = \sum_{n\in Z} e^{in\theta_1} R(\theta_1)\begin{pmatrix}a_n\\ b_n\end{pmatrix} \tag{8.5}$$

If we define a_n, b_n as

$$\begin{pmatrix}a_n\\ b_n\end{pmatrix} = \frac{1}{2\pi}\int_{S^1} d\theta_1{}'\ e^{-in\theta_1{}'} R^\dagger(\theta_1{}')\begin{pmatrix}\mu(\theta_1{}')\\ \nu(\theta_1{}')\end{pmatrix} \tag{8.6}$$

we verify by insertion that (8.5) is identically satisfied:

$$\begin{aligned}\begin{pmatrix}\mu(\theta_1)\\ \nu(\theta_1)\end{pmatrix} &= \frac{1}{2\pi}\sum_{n\in Z}\int_{S^1} d\theta_1{}'\ e^{in(\theta_1-\theta_1{}')} R(\theta_1)R^\dagger(\theta_1{}')\begin{pmatrix}\mu(\theta_1{}')\\ \nu(\theta_1{}')\end{pmatrix} \\ &= \int_{S^1} d\theta_1{}'\ \delta(\theta_1-\theta_1{}')R(\theta_1)R^\dagger(\theta_1{}')\begin{pmatrix}\mu(\theta_1{}')\\ \nu(\theta_1{}')\end{pmatrix} = \begin{pmatrix}\mu(\theta_1)\\ \nu(\theta_1)\end{pmatrix}\end{aligned} \tag{8.7}$$

3) This gives us an insight about the composition of the set of eigenfunctions of U_F. For $\mathcal{H} = C^N$ the set of eigenfunctions has the structure $\{e^{in\theta_1}\phi_j(\theta_1)\}$, $n \in Z, j = 1,\ldots,N$, where ϕ_j are N functions that are mutually orthogonal in $\mathcal{H}$ for almost all θ_1 (and therefore form a basis of $\mathcal{H}$ for a.a. θ_1). The corresponding eigenvalues are of the form $exp - i(\lambda_j T_2 + n_1 2\pi\alpha)$, and thus they form a dense set.

8.2. Superconvergent Perturbation Theory

As a consequence of the results of the preceding properties, when the spectrum of the quasienergy operator is pure point it has the structure

$$\lambda_{n,m} = \bar{\lambda}_m + n_1\omega_1 + n_2\omega_2, \qquad m = 1,\ldots N, \quad n \in Z^2 \tag{8.8}$$

i.e. it is a dense subset of R. Therefore one cannot apply the usual perturbation theory [107], since already in the first order terms there appear sums whose convergence is not clear due to the presence of small denominators. Under suitable diophantine conditions on the frequencies one can control the convergence of the perturbation series by using a KAM-type technique introduced by Bellissard [84–85].

Theorem 8.1: Let

$$K = -i\omega_1\frac{\partial}{\partial\theta_1} - i\omega_2\frac{\partial}{\partial\theta_2} + H_0 + \varepsilon V(\theta_1,\theta_2) \equiv K_0 + \varepsilon V(\theta_1,\theta_2) \tag{8.9}$$

where H_0 and $V(\theta_1,\theta_2)$ are hermitean 2×2 matrices, H_0 being constant and $V(\theta_1,\theta_2)$ such that each component is an analytic function in the strip $\{\theta \mid \mathcal{I}m\theta_j < r_0\}$. Assume e.g. that $\bar{\alpha} = \omega_2/\omega_1 > 1$, and that $(2\beta/\omega_1)_{mod\ 1} > 0$, where $2\beta > 0$ is the difference between the two eigenvalues of H_0. Then, for any given $\eta > 0$ and fixed ω_1 there is a set of $\bar{\alpha}$'s $S_\eta \subset (1,\infty)$ of Lebesgue measure $|S_\eta| < \eta$ and a value $\varepsilon_c(\eta)$, such that if $\bar{\alpha} \in (1,\infty)\setminus S_\eta$ and $\varepsilon < \varepsilon_c$, the spectrum of K is pure point.

8.3. Sketch of the Proof

The main ingredients of the construction can be described as follows. Without loss of generality we set

$$H_0 = \beta\begin{pmatrix} 1 & 0 \\ 0 & -1 \end{pmatrix} \tag{8.10}$$

We construct a unitary transformation $R(\bar{\alpha},\varepsilon)$ such that

$$RKR^{-1} = K^{(0)} + \begin{pmatrix} g_+(\bar{\alpha},\varepsilon) & 0 \\ 0 & g_-(\bar{\alpha},\varepsilon) \end{pmatrix} \tag{8.11}$$

where $g_\pm$ are independent of $\theta \equiv (\theta_1,\theta_2)$. Stated equivalently, R transforms K into an operator that is diagonal in the basis of eigenfunctions of the unperturbed $K^{(0)}$, which are of the form

$$\psi_{n,m} = \begin{cases} e^{in\cdot\theta}\binom{1}{0}, & \text{if } m = +1 \\ e^{in\cdot\theta}\binom{0}{1}, & \text{if } m = -1 \end{cases}; \qquad n \equiv (n_1,n_2) \in Z^2; m \in \{+1,-1\}. \tag{8.12}$$

corresponding to eigenvalues $\omega \cdot n \pm \beta$, where $\omega = (\omega_1, \omega_2)$. The transformation R is constructed by iteration $R = \ldots R_k \ldots R_2 R_1$; at each step the order of the θ-dependent perturbation is reduced from order ε^j to order ε^{2j}.

The kth iteration step is defined as follows: We start with an operator of the form

$$K_k = D_k + V_k; \qquad D_k \equiv K^{(0)} + g_k \tag{8.13}$$

where g_k and D_k are diagonal in the basis $\{\psi_{n,m}\}$ and V_k is hermitean. g_k has been generated in the previous iterations and can depend explicitly on $\bar{\alpha}$, but it has no dependence on θ. We represent the transformation as $R_{k+1} = e^{W_{k+1}}$, with $W_{k+1}^{\dagger} = -W_{k+1}$.

Expanding the exponentials and regrouping the terms that we expect to be of the same order as V_k (i.e. making the hypothesis that will be justified a posteriori that W_{k+1} is of the same order as V_k) we obtain

$$\begin{aligned} K_{k+1} \equiv e^{W_{k+1}} K_k e^{-W_{k+1}} &= K_k + [W_{k+1}, K_k] + \frac{1}{2!}[W_{k+1}, [W_{k+1}, K_k]] \\ &\quad + \frac{1}{3!}[W_{k+1}, [W_{k+1}, [W_{k+1}, K_k]] + \ldots \\ &= D_k + V_k + [W_{k+1}, D_k] + V_{k+1} \end{aligned} \tag{8.14}$$

where we have selected

$$V_{k+1} = [W_{k+1}, \{[W_{k+1}, D_k]\frac{1}{2!} + V_k\}] + [W_{k+1}, [W_{k+1}, \{[W_{k+1}, D_k]\frac{1}{3!} + V_k\frac{1}{2!}\}]] + \ldots \tag{8.15}$$

We will work with the matrix representation corresponding to the basis $\{\psi_{n,m}\}$, using the notation

$$W_{k+1}(m, m', n - n') \equiv < \psi_{n,m} \ , \ W_{k+1}\psi_{n',m'} > \tag{8.16}$$

The matrix elements depend only on the difference $n - n'$ because of the special form (8.12) of the eigenfunctions $\psi_{n,m}$. For the diagonal operators D we use the simplified notation

$$D(m, n) \equiv < \psi_{n,m} \ , \ D\psi_{n,m} > \tag{8.17}$$

and

$$g_k(m) \equiv < \psi_{n,m} \ , \ g_k\psi_{n,m} > \tag{8.18}$$

which is independent n since g_k has no dependence on θ.

Motivated by the decomposition (8.14) we proceed as follows:

1) Determine W_{k+1} and a diagonal δg such that

$$e^{W_{k+1}} K_k e^{-W_{k+1}} = D_k + \delta g + V_{k+1} \tag{8.19}$$

This will be satisfied if W_{k+1} and δg satisfy the equation

$$[W_{k+1}, D_k] + V_k = \delta g \tag{8.20}$$

which one can solve explicitly: The diagonal terms of (8.20) yield

$$\delta g(m) = V_k(m, m, \underline{0}) \tag{8.21}$$

and one can choose $W_{k+1}(m, m, \underline{0}) = 0$. The off-diagonal terms yield

$$W_{k+1}(m, m', n - n') = \frac{V_k(m, m', n)}{D_k(m, n) - D_k(m', n')} \tag{8.22}$$

The denominator

$$D_k(m, n) - D_k(m', n') = \omega \cdot (n - n') + (m - m')\beta + g_k(m) - g_k(m') \tag{8.23}$$

can be zero or arbitrarily close to zero for infinitely many indices $m, m', n - n'$. To guarantee the convergence of the series (8.22) and to obtain suitable estimates, we fix ω_1 and restrict the values of $\bar{\alpha}$ to the ones belonging to the set characterized by the following diophantine condition:

$$\begin{aligned}\Omega_{k+1}(\gamma_{k+1}) = &\{\bar{\alpha} \in \Omega_k(\gamma_k); \textit{such that } \forall n \in Z^2, \textit{ and } m, m' \in \{+1, -1\} \\ &|\omega \cdot n + (m - m')\beta + g_k(m) - g_k(m')| \geq \frac{\gamma_{k+1}}{(1 + |n|)^\sigma}\}\end{aligned} \tag{8.24}$$

where n and $m - m'$ are not simultaneously zero, $\Omega_0 = (1, \infty)$, $\sigma > 2$ is fixed and γ_{k+1} is a constant that we will choose for each step. It can be shown that the Lebesgue measure of the complement of this set is small. Since the g_k in the denominator depend on $\bar{\alpha}$, in order to get estimates of diophantine type we need to keep control on the size of g_k as well as on its variation as a function of $\bar{\alpha}$. This is achieved by choosing an appropriate norm [85]. The details of the proof are given in Ref. 104.

9. Quantum Perturbation Theory in Differential Form: Relation with Classical Integrable Many-Particle Systems

A new formulation of quantum perturbation theory has been recently developed [109-111] for Hamiltonians of the form

$$H = H_0 + \gamma V$$

Instead of working with a series in powers of the perturbation parameter γ, one constructs a differential equation for the eigenvalues $E_n(\gamma)$ as a function of γ. The terms appearing in the equation depend on the other eigenvalues $E_{n'}(\gamma)$ but also on the matrix elements of the perturbation $V_{n,n'} = < \varphi_n(\gamma)\ ,\ V\ \varphi_{n'}(\gamma) >$. Therefore, in order to get a closed system of differential equations one has to write also equations for the matrix elements. The solution of the eigenvalue problem for arbitrary γ is then equivalent to solving this system of differential equations. It turns out that if we reinterpret γ as a time variable, this system of equations has the form of the *classical* Hamilton equations for a system consisting of an infinite number of particles moving in one dimension. The system is not only Hamiltonian but it is even completely integrable, in the sense that it has an infinite number of conserved quantities. Similar models had in fact been studied from a mathematical point of view, precisely because of their integrability properties. They are of the same type as e.g. the Toda lattice, or the Korteweg-deVries and sine-Gordon equations, which are prototypes of soliton behavior. The concrete system that arises from the quantum perturbation theory can be considered as a generalization [112, 113] of a model first treated by Calogero [114] and Moser [115].

The complete integrability, which at first might seem surprising, is quite natural in hindsight. It means that the dynamics of the system of classical particles can be determined by algebraic methods. This is precisely what we started with, since the positions of the particles at time γ are just the eigenvalues, that in principle can be determined by diagonalizing an infinite matrix, i.e by an algebraic operation.

The formalism was first developed for eigenvalues of time independent Hamiltonians. It was then extended to the treatment of the eigenvalues of the Floquet and the quasienergy operators [116, 48, 117]. This formalism provides a new method to calculate eigenvalues numerically, by solving a (truncated) system of ordinary Hamilton equations, instead of diagonalizing truncated matrix representations. Explicit examples in the time independent case have been treated using traditional molecular dynamics techniques [118, 119].

Acknowledgements:

I would like to thank P. Bleher, L. Bunimovich, M. Combescure, G. Gallavotti, J.L. Lebowitz, M. Nerurkar, A. Pellegrinotti, and P. Seba for useful discussions and contributions. This work has been partially supported by a grant of the Swiss National Science Foundation and AFOSR Grant 87-0010.

References

1. G. Gallavotti; *The elements of mechanics*, Springer Vlg. 1983.
2. V.I. Arnold; *Mathematical Methods of Classical Mechanics*, Springer Vlg. Berlin 1978.
3. H. Goldstein; *Classical Mechanics*, Addison-Wesley, Reading, MA, 1980.
4. R. Tabor; *Chaos and Integrability in Nonlinear Dynamics*, J. Wiley 1989.
5. R.D. Ruth; A canonical integration technique, IEEE Tran. Nucl. Sci. **NS-30** (1983) 2669.
6. E. Forest, R.D. Ruth; Fourth order symplectic integration, Physica **D 43** (1990) 105.
7. C.R. Menyuk; Some properties of the discrete Hamiltonian method, Physica **11 D** (1984) 109.
8. R.I. McLachlan, P. Atela; The accuracy of symplectic integrators, Nonlinearity **5** (1992) 541.

9. Feng Kang, Qin Meng-zhao; Hamiltonian algorithms for Hamiltonian systems and a comparative numerical study, Comput. Phys. Commun. **65** (1991) 173.
10. P.J. Channel, C. Scovel; Symplectic integration of Hamiltonian systems, Nonlinearity **3** (1990) 231.
11. J.M. Sanz-Serna; Symplectic integrators for Hamiltonian problems: an overview, Acta Numerica **1** (1992) 243.
12. B. Chirikov, Phys. Rep. **52** (1979) 263.
13. J. Candy, W. Rozmus; A symplectic integration algorithm for separable Hamiltonian functions, J. Comput. Phys. **92** (1991) 230.
14. F. Rannou; Numerical study of discrete plane area-preserving mappings, Astron. and Astrophys. **31** (1974) 289.
15. D.J.D. Earn, S. Tremaine; Exact numerical studies of Hamiltonian maps: iterating without roundoff error, Physica **D** (1992) to appear.
16. D.J.D. Earn, S. Tremaine; Exact solutions for Hamiltonian systems using integer arithmetic, in B. Sundelius (ed.), *Dynamics of Disc Galaxies (1991)*, Göteborg, Sweden.
17. C.F.F. Karney; Long-time correlations in the stochastic regime, Physica **D8** (1983) 360.
18. L.A. Smith; Strange accumulators, Ann. NY Acad. Sci. **497** (1987) 61.
19. C. Scovel; On symplectic lattice maps, Phys. Let. **A 159** (1991) 396.
20. R.S. MacKay, J.D. Meiss;*Hamiltonian dynamical systems*, Adam Hilger, Bristol, 1987.
21. J.D. Meiss; Symplectic maps, variational principles, and transport, Rev. Mod. Phys. **64** (1992) 795.
22. D.F. Escande; Stochasticity in classical Hamiltonian systems: universal aspects, Phys.Rep.**121** (1985) 165.
23. D.F. Escande, F. Doveil; J. Stat. Phys. **26** (1981) 257.
24. M. Month, J.C. Herrera; *Nonlinear dynamics and the beam-beam interactions*, AIP, New York, 1979.
25. J.M. Jowett, M. Month, S. Turner; *Nonlinear Dynamics Aspects of Particle Acceleratoes*, Lecture Notes in Physics Vol. 247, Springer Vlg., Berlin 1986.
26. S. Wiggins; *Introduction to Applied Nonlinear Dynamical Systems and Chaos*, Springer Vlg., Berlin 1990.
27. A.J. Lichtenberg, M.A. Lieberman; *Regular and Stochastic Motion*, Springer Vlg., Berlin 1983.
28. J. Moser; Recent developments in the theory of Hamiltonian systems, SIAM Review, **28** (1986) 459.
29. J. Denzler; Mather sets for plane Hamiltonian systems, J. Appl. Math. and Phys. (ZAMP) **38** (1987) 791.
30. G. Gallavotti; Quasi-integrable mechanical systems, in Les Houches Lectures XLIII: *Critical phenomena, random systems, gauge theories*, edited by K. Osterwalder and R. Stora, Elsevier Publ., 1986.
31. W. Thirring; *A course in mathematical physics I: classical dynamical systems*, Springer Vlg., Berlin 1992.
32. V.I. Arnold, A. Avez; *Problèmes ergodiques de la mécanique classique*, Gauthier-Villars, Paris 1967.
33. J. Guckenheimer, P. Holmes; *Nonlinear Oscillations, Dynamical Systems, and Bifurcations of Vector Fields*, Springer Vlg., Berlin 1983.

34. H.R. Jauslin; Melnikov's criterion for nondifferentiable weak-noise potentials, J. Stat.Phys. **42** (1986) 573, and Physica **A 144** (1987) 179.
35. J. Moser; *Stable and random motions in dynamical systems*, Princeton U. Press, 1973.
36. A.N. Kolmogorov, S.V. Fomin; *Introductory Real Analysis*, Dover Publ, New York 1970.
37. S. Aubry, P.Y. Le Daeron; The discrete Frenkel-Kontorova model and its extensions, Physica **8D** (1983) 381.
38. J. Mather; Existence of quasiperiodic orbits for twist homeomorphisms of the annulus, Topology **21** (1982) 457.
39. R.B. Griffiths, H.J. Schnellnhuber, H. Urbschat; Exactly solvable model for cantorus phase transitions, Phys. Rev. Lett. **65** (1990) 2551.
40. O. Bolza; *Variationsrechnung*, Chelsea Publ. , New York, 1961.
41. I.M. Gelfand, S.V. Fomin; *Calculus of Variations*; Prentice-Hall, NJ, 1963.
42. G.A. Bliss; *Lecture on the Calculus of Variations*, Univ. of Chicago Press, Chicago, 1946.
43. G.M. Ewing; *Calculus of Variations*, Dover Publ. New York, 1985.
44. J. Hietarinta; Phys. Lett. **A93** (1982) 55, and J. Math. Phys. **25** (1984) 1833.
45. D. Dürr, S. Goldstein, N. Zanghi; J. Stat. Phys. **67** (1992) 843.
46. M.C. Gutzwiller; *Chaos in Classical and Quantum Mechanics*, Springer Vlg., Berlin 1990.
47. L.E. Reichl; *The Transition to Chaos in Conservative Classical Systems: Quantum Manifestations*, Springer Vlg., Berlin 1992.
48. F. Haake; *Quantum Signatures of Chaos*, Springer Vlg., Berlin 1990.
49. A.M. Ozorio de Almeida;*Hamiltonian Systems: Chaos and Quantization*, Cambridge Univ. Pr., Cambridge, 1988.
50. *Quantum Chaos*, Les Houches Lectures 1989.
51. H.A. Cerdeira, R. Ramaswamy, M.C. Gutzwiller, G. Casati (eds.); *Quantum Chaos*, World Scientific, Singapore 1991.
52. G. Casati, L. Molinari; Progr. Theo. Phys. Suppl. **98** (1989) 287.
53. M. Reed, B. Simon; *Methods of Modern Mathematical Physics*, Vol IV, Thm XIII.76; Academic Press, New York 1979.
54. C. Piron, *Foundations of Quantum Physics*, Benjamin, Reading 1976.
C. Flesia, C. Piron; Helv. Phys. Acta **57** (1984) 697.
55. J.S. Howland; Math. Ann. **207** (1974) 315,
and Indiana Univ. Math. J. **28** (1979) 471.
56. J.E. Bayfield, P.M. Koch; Phys. Rev. Lett. **33** (1974) 258.
57. G. Casati, B.V. Chirikov, F.M. Izrailev, J. Ford; Lecture Notes in Physics Vol. 93, Springer Vlg., Berlin 1979.
58. G. Casati, I. Guarneri, J. Ford, F. Vivaldi; Phys. Rev. **A 34** (1986) 1413.
59. B.V. Chirikov, F.M. Izrailev, D.L. Shepelyansky;
Sov. Sci. Rev. Sec. C 2 (1981) 209.
60. F.M. Izrailev, D.L. Shepelyansky; Teor. Mat. Phys. **95** (1980) 417.
61. G. Casati, I. Guarneri; Comm. Math. Phys. **95** (1984) 121.
62. G. Casati, B.V. Chirikov, D.L. Shepelyansky; Phys. Rev. Lett. **53** (1984) 2525.
63. G. Casati, B.V. Chirikov, I. Guarneri, D.L. Shepelyansky;
Phys. Rep. 154 (1987) 77.

64. M. Combescure; Crystallization of trapped ions - a quantum approach, Ann. Phys. **204** (1990) 113.
65. S. Yücel, E. Andrei; Phys. Rev. **B 42** (1990) 2088, and **B 43** (1991) 12029.
66. R. Jensen; Phys. Rev. Lett. **49** (1982) 1365.
67. J.S. Howland; in *Mathematical methods and applications of scattering theory*, J.A. de Santo (ed.), Lecture Notes in Physics **130**, Springer Vlg. Berlin 1980.
68. C. Piron; *Mécanique quantique*, Presses polytechniques et universitaires romandes, Lausanne 1990.
69. J. Bellissard, in *Stochastic Processes in Classical and Quantum Systems* edited by S. Albeverio, G. Casati and D. Merlini, Springer 1986.
70. K. Yajima; Comm. Math. Phys. **87** (1982) 331.
71. V. Enss, K. Veselic; Ann. Inst. H. Poincaré **A 39** (1983) 159.
72. J.-M. Combes; Connections between quantum dynamics and spectral properties of time-evolution operators, preprint, Marseille 1992.
73. R. Ketzmerick, G. Petschel, T. Geisel; Slow decay of temporal correlations in quantum systems with Cantor spectra, Phys. Rev. Lett. **69** (1992) 695.
74. S. Fishman, D.R. Grempel, R.E. Prange; Phys. Rev. Lett. **49** (1982) 509.
75. J. Moser; Quasi-periodic solutions of nonlinear elliptic partial differential equations, Bol. Soc. Bras. Mat. **20** (1989) 29.
76. M. Levi; KAM theory for particles in periodic potentials, Ergod.Th.Dynam.Sys. **10** (1990) 777.
77. M. Levi; On Littlewood's counterexample of unbounded motions in superquadratic potentials, Dynamics Reported, C. Jones, U. Kirchgraber, H. Walther (eds.) **1** (1991) 113.
78. K. Yajima, H. Kitada; Ann. Inst. H. Poincaré **A 39** (1983) 145.
79. L. Bunimovich, H.R. Jauslin, J.L. Lebowitz, A. Pellegrinotti, P. Nielaba; J. Stat.Phys. **62** (1991) 793.
80. D. Ruelle, N. Cimento, **59A** (1969) 655; W.O. Amrein, V. Georgescu, Helv.Phys. Acta **46** (1973) 635.
81. D.J. Moore; The calculation of nonadiabatic Berry phases, Phys. Rep. **210** (1991) 1.
82. G.A. Hagedorn, M. Loss, J. Slawny; J. Phys. **A19** (1986) 521.
83. M. Combescure; Ann. Inst. H. Poincaré **44** (1986) 293, and Ann. Phys. **173** (1987) 210.
84. J. Bellissard; in *Trends in the Eighties*, edited by S. Albeverio and Ph. Blanchard, World Scientific, Singapore 1985.
85. M. Combescure; Ann. Inst. H. Poincaré **47** (1987) 63, and Ann. Phys. **185** (1988) 86.
86. J.S. Howland; Ann. Inst. H. Poincaré, **49** (1989) 309 and **49** (1989) 325.
87. G. Nenciu; private communication.
88. M. Toda, S. Adachi, K. Ikeda; Progr. Theo. Phys. Suppl. **98** (1989) 323.
89. M. Samuelides, R. Fleckinger, L. Touzillier, J. Bellissard; Europhys. Lett. **1** (1986) 203.
90. J.V. José, R. Cordery; Phys. Rev. Lett. **56** (1986) 290.
91. W.M. Visscher; Nucl. Phys. B, Proc. Suppl. **2** (1987) 619, and Phys. Rev. **A 36** (1987) 5031.

92. P. Seba; Quantum chaos in the Fermi-accelerator model, Phys. Rev. **A41** (1990) 2306.
93. J.M. Luck, H. Orland, U. Smilansky; J. Stat.Phys. **53** (1988) 551.
94. Y. Pomeau, B. Dorizzi, B. Grammaticos; Phys. Rev. Lett. **56** (1986) 681.
95. R. Badii, P.F. Meier; Phys. Rev. Lett. **58** (1987) 1045.
96. P.W. Miloni, J.R. Ackerhalt, M.E. Goggin; Phys. Rev. **A 35** (1987) 1714.
97. M. Combescure; *Recurent versus diffusive dynamics for a kicked quantum system*, J. Stat.Phys. **62** (1991) 779.
98. M. Combescure; *Recurrent versus diffusive dynamics for a kicked quantum oscillator*, preprint 1991.
99. I. Guarneri; Lett. Nuovo. Cim. **40** (1984) 171.
100. C.-A. Pillet; Comm. Math. Phys. **102** (1985) 237, and **105** (1986) 259.
101. C.-A. Pillet, in *Stochastic Processes in Classical and Quantum Systems* edited by S. Albeverio, G. Casati and D. Merlini, SpringerVlg., Berlin 1986.
102. H.R. Jauslin, J.L. Lebowitz; Chaos **1** (1991) 114.
103. L. Amerio, G. Prouse; *Almost-Periodic Functions and Functional Equations*, van Nostrand Reinhold Co., New York 1971.
104. P. Blekher,H.R. Jauslin, J.L. Lebowitz; Floquet spectrum for two-level systems in quasiperiodic time dependent fields, J. Stat.Phys. **68** (1992) 271.
105. B.M. Levitan, V.V. Zhikov; *Almost-periodic functions and differential equations*, Cambridge University Press, Cambridge 1982.
106. M. Rychlik; *Renormalization of cocycles and linear ODE with with almost-periodic coefficients*, Inventiones Math., to appear 1992.
107. T. Kato; *Perturbation Theory for Linear Operators*, Springer, Berlin 1980.
108. J. Bellissard; in *Chaotic Behavior in Quantum Systems*, edited by G. Casati, Plenum Press 1985.
109. Phys. Rev. Lett. **51** (1983) 943.
110. T. Yukawa; Phys. Rev. Lett. **54** (1985) 1883, Phys. Lett. **A 116** (1986) 227.
111. K. Nakamura, M. Lakshmanan; Phys. Rev. Lett. **57** (1986) 1661
112. J. Gibbons, T. Hermsen; Physica **D 11** (1984) 337.
113. S. Wojciechowski; Phys. Lett. **A 111** (1985) 101, Phys. Scripta **34** (1986) 304.
114. F. Calogero; J. Math. Phys. **12** (1971) 419.
115. J. Moser; Adv. Math. **16** (1975) 197.
116. K. Nakamura, H.J. Mikeska; Phys. Rev. **A 35** (1987) 5294.
117. R. Scharf, M. Kuś; Level dynamics for a periodically driven quantum system, Phys. Rev. **A 40** (1989) 4774.
118. Xiazhou Yang, J. Burgdörfer; Molecular dynamics approach to the statistical properties of energy levels, Phys. Rev. Lett. **66** (1991) 982.
119. H. Ishio, K. Nakamura; Phys. Rev. **A 46** (1992) R2193.

NONEQUILIBRIUM ORDERING AND CRITICAL PHENOMENA IN LATTICE SYSTEMS[a]

J. MARRO,[b] A. ACHAHBAR and J. J. ALONSO

Departamento de Física Aplicada, Facultad de Ciencias,
Universidad de Granada, Campus de Fuentenueva,
E-18071-Granada, España

ABSTRACT

The computer implementation of four kinetic lattice model systems in which the condition of detailed balance is frustrated by the presence of either competing tendencies or else driven dissipative fields is illustrated in Section 2. The systems exhibit interesting nonequilibrium ordering and critical phenomena. Some of the most outstanding facts concerning this behaviour are reviewed in Section 3.

1. Introduction

The onset of ordering in nonequilibrium conditions is related to several interesting phenomena in nature. Two main kinds of macroscopic ordering have been described within the realm of statistical physics in systems that are separate from thermodynamic equilibrium. One belongs to the field known as *kinetics of phase transitions*. It deals basically with time evolution properties as the system relaxes towards an ordered state, e.g., one in which two thermodynamic phases coexist. We shall not refer to this case hereafter.[c] Instead, our interest in the following is in the kind of ordering that occurs under suitable experimental *steady* nonequilibrium conditions. It has been described variously as dissipative structures, self organization, broken symmetries, pattern formation, etc. The more canonical term *nonequilibrium phase transitions and critical phenomena* is also applicable given that, in principle, some apparent similarity exists with the equilibrium counterpart.[1] The similarity is not simple: the precise extension to nonequilibrium phenomena of concepts such as symmetry breaking, thermodynamic instability and universality, for example, remains an open problem. It reflects the relatively poor development of nonequilibrium theory which needs to deal, almost by definition, with much more varied and complex situations than equilibrium theory.

As a matter of fact, the study of macroscopic, observable phenomena in physics and

---

[a] Supported by the Spanish *DGICYT*, Project PB91-0709, and *PAI* (*Junta de Andalucía*), and by the Commission of the European Communities, Contract CI1.0409.

[b] Electronic mail address: *JMARRO @ UGR . ES*

[c] See, for instance, the paper by A. M. Lacasta *et al* in this volume, and references therein.

biology, for instance, reveals that most systems of interest cannot be assumed to be in the equilibrium state described by thermodynamics. In practice, one of the simplest situations one observes in nature is stationary, but it involves systems that are not in equilibrium due to their relation with the environment or with an external agent. Thus, system and environment are not in equilibrium among themselves, and a net (e.g.) steady current of some extensive quantity, such as matter and/or energy, occurs with the exterior. The system is then said to be in a *steady nonequilibrium state.*[a] A more precise and general definition of the latter would be useful. Following the methods of equilibrium theory, steady nonequilibrium states have been associated sometimes to an extremum of a potential function which generalizes more or less naturally the concept of entropy. A similar strategy leads to the search for prescriptions to specify appropriate ensembles as in Gibbs theory. None of those approaches has so far provided a systematic method to obtain the required information about any steady condition, however. In fact, the present investigation of steady nonequilibrium states is still based rather on a collection of methods, most of them approximate, each developed to deal with a particular situation.[2-4]

It is our aim in the present lecture notes to illustrate how the methods of *computational physics* may help the understanding of nonequilibrium ordering given the situation sketched above. More precisely, Ising-like, lattice models, which can be implemented very efficiently in a computer, may be defined. They capture some of the essence in natural steady nonequilibrium conditions that are characterized by the presence of some ordering, cooperative phenomena.[b] Significant information concerning specific situations may be obtained by means of computer simulations of the behaviour of such mathematically well-defined systems. Furthermore, a systematic study may lead to extract some general facts about nonequilibrium phenomena and, one hopes, it may help the development of theory given the versatility of the models.

2. The Models

Consider a periodic lattice in d dimensions with N sites and toroidal boundary conditions; say, $\Lambda=\{x\in Z^d\}$. Let $s = \{s_x=\pm 1\}$ be a *configuration*. The two possible values of the occupation or spin variable s_x represent, respectively, either particle-hole or spin *up-down* states at site x . The models of interest have in common also that s evolves in time due to a stochastic process which is characterized by the following *master equation* for the probability of s at time t :

$$\partial P(s;t)/\partial t = L\,P(s;t) \equiv \sum_{s'} [c(s'\to s)\, P(s';t) - c(s\to s')\, P(s;t)]$$

---

[a] For some general facts about steady nonequilibrium states see Refs. 2-4, for instance.

[b] Ref. 5 contains a detailed description of several irreversible Ising-like model systems, and some related analytical results.

Here, $c(s \rightarrow s')$ denotes the probability per unit time for a transition from s to a new configuration s'. Such stochastic transitions are assumed to be induced by the contact with a thermal bath.[6-9]

In practice, different models correspond to different realizations of L. Two main cases have been described in the literature:

(*i*) The process is implemented in a computer choosing an arbitrary initial configuration and performing the following Monte Carlo step reiteratively: First, a site x is selected at random; then, one attempts to *flip spin* s_x, i.e., one attempts the change $s_x \rightarrow -s_x$, according to some given criterion (cf. later on). The corresponding operator L is called the *Glauber generator*,[6] to be denoted L_G, and s' is specifically written as s^x.

(*ii*) Alternatively, the initial arbitrary configuration is changed reiteratively by the following Monte Carlo step: First, two nearest-neighbour (*NN*) sites, x and y, are selected at random; then, one attempts the interchange, $s_x \rightleftarrows s_y$, of the corresponding occupation variables (only) if $s_x s_y = -1$ (i.e., if any one of the sites is occupied by a particle and the other is empty) according to some given criterion. The corresponding operator is called the *Kawasaki generator*,[6] to be denoted L_K, and s' is specifically written as s^{xy}.

The criterion one uses in (*i*) and (*ii*) to decide whether or not the attempted move should be performed is crucial to determine the nature of the asymptotic steady state to which the system evolves as $t \rightarrow \infty$. Consider the following one:

(*iii*) Define a configurational energy given (e.g.) by the *NN* Ising Hamiltonian, namely, $H_J(s) = -J \sum_{NN} s_x s_y$ using some standard notation.[7] Then, compute the energy cost associated to the move, i.e., $\Delta H_J \equiv H_J(s') - H_J(s)$. The move is performed with probability $\phi(\beta \Delta H_J) = min\{1, exp(-\beta \Delta H_J)\}$, where $\beta = (k_B T)^{-1}$ is the inverse temperature of the involved bath. Such a procedure corresponds to the so-called *Metropolis algorithm.*[8]

Further choices are allowed for the function $\phi(X)$, in principle.[5] It is critical in determining the asymptotic steady state, however. The simplest case corresponds to any arbitrary positive function such that

$$\phi(X) = exp(-X)\, \phi(-X) ,$$

in addition to $\phi(0) = 1$, and $\phi(X) \rightarrow 0$ as $X \rightarrow \infty$. The latter assure proper boundary conditions. The first mentioned condition on $\phi(X)$ corresponds to the so-called *detailed balance* which is sufficient (but not necessary) to guarantee that the system will evolve in time towards the canonical equilibrium state. That is, in as long as the detailed balance condition is satisfied, which is the case of the Metropolis algorithm of interest, the stationary solution of the master equation we have introduced above is $P_{st}(s) \propto exp[-\beta H_J(s)]$, as one may convince oneself. Consequently, the kinetic Ising magnetic model[6] and the Kawasaki binary alloy model[6] correspond, respectively, to the generators L_G, as defined in (*i*), and L_K, as defined in (*ii*), if they are complemented with any criterion, such as (*iii*), that satisfies detailed balance.[9] Otherwise, if the latter is not fulfilled, the master equation may imply tendency towards a steady nonequilibrium state that may differ essentially (from a conceptual point of view, at least) from an equilibrium state. It occurs, for instance, when the stochastic process defined by L corresponds to one of the cases defined below. In fact, none of them has a proper Hamiltonian, in which they also differ from the ordinary, equilibrium lattice models. A configurational function is still involved in some of the cases, however, which for simplicity is taken equal to the function $H_J(s)$ defined above, i.e., the *NN* Ising Hamiltonian.

2.1. *Driven Diffusive System Near Dimension Two* [10]

Consider a lattice, Λ , consisting of two twin square lattices, $\lambda_1=\lambda_2=\lambda$, such that any site has five *NN*, with one of them in the other plane; $\Lambda \equiv \lambda_1 \bigcup \lambda_2$ and $\lambda_1 \bigcap \lambda_2 = \varnothing$. Kinetics proceed as a consequence of the Kawasaki generator L_K ; it implies, in particular, that $s_x = \pm 1$ correspond to particle-hole states, respectively, and that the density of particles, say ρ , remains constant during the time evolution, as in the familiar lattice-gas model.[7,9] There is no restriction on the possible configurations of Λ , e.g., any given particle has access to any of the two planes. Any two particles attract each other only when they are located at *NN* sites such that both belong either to λ_1 or to λ_2 ; i.e., $J>0$ anywhere except that all bonds between the planes are broken, with $J=0$.

When L_K is complemented with criterion *(iii)*, so that tendency towards canonical equilibrium exists, the phase diagram of the system in Λ differs essentially from the one for the ordinary two-dimensional lattice gas. For example, the system exhibits a phase transition which is of second order for any ρ (while it is of first order for $\rho < ½$ in the ordinary system) but Onsager critical exponents occur only for $\rho = ½$. This is consistent with the known fact that the nature of a phase transition depends on the dimensions and symmetries characterizing the system and its order parameter. In fact, the phase transition seems strongly influenced by the existence of an additional degree of freedom in Λ : the particles can hop to other plane (which justifies the present sub-section heading). The latter peculiarity of Λ is important from a practical point of view also: the relaxation towards ordered states via diffusion is faster in Λ than in the case of a single plane, and the phase transition is sharper.

The presence of steady nonequilibrium states may be induced by frustrating the condition of detailed balance, as indicated above. It occurs, for instance, if the particles are assumed to be driven by a constant external *electric* field $\boldsymbol{E}$ that points along one of the principal lattice directions. For simplicity, $E \rightarrow \infty$, so that no particle can jump against the field direction under the influence of thermal fluctuations. Then, the time evolution of $\boldsymbol{s}$ consists of stochastic jumps of particles to *NN* empty sites, including jumps from one plane to the other, biased by the existence of the field. That is, one implements case *(ii)* in the computer, and criterion *(iii)* is modified in the sense that one adds the work done by the field to the energy cost. Consequently,

$$\Delta H_J \equiv H_J(\boldsymbol{s}^{xy}) - H_J(\boldsymbol{s}) - \boldsymbol{E} \bullet (\boldsymbol{x}-\boldsymbol{y})(s_x - s_y) \ ,$$

where $H_J(\boldsymbol{s})$ does not involve any interaction between planes, i.e., it is the sum of two independent NN Ising Hamiltonians. Thus, the actual action depends on the direction of the vector x-y , e.g., $\phi=1$ or 0 (for $E=\infty$) when one attempts to move the particle in the directions of $+E$ or $-E$, respectively. We remark that, in addition to the fact that detailed balance only holds locally here, due to the last term in the expression for ΔH_J above, E is not the gradient of a potential, so that no obvious Hamiltonian including the electric energy exists, and a net steady *dissipative* current sets in for periodic boundary conditions. The resulting irreversible system has some practical interest; namely, it is a sort of *driven diffusive system* that may model low-dimensional solid electrolytes or fast ionic conductors.[a] The main facts concerning ordering in this system are briefly reviewed in Section 3.3.

2.2. *Generalized Reaction Diffusion Ising System* [13]

Consider a lattice gas (d = 1 or 2 in practice) with attracting interactions again. The stochastic changes in configuration s are a consequence now of a competition between *reaction processes* activated by a thermal bath at temperature T , and *diffusion processes* consisting of NN exchanges driven by a thermal bath at temperature T' . The two processes are independent in continuous time, with p the *a priori* probability of exchanges per bond and 1-p that of reactions per site. That is, the generator characterizing the master equation is

$$L = p\,L_K + (1\text{-}p)\,L_G$$

in the present case. Here, L_G and L_K are implemented in the computer by using criterion (*iii*) with probabilities $\phi[\Delta H_J(s^x)/k_BT]$ and $\psi[\Delta H_J(s^{xy})/k_BT']$, respectively, where $\Delta H_J(s')$ $\equiv H_J(s') - H_J(s)$ and $H_J(s)$ is the NN Ising Hamiltonian; s' represents either s^{xy} or s^x , depending on the process in consideration. Both ϕ and ψ are assumed to satisfy detailed balance, but their combination does not; in addition, one has $T \neq T'$. Consequently, the familiar systems with equilibrium states (e.g., the Ising magnetic system and the binary alloy model) only follow when one has either $p=0$ or $p=1$ strictly, while steady nonequilibrium states occur in general otherwise. The resulting behaviour is very interesting[b]; some details are given in Section 3.2.

2.3. *A Nonequilibrium Version of the Spin Glass Problem* [15]

Next, consider the ordinary magnetic Ising model on a square lattice. Let us modify the familiar kinetics, however, in the sense that the basic Monte Carlo step corresponds to a combination of two different, independent Glauber generators. Namely,

$$L = p\,L_G^{+} + (1\text{-}p)\,L_G^{-} ,$$

---

[a] A discussion on the relation between the model and real materials may be found in Ref. 11. We refer to Refs. 5 and 12, and references therein, for some related theory.

[b] Some theory concerning the reaction-diffusion system may be found in Ref. 14, and references therein.

where a mention to the sign of the exchange interaction J in the configurational function $H_J(s)$ is made. More explicitly, after a site x is selected at random, one attempts to flip spin s_x according to the following criterion: all the exchange interactions between s_x and its four *NN* are assumed to have the value $J_o>0$ with probability p, and $-J_o$ with probability $1\text{-}p$. The energy cost of the attempted flip is computed then by using H_J with J equal to the selected value. Thus, one gets either $\Delta H_{J_o} = 2J_o s_x \sum_{NN \text{ of } x} s_y$ or $\Delta H_{-J_o} = -2J_o s_x \sum_{NN \text{ of } x} s_y$, where the sums extend to all *NN* of site x. Following the Metropolis algorithm, the flip is performed with probability either $min\{1, exp(-\beta\Delta H_{J_o})\}$ or $min\{1, exp(-\beta\Delta H_{-J_o})\}$, respectively. Several facts concerning both the significance and the behaviour of this system are discussed in Section 3.1.[a]

2.4. *A Probabilistic Cellular Automaton* [19]

As a final example of a nonequilibrium lattice model, consider a lattice in d dimensions with probabilistic *cellular-automaton* kinetics. Namely, the thermal fluctuations which are represented by the master equation are absent, and one uses instead the following rule to attempt the flip of spin s_x : Compute

$$\alpha_x \equiv s_x \sum_{NN \text{ of } x} s_y ,$$

where the sum extends to the $2d$ *NN* of site x ; the flip is performed with probability p or $1\text{-}p$ according to whether α_x is negative or positive, respectively; the flip is performed anyway if α_x is zero. This criterion is similar (but not identical) to the majority rule that characterizes the so-called *voter model*.[20] It is also remarkable that it corresponds to the system in Section 2.3 when $T=0$. Thus, the ground state of that system may be studied by using the present rule which is very convenient for numerical analysis, e.g., thermal noise is absent. The resulting steady states exhibit a sort of nonequilibrium percolation phenomena whose critical exponents are the ones for the ordinary Ising model. Further details, and a discussion of these effects will be published elsewhere.[19]

3. Nonequilibrium Phase Transitions

3.1. *Competition of Glauber Generators*

We describe in this section some details concerning the nature and the behaviour of the model defined in Section 2.3 when it is implemented by a two-dimensional lattice.[b] An outstanding fact is that the system may be interpreted as a *nonequilibrium impure model*:

---

[a] Some theory for the nonequilibrium model in section 2.3 may be found in Refs. 16-18.

[b] Cf. Ref. 15 for a detailed description of Monte Carlo results for both $d = 2$ and 3 in which this section is based, Ref. 16 for some exact results, mainly concerning $d = 1$, and Ref. 17 for field theory.

Kinetics involve the simultaneous action of two independent spin-flip mechanisms, each satisfying individually detailed balance, which occur *as if* the exchange interaction between the involved spins has a different value. It may be proved[a] that such a competition is equivalent to the existence of fast and random, spin-configuration-independent diffusion of microscopic disorder, of the kind that takes place in spin glasses when magnetic ions diffuse.

The kinetic competing process drives asymptotically the system towards pure ferromagnetic and antiferromagnetic steady states for $p = 1$ and 0, respectively, while the situation is expected to be less simple otherwise. Namely, the competition generates steady nonequilibrium states in general. In fact, detailed balance is not satisfied, and no single Hamiltonian exist that characterizes microscopically the system. Note that the consideration of a *nonequilibrium* impure model is appealing because some *unusual* observations in disordered materials are consistent with the existence of nonequilibrium effects, e.g., a dependence of the steady state on the past history has been reported. On the other hand, the model, which one may expect *a priori* to behave differently from the (equilibrium) quenched spin-glass model, essentially differs also from the annealed version of the latter which involves an unrealistic representation (e.g.) of impurity diffusion towards a highly correlated distribution of bonds.

The Monte Carlo study of square lattices (the situation for $d=3$ is qualitatively similar [15]) reveals the following main facts:

(*A*) A ferromagnetic (antiferromagnetic) -like phase occurs at temperatures $T<T_C(p)$ for p large (small) enough. The system is symmetrical around $p=½$, so that only the case $p>½$ needs to be considered.

(*B*) The parameter p has three basic effects on the phase transition: (*a*) the degree of saturation revealed by the $m(T)$ curves as T is lowered tends to decrease with p; (*b*) the transition temperature $T_C(p)$ also decreases with p; and (*c*) there is no low-temperature ferromagnetic phase for any $p\leq 0.91$, approximately.

(*C*) The indicated phase transition is similar to the one in the pure system, i.e., to the one for the equilibrium case $p=1$. In particular, it is of second order, and data give consistency with the Onsager critical exponents.

Facts (*A*)-(*C*) correspond to a rather *conservative* picture, i.e., to the simplest one we might expect given the $\pm J$ competition that characterizes the system. This probably reflects the fact that the symmetries of the original Ising model are preserved. The (preliminary) observations that follow are thus intriguing.

(*D*) The ground $T=0$ state exhibits a sharp, second order phase transition for $p=p_o\approx 0.928$, as indicated in Section 2.4. It probably corresponds to a sort of *nonequilibrium percolation phenomena* of the kind that takes place in other lattice systems.[20] It is interesting that the associated critical behaviour, which is induced by a mechanism other than thermal fluctuations, is precisely the same as in the ordinary Ising model at finite temperatures. It agrees with the situation in other Ising-like systems[21-23], in accordance with

[a] Cf. Ref. 14 and references therein.

some expectations, unlike some critical behaviour described in Sections 2.1, 3.2 and 3.3.

(*E*) The data at finite temperature for $p=0.93$ suggest that critical exponents might differ from the Onsager values for $p_o<p\approx 0.93$ at least. This is not surprising in the light of the exact results found exactly for the one-dimensional system, which exhibits a rich, involved critical behaviour (near $T=0$),[16] but it is puzzling in the light of observations (*C*) and (*D*). One may argue that the percolation-like phenomena described in (*D*) influences the behaviour of the system at low T also, so that a sort of crossover is observed for $p\approx 0.93$.

(*F*) The data, e.g., direct inspection of spin configurations, also suggest the presence of a sort of order (with zero total and sublattice magnetizations, however) for $p<p_o$ at low temperature. No appropriate order parameter has been found to describe this effect quantitatively, however, the precise nature of which is uncertain at the time of writing.

3.2. *Competition of Glauber and Kawasaki Generators*

Monte Carlo results for the reaction-diffusion system in Section 2.2 have been reported elsewhere.[13,14] Here we describe the main facts from a mean-field theory[14] that seems to contain the essence of the system behaviour, and compare with existing Monte Carlo results. For $d>1$, the latter refer to the case in which the *reaction rate*, denoted $\phi[\Delta H_J(s^x)/k_BT]$, and the *diffusion rate*, denoted $\psi[\Delta H_J(s^{xy})/k_BT']$, both have been implemented by means of the Metropolis algorithm.

First, we remark that the mean-field theory clearly reveals the distinct role played by the two rates. Namely, while different realizations of the diffusion rate just induce minor quantitative differences, which may be interpreted as due to a change in the effective value for p (cf. the expression for L in Section 2.2), the choice of reaction rate may have a decisive, qualitative influence on the nature of the steady state. This is quite evident for the one-dimensional lattice where long range order below a finite temperature occurs for a certain family of reaction rates only. This family, that includes the case $\phi(X)=exp[\Delta H_J(s^x)/2k_BT]$, has in common that it favours low-energy states rather atypically (in which the Metropolis algorithm is excluded). Furthermore, it follows that the phase transition (which is of second order, by the way) occurs for any $p>0$. This result is consistent with the known exact results for $T'=\infty$ and $p\to 1$, and with a Monte Carlo simulation for $T'=\infty$ that uses the same rates as for the mean-field equations. The Monte Carlo data indicate the existence of long range order for $p\geq ½$ *at least* (the "experimental" situation for $p<½$ is not so clear cut due to the very slow evolution of the system that one should probably associate with a peculiarity of the relaxation towards an ordered state via diffusion).

The combined picture that ensues from exact, Monte Carlo and mean-field results, as depicted in Fig.1, for instance, is very interesting. That is, given that one may argue[14] that the present mean-field theory provides an accurate description of this problem, the chances are that the system has indeed a phase transition at finite temperature when $d=1$, which is a rare case, for the realization of rates indicated above.

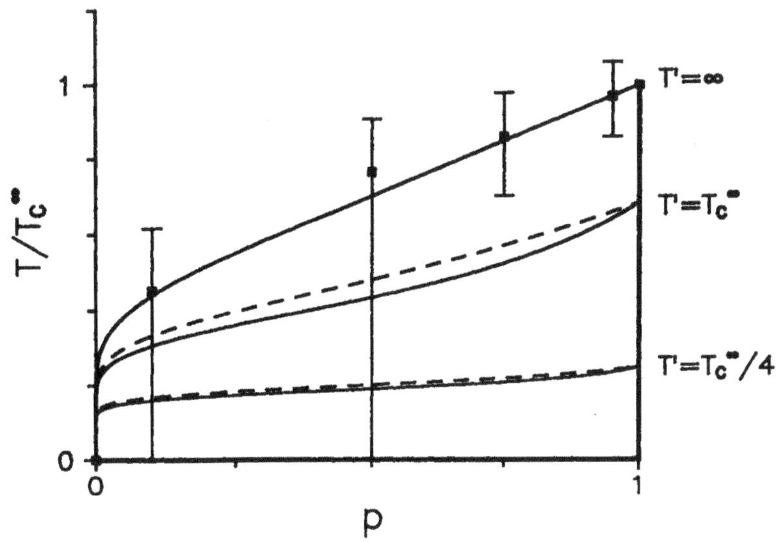

Figure 1: The phase diagram for one of the versions (cf. the main text) of the one-dimensional system in which long range order exists at finite temperature. The horizontal line corresponds to the equilibrium zero-T critical point. The right-most vertical line represents critical points for $p \to 1$ and any $T' > T$. The solid and dashed curves are, respectively, for diffusion rates of the Metropolis and Kawasaki types for different values of T', as indicated. The symbols represent Monte Carlo data whose error bars are also shown. $T_C^\infty \equiv 2J(k_B \log 3)^{-1}$ is the exact critical temperature for $p \to 1$ and $T' = \infty$.

The two-dimensional system with rates of the Metropolis type (cf. Fig.2) exhibits a phase transition whose order may vary from second to first (thus crossing a *tricritical point*) as T' is increased, depending on the value of p: For $p \to 1$, the transition is of second order for $T^B < T' < 2.57T^B$, where T^B is the Bethe temperature, the transition is standard of first order for $T' > 2.57T^B$, and it has weak discontinuities, and very slow relaxations (apparently due to existing coexistence of phases) for $T' << T^B$. The indicated ranges of T' values become narrower and tend, finally, to disappear as p is decreased; e.g., only transitions of second order occur for $p < 0.37$. The Monte Carlo analysis reveals the same qualitative picture. The mean-field theory demonstrates also that the use of the choice by Kawasaki[14] for both the reaction and diffusion rates (though only the change in the latter is relevant in practice) produces a qualitatively different picture: only second order phase transitions occur, and the critical temperature is an increasing function of both p and T'.

3.3. Kawasaki Generator with a Driving Field

The irreversible quasi two-dimensional lattice gas in Section 2.1 has been studied by the Monte Carlo method.[10] When the temperature T is high enough, the particles are always observed to distribute evenly between the two planes. The particle distribution within each plane is apparently homogeneous. The system exhibits two kinds of phase transitions as the temperature T is lowered, however. For $T<T_\infty(\rho)$, there is segregation within each plane into a particle-rich, say *liquid* phase and a particle-poor, *gas* phase. The liquid phase is highly anisotropical; namely, two approximately equal strips form along the field direction, one on top of the other. This case is illustrated in Fig.3.

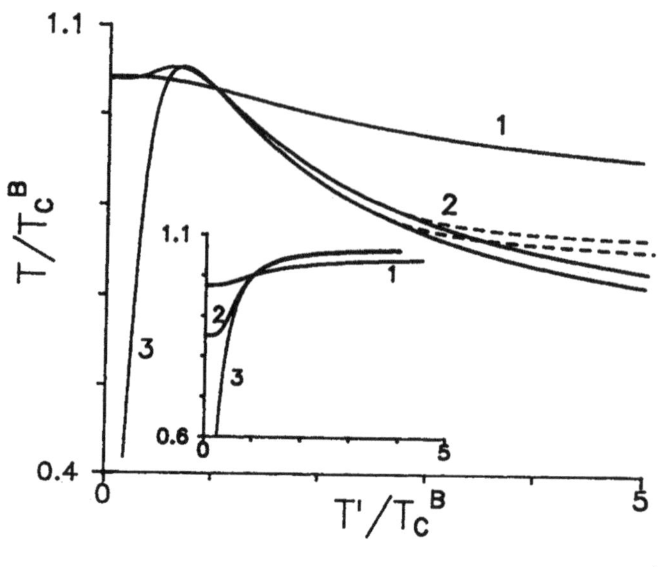

Figure 2: The phase diagram for the two-dimensional system as implied by the mean-field theory. The solid (dashed) lines represent the lower (upper) limits for the stability of the disordered (ordered) phase. The transition is of second order when both lines coincide, and of first order otherwise. T^B is the Bethe critical temperature. The main graph corresponds to the case in which the system evolves according to Metropolis reaction and diffusion rates, and different values of p : curves 1, 2 and 3 are for $p=0.5$, $p=0.95$ and $p\to 1$, respectively. The inset corresponds to a reaction rate of the Kawasaki type.

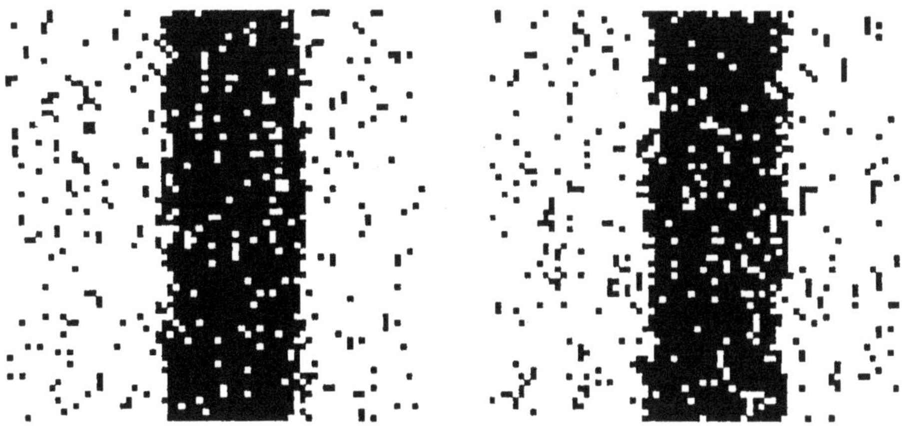

Figure 3: Typical steady state configurations for $T_\infty^*(\rho) < T < T_\infty(\rho)$ corresponding to the two planes of the "quasi two-dimensional" system with a driving dissipative field. The graphs have been obtained by the Monte Carlo method[10] for T equal to the Onsager critical temperature and $\rho = 0.35$.

In order to monitor the underlying anisotropy, one may computed, in particular, the *morphological order parameter* $m = \frac{1}{2}(m_1 + m_2)$, where $m_i = | <M_h^2> - <M_v^2> |^{\frac{1}{2}}$ and $M_{h(v)}^2 = | \lambda |^{-3/2} \sum_{h(v)} [\sum_{v(h)} (1 - 2\sigma_x)]^2$. Here, h (v) indicates summation along the horizontal (vertical) directions on each plane. (It seems that different realizations of the order parameter lead to the same conclusions but their use is less convenient in practice.) It follows that the phase transition at $T_\infty(\rho)$ is continuous for $\rho = \frac{1}{2}$, while it is of first order when $\rho < \frac{1}{2}$. The discontinuities are quite evident for $\rho << \frac{1}{2}$, and become smaller as one approaches $\rho = \frac{1}{2}$.

In addition to the phase transition revealed by m , there is another phase transition which is revealed, for instance, by the order parameter $\Delta\rho(T) \equiv | \rho_1(T) - \rho_2(T) |$, where $\rho_i(T)$ is the actual density in plane i at T . Fig.4 illustrates the behaviour of both $m(T)$ and $\Delta\rho(T)$ for a given density.

It is observed that $\Delta\rho$ increases sharply at $T_\infty^*(\rho)$; the situation is similar to the one in an equilibrium first-order phase transition for any ρ . The anisotropic liquid phase lies in one of the planes only below $T_\infty^*(\rho)$, while the gas phase fills out the rest, as illustrated in Fig.5.

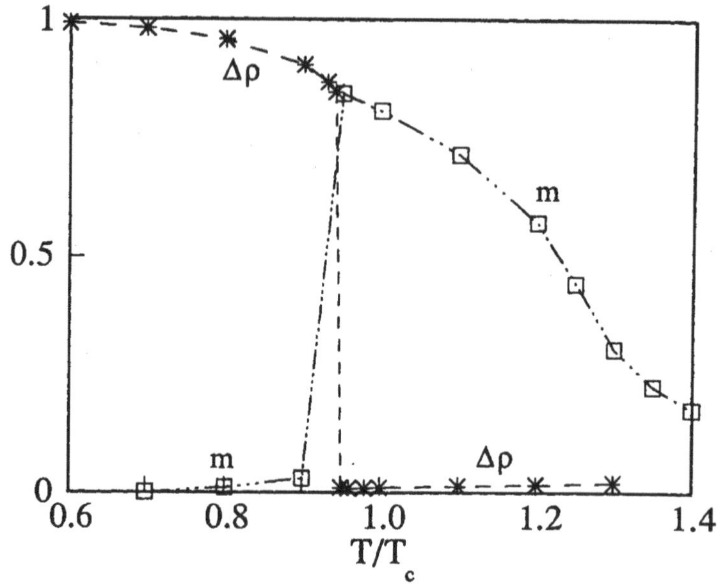

Figure 4: The variation with temperature of the two order parameters (cf. the main text) for $\rho = \frac{1}{2}$.

Figure 5: Typical steady state configurations for $T < T_\infty^*(\rho)$ corresponding to the same system. The graphs have been obtained by the Monte Carlo method[10] for $T = 0.8 T_C^{\mathrm{Onsager}}$ and $\rho = 0.35$.

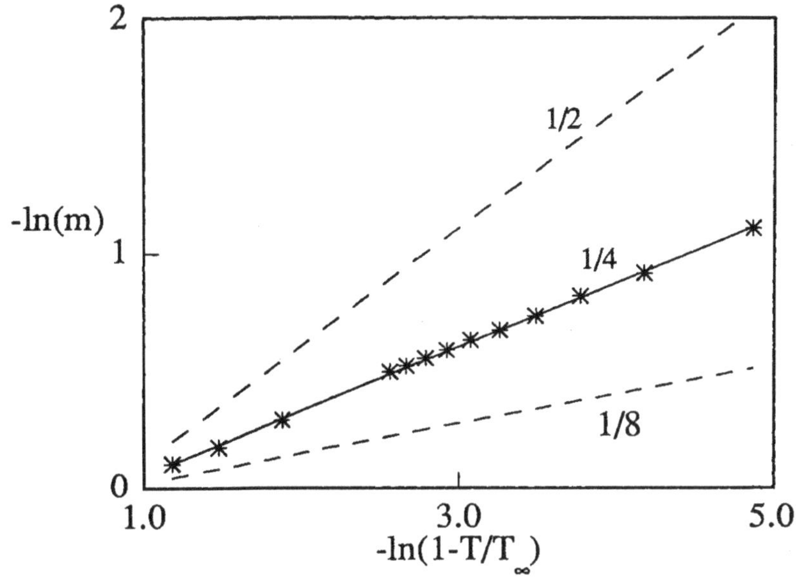

Figure 6: The log-log plot that is customary to obtain values for critical exponents is used here to show that the Monte Carlo data for m as $T \to T_\infty$ is well represented by the value $b \approx ¼$. For comparison purposses, the slopes corresponding to the classical and equilibrium values are also shown.

The critical behaviour $m \sim |\, T\text{-}T_\infty(\rho = ½)\, |^{\,b}$ as $T \to T_\infty$ is a specific question here. The data seems to exclude the classical Landau case, $b = ½$, and deviates clearly also from the Onsager equilibrium value, $b = ⅛$. Data are consistent with $b \approx ¼$ instead; an illustration of this fact is in Fig.5.

4. Conclusion

Finally, we remark that, lacking a general theory for steady nonequilibrium states and nonequilibrium ordering, the study of suitable lattice systems by means of computational statistical physics is most convenient to collect relevant information. The following procedure seems helpful. One considers any of the (now standard) kinetic lattice models, such as the lattice gas, the binary alloy, or the magnetic Ising model, for example. They have in common that their (particle, spin, etc.) configurations evolve in time by a stochastic process that satisfies detailed balance which implies an asymptotic tendency to canonical equilibrium. Next, detailed balance is frustrated by modifying kinetics appropriately, e.g., by introducing driving dissipative fields (Section 2.1) or competing tendencies (Sections 2.2 and 2.3), or by replacing the master equation by an irreversible rule (Section 2.4). It induces nonequilibrium conditions that are similar in a sense to the case in which the (particle, spin, etc.) configuration is acted on by an external, non-Hamiltonian agent. The resulting models may be somewhat arbitrary, but they may be shown to have a practical relevance in some occasions as a representation of certain situations in nature. As a matter of fact, some of them are known to model fast ionic conductors, chemically reacting systems and disordered materials, for instance. In any case, they usually exhibit and exceptionally rich behaviour whose systematic study may be very helpful at the present stage of development of theory.

Acknowledgments

We are indebted to Julio F. Fernández, Pedro L. Garrido, Jesús M. González-Miranda, Amilcar Labarta, and Marcello Puma for collaboration in some of the results mentioned in the present lecture notes, and for many useful discussions.

References

1. See, for instance, R. Graham and H. Haken, *Z.Phys.* **213** (1968) 420; **237** (1970) 31.
2. S. R. de Groot and P. Mazur, *Nonequilibrium Thermodynamics* (Dover, New York, 1984).
3. H. Haken, *Synergetics* (Springer-Verlag, Berlin, 1978).
4. I. Prigogine, *Introduction to Thermodynamics of Irreversible Processes* (Thomas, New York 1955); G. Nicolis and I. Prigogine, *Self-organization in Nonequilibrium Systems* (Wiley Interscience, New York, 1977).
5. J. Marro, *Nonequilibrium Statistical Physics* (Lecture Notes, *Troisième Cycle de la Physique en Suisse Romande, Physique Théorique*, Lausanne, 1992).
6. R. J. Glauber, *J. Math. Phys.* **4** (1963) 294; K. Kawasaki, *Phys. Rev.* **145** (1966) 224; see also K. Kawasaki, in *Phase Transitions and Critical Phenomena*, vol. 4, ed. C. Domb and M. S. Green (Academic Press, London, 1972).
7. See, for instance, C. J. Thompson, *Mathematical Statistical Mechanics* (Princeton Univ. Press, N. J., 1972).
8. N. Metropolis, A. W. Rosenbluth, M. M. Rosenbluth, A. H. Teller and E. Teller, *J. Chem. Phys.* **21** (1953) 1087.
9. H. E. Stanley, *Introduction to Phase Transitions and Critical Phenomena* (Oxford Univ. Press, New York 1971).
10. A. Achahbar, P. L. Garrido and J. Marro, *Phys.Lett.* **A**, in press.
11. J. Marro, P. L. Garrido and J. L. Vallés, *Phase Transitions* **29** (1991) 129, and references therein.
12. P. L. Garrido, J. Marro and R. Dickman, *Ann. Phys.* (N. Y.), **199** (1990) 366.
13. J. M. González-Miranda and J. Marro, *J. Stat. Phys.* **61** (1990) 1285.
14. J. J. Alonso, J. Marro and J. M. González-Miranda, *Phys. Rev.* **A**, in press.
15. J. M. González-Miranda, A. Labarta, M. Puma, J. F. Fernández, P. L. Garrido and J. Marro, to be published.

16. P. L. Garrido and J. Marro, *Europhys. Lett.* **15** (1991) 375.
17. J. J. Alonso and J. Marro, *Phys. Rev.* **B45** (1992) 10408.
18. A. I. López-Lacomba and J. Marro, *Phys. Rev.* **B46**, 8244 (1992).
19. J. Marro, J. F. Fernández, J. M. González-Miranda and M. Puma, to be published.
20. See, for instance, R. Durret, *Particle Systems and Percolation* (Wadsworth, California, 1988), and references therein.
21. D. Stauffer, in *Proceedings of Physics Computing '92* (World Scientific, Singapore, scheduled to appear in 1992); see also D. Stauffer and R. B. Pandey, *Computers in Physics* *6* (1992) 404.
22. T. Tomé and R. Dickman, *The Ziff-Gulari-Barshad Model with Desorption*, preprint, and references therein.
23. G. Grinstein, C. Jayaprakash and Y. He, *Phys.Rev.Lett.* **55** (1985) 2527.

LAPLACIAN GROWTH
AND
DIFFUSION LIMITED AGGREGATION

JULIO F. FERNÁNDEZ
Instituto Venezolano de Investigaciones Científicas
Apartado 21827, Caracas 1020-A, Venezuela

ABSTRACT

Laplace's equation is essential to the theory of a wide variety of seemingly different phenomena, such as electrodeposition, dielectric breakdown, and fluid-fluid displacement. For this reason, these effects are all grouped under the heading of *Laplacian growth*. Smooth geometrical shapes are unstable in this kind of growth; consequently, extremely irregular patterns, known as *fractals*, are formed. Diffusion Limited Aggregation (DLA) is a stochastic algorithm that models the physics of such growth. All of this is the subject of these three lectures, which are organized as follows: 1. Growth models: Eden's model, DLA, and invasion percolation. Fractals. Laplacian growth. 2. Instabilities and similarity in Laplacian growth. DLA and Laplacian growth: their relationship. 3. DLA generalizations: (a) non-vanishing surface tension; (b) fluid-fluid displacement in porous media; (c) Poisson growth.

1. Growth Models, Fractals and Laplacian Growth

This lecture is about *growth models* (the relevant physics is discussed in the second lecture), about the growth forms (fractals) which they generate, and about Laplacian growth.

1.1. Eden's Model

The original aim of this model[1] was to simulate growth in some biological systems. It is illustrated in Fig. 1.1. Given a growth seed (the black area in Fig. 1.1), and its perimeter (the shaded squares in Fig. 1.1), growth takes place as follows: a perimeter square, chosen at random (without bias), becomes aggregated to the growth seed; the perimeter is then redefined; a new perimeter square is again chosen at random and aggregated, and so on. Growth turns out to be compact, but surfaces are rough. In spite of its simplicity, this model has not been solved exactly.

1.2. Diffusion Limited Aggregation (DLA)

Witten and Sander first proposed this model[2] in order to simulate soot for-

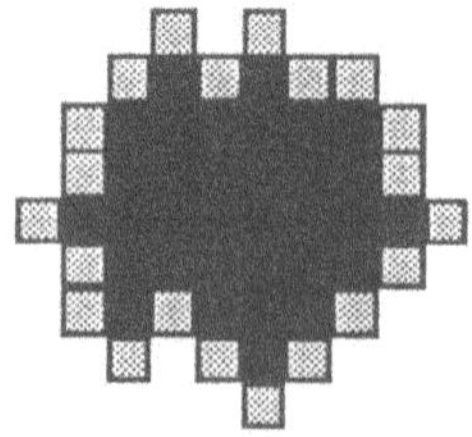

Figure 1.1: Shows a growth seed (black squares), and its perimeter (grey squares). All squares on the perimeter are equally likely to become aggregated in Eden's model.

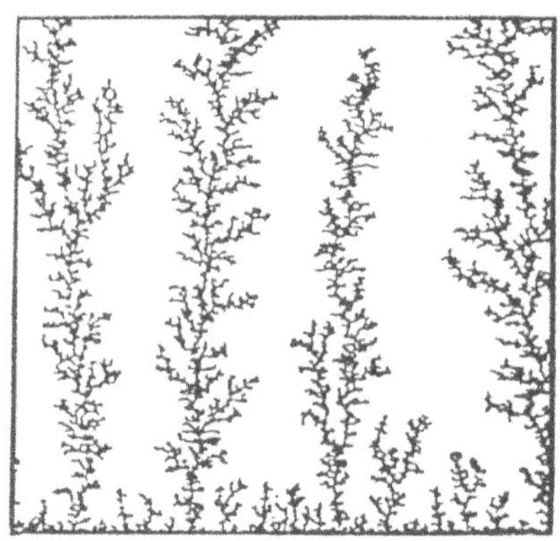

Figure 1.2: Outcome of a DLA run on a square lattice of 1024×1024 sites.

mation. The relation between DLA and Laplacian growth, which is a central theme in these lectures, was realized a few years later[3]. Growth usually takes place in a circular geometry, or in a rectangular one. A random walker, first launched with equal probability from any point on a circular boundary encompassing the aggregate (in rectangular cells, random walkers start from the top), moves randomly in the open space until it either comes back to the outer circular boundary (in which case a new walker is launched), or it lands on a site belonging to the aggregate; the last site visited by the random walker before landing on the aggregate then becomes part of it[4]. Another random walker is then launched and so on. Figure 1.2 was obtained this way.

1.3. Invasion Percolation

Consider two immiscible fluids moving through a porous medium, as, for instance, when water is injected into an oil reservoir in order to dislodge oil which will not flow out on its own. When the rate of injection is slow enough, viscous forces become negligible and motion is then determined mostly by capillary pressure.

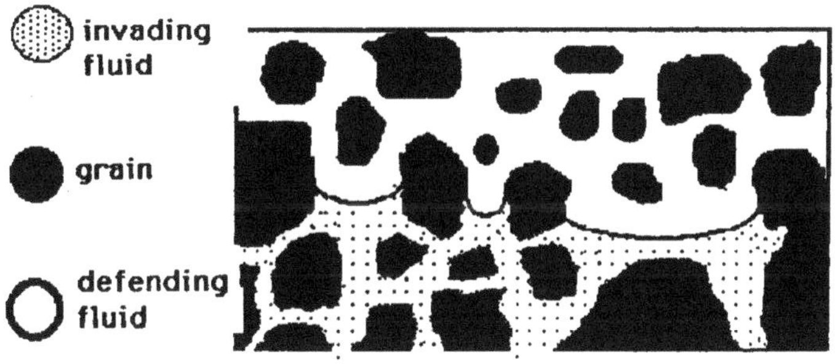

Figure 1.3: An *invading* fluid (assumed to be non-viscous) pushing upwards through some fixed barriers (*grains*) displaces a viscous liquid (shown as white).

The invasion percolation (IP) model[5] was designed to simulate the time evolution of interfaces between two fluids moving *very slowly* (just how slowly it must move is part of a wider issue to be discussed in the third lecture) through porous media. Consider, for instance, the conditions exhibited in Fig. 1.3, where an invading fluid (below) wets rocks more readily than the defending fluid (above). The largest pressure drop occurs across the piece of interface where the curvature is largest, which will, assuming negligible viscous forces, move forward faster than the rest of the interface. Accordingly, in IP, a piece of the interface is advanced wherever the capillary pressure along the interface is most favorable for it. Figure 1.4 shows the area invaded (in black) in an IP computer simulation [6].

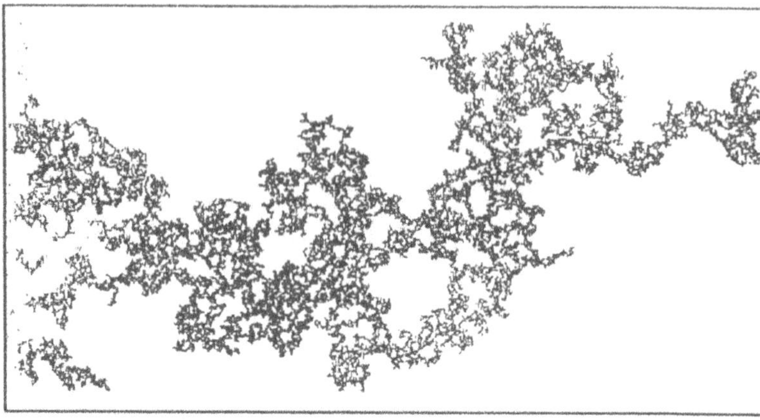

Figure 1.4: Invasion percolation grown pattern. (After Furuberg et al[6].)

1.4. Fractals

Growth forms generated by DLA and IP algorithms, Figures 1.2 and 1.4, respectively, have something in common: both seem to exhibit empty spaces on all length scales shown. That is a characteristic feature of geometrical objects, of widespread occurrence in nature, which have come to be known as *fractals* -after Mandelbrot [7]. A very brief introduction for beginners follows. We start with an example. Consider Koch's curve shown in Fig. 1.5.

The mass $M(\varepsilon)$ between two points increases with the straight line distance ε between them as follows,

$$\frac{M(3\varepsilon)}{M(\varepsilon)} = 4. \tag{1.1}$$

Iterating this equation n times gives $M(3^n\varepsilon)/M(\varepsilon) = 4^n$. To define a linear density, let $\lambda = M(3^n\varepsilon)/(3^n\varepsilon)$; substitution of the previous equation yields $\lambda = [M(\varepsilon)/\varepsilon](4^n/3^n)$. Obviously, $\lambda \to \infty$ as $n \to \infty$. Such a definition of λ is therefore meaningless for

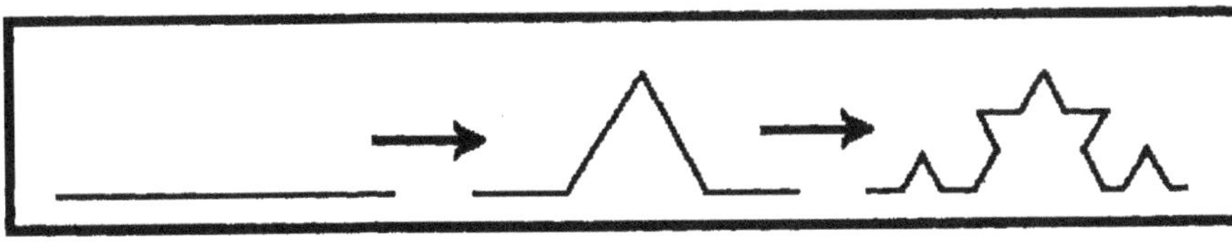

Figure 1.5: Infinite iteration of the indicated sequence of operations produces Koch's curve.

Koch's curve. That suggests the definition of,

$$\lambda = \frac{M(3^n \varepsilon)}{(3^n \varepsilon)^D} \tag{1.2}$$

instead of a *linear* density. D plays the role of the dimension here. Again, this expression can be cast into,

$$\lambda = \frac{M(\varepsilon)}{\varepsilon^D} \frac{4^n}{3^{nD}}, \tag{1.3}$$

which becomes conveniently independent of n if we set

$$3^D = 4. \tag{1.4}$$

Mandelbrot[7] has coined the term *fractal dimension* for quantities such as D [see Eq. (1.2)], that are effective dimensionalities but do not have integral values. There is no universally accepted definition of *fractals*. The requirement that the topological dimension D_T (= 1 for Koch's curve) fulfill $D_T < D$ is sometimes used to define them. Having suitably defined λ and D in Eqs. (1.2) and (1.4), one can write

$$M(\varepsilon) = \lambda \varepsilon^D. \tag{1.5}$$

The patterns shown in Figs. 1.2 and 1.4 are *random* fractals. Figure 1.6 shows a plot of $ln[M(\varepsilon)]$ versus $ln(\varepsilon)$ corresponding to the DLA fractal shown in Fig. 1.2; the slope of the straight line fit gives the value (1.7) of D.

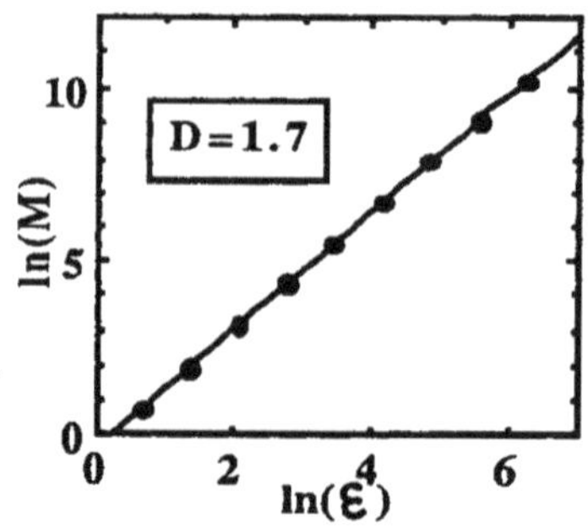

Figure 1.6: Natural logarithm of the average mass M (i.e., number of occupied sites) in a box of side ε versus the natural logarithm of ε for the DLA pattern shown in Fig. 1.2. D is given by the slope.

Its *universality* makes fractal dimensionality even more interesting than it would otherwise be. It is not that all random fractals have the same dimensionality; the fractal dimensions for Figs. 1.2 and 1.4 have different values: $D = 1.7$ and 1.8, respectively. Rather, D seems to depend on the type of growth mechanism but not on the boundary conditions. A DLA pattern grown with circular geometry has dimensionality no different from that of Fig. 1.2. (For some practical consequences, see exercice 2 at the end of this lecture.)

A fractal is not completely characterized by its dimensionality. The fractal dimension for the pattern shown in Fig. 1.7 is $log(16)/log(5) \approx 1.72$ (see exercice 1); it differs little from the value of D for a DLA growth, even though the two patterns are quite different.

Other concepts, such as texture[7], and growth measures[8] help to characterize fractals. No scheme that completely characterizes fractals is known to exist, however.

Finally, we mention a basic feature of ideal fractals: selfsimilarity. It is illustrated in Fig.1.8 for a statistical fractal. It conveys the message that even statistical fractals are not completely random objects: the whole is represented, on smaller scales, in every one of its small non-empty regions.

Figure 1.7: To generate this fractal, cut out a $3/5 \times 3/5$ square (let the side of the square boundary be 1) as shown; then, out of the remaining 16 squares ($1/5 \times 1/5$ each) cut out the corresponding squares as shown, and so on.

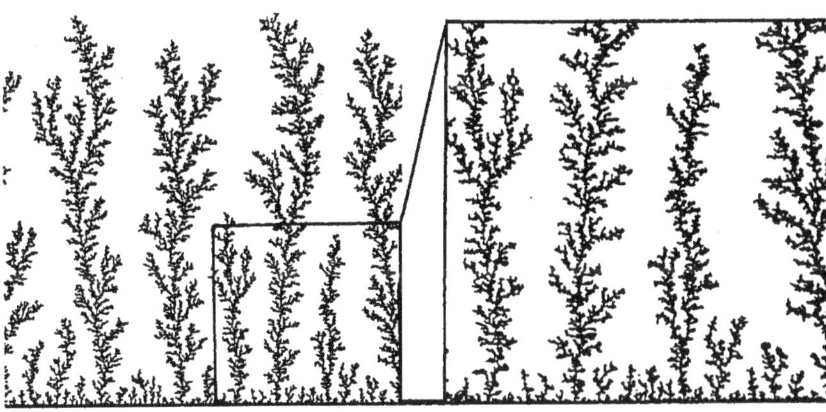

Figure 1.8: A DLA grown fractal (on the left) and an enlarged portion of it (on the right) to illustrate statistical self-similarity.

1.5. Laplacian Growth

We have thus far only mentioned growth *models*. We now turn our attention to growth elevenit experiments that fall under the heading of *Laplacian growth*, and to a bit of the underlying physics. The connection between Laplacian growth and the DLA model is established in the next lecture. Two specific instances of Laplacian growth will be described below: *viscous fingering* and electrodeposition.

The best known instance of viscous fingering is the Saffman-Taylor[9] finger. The experimental set up is a Hele-Shaw cell, which is essentially made up of two parallel glass plates separated by a narrow (ususally a fraction of a millimiter) gap as shown in Fig.1.9.

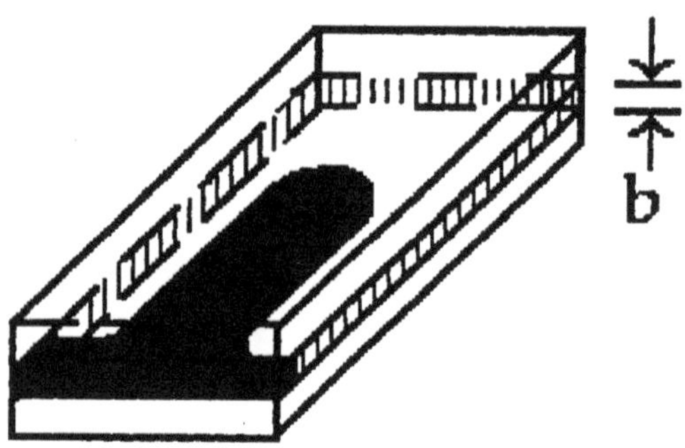

Figure 1.9: Hele-Shaw cell showing a finger (black) of non-viscous fluid invading a the gap between two plates -originally filled with a viscous fluid (shown as white).

The space between the two plates is initially filled with some viscous fluid. During the experiment, a nearly non-viscous fluid (such as water or air) which does not mix with the viscous one is injected through the lower end of the cell, as shown in Fig. 1.9. Furthermore,the injected fluid does not displace the viscous one in a piston like fashion; instead, it advances as shown in Fig. 1.9 for a wide range of injection rates. λ/W (λ and W are the finger and cell width, respectively) extrapolates to a value of $1/2$ as the injection rate increases to infinity. This interesting result has finally been accounted for not long ago[10]. At sufficiently large injection rates, given any small amount of noise or inhomogeneities, the Saffman-Taylor finger breaks up into ramified patterns such as the one shown in Fig. 1.10 obtained by Arnéodo et al[11].

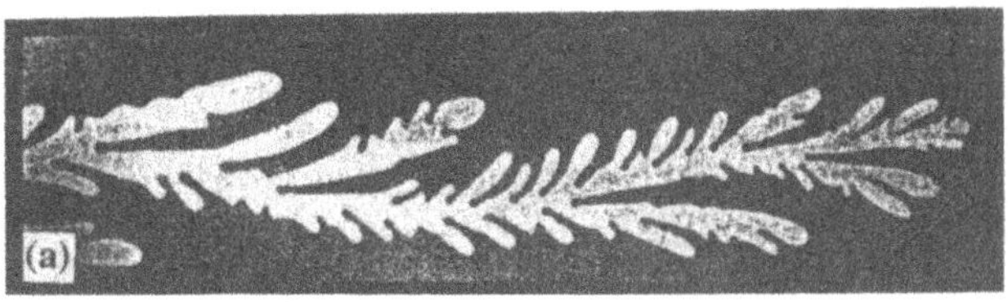

Figure 1.10: After Arnéodo et al, Ref. 11.

What is the physics behind this phenomenon? The velocity field **v** for the viscous fluid fulfills

$$\mathbf{v} = -(\kappa/\mu)\nabla p, \tag{1.6}$$

where κ is the medium permeability, given in this case by $\kappa = b^2/12$, b is the Hele-Shaw gap width, μ is the viscosity and p is the pressure field. Equation (1.6) is a two dimensional equation which merely states that the dominant force controlling viscous flow is friction with the plates in Hele-Shaw's cell. In order to obtain the proportionality constant (κ/μ), one may easily derive Eq. (1.6) (for length scales much longer than the gap distance b in the Hele-Shaw cell) from the quasi steady state Navier-Stokes equation[12] for incompressible fluids. Equation (1.6) has the same form as Darcy's equation for viscous flow in porous media, and, for that reason, it is sometimes referred to by that name. It follows from Eq. (1.6) that if μ is negligible for the injected fluid, then p is constant over the space invaded by it; furthermore, incompressibility (that is, $\nabla . v = 0$) of the viscous fluid allows one to write Laplace's equation,

$$\nabla^2 p = 0. \tag{1.7}$$

In order to determine how the interface between the two fluids moves, one must first solve Laplace's equation over the region of space R filled by the viscous fluid; equation (1.6) then gives **v** everywhere in R, including **v** near the interface, which is what one needs to be able to move the interface. Equation (1.7) must be solved again after the interface is moved, and so on. The ensuing time evolution is, for that reason, known as *Laplacian growth.* In order to specify the boundary conditions, let the invading fluid be injected at pressure p_o, and let the viscous fluid emerge at the top of the cell under no pressure (i.e., with $p = 0$). Then, if there is no discontinuity across the interface (e.g., negligible surface tension), $p = p_o$ thereon, ∇p is parallel to the sides of the cell (assuming they are sealed). Having specified the boundary conditions, Laplace's equation may be solved (at least in principle).

Electrodeposition is another instance of Laplacian growth. Consider some ionic solution and two electrodes. For very dilute (with negligible self-diffusion) solutions,

$$\mathbf{j} = -\sigma \nabla \varphi, \tag{1.8}$$

where j is the ionic current, σ is the conductivity and φ is the electrostatic potential. In quasi steady state conditions, $\nabla . j = 0$, and therefore,

$$\nabla^2 \varphi = 0. \tag{1.9}$$

This is another case of Laplacian growth: the cathode *grows* (as metallic ions deposit on it, at a local rate proportinal to $\mathbf{j}$). It was first observed experimentally by Brady and Ball[13]. Figure 1.11 exhibits some electrodeposits.

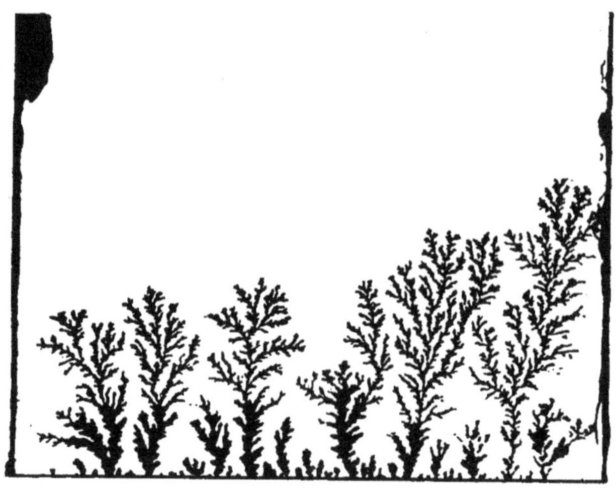

Figure 1.11: Silver deposited from a 0.1 mole solution of $AgNO_3$ with gold (after Ref. 14) electrodes.

Many other instances of Laplacian (or nearly Laplacian -such as propagation of sparks or lightning) growth exist. Consider an electrode (or cloud) from which a spark is about to *jump* into a dielectric medium, where $\nabla^2\varphi = 0$. Once again, this equation must be solved -subject to the boundary condition $\varphi =$ *constant* over the electrode. Since molecular ionization rates s depend on $\nabla\varphi$, growth of the ionized region of space is determined by the value of $\nabla\varphi$ on its surface. If $s \sim \nabla\varphi$ were to hold, then the ionized region of space would grow by precisely the same rule as viscous fingers and electrodeposition do. In the *dielectric breakdown model* [13,15], $s \sim (\nabla\varphi)^{\eta}$, where the value of η is unknown.

Exercices

1. To find the fractal dimension of the square pattern shown in Fig. 1.7, cover it with 25 squares (each 1/5 as large on the side as the large square). Use $M(5\varepsilon)/M(\varepsilon) = 5^D$, which follows from Eq. (1.5), to obtain $D = log(16)/log(5)$.

2. After oil wells *run dry*, that is, oil stops coming out under its own pressure, oil can be forced out by the so called *secondary recovery* technique: Water (or some other nearly non-viscous fluid) is injected into the oil reservoir at some points in order to displace oil towards extraction wells. Assume water does not displace oil piston like, but fills space as a fractal of dimensionality D and that D has a *universal value* (it is independent of how the wells are arranged spatially). By how much does the fraction of oil recovered decrease if all distances between wells are increased by a factor of α ?

3. Show that a set of points placed uniformly at random within a given rectangle is not self-similar, and, furthermore, that $D = 2$ for it.

4. Design a fractal of your own and find its dimension.

2. Instabilities, Similarity, and DLA

This lecture provides a brief introduction to instabilities and similarity in Laplacian growth, and to how this type of growth is related to DLA. Instabilities of smooth forms under Laplacian growth are essential to the development of Laplacian fractals. Similarity is a revealing and fruitful concept. Finally, DLA is shown to be the right stochastic algorithm for simulations of Laplacian fractal growth phenomenon.

2.1. Instabilities

We look only at instabilities of moving *flat* interfaces between two inmiscible fluids. Consider Laplacian growth, that is, $\nabla^2 p = 0$ and an advancing interface with velocity $(-v) \propto \nabla p$. We shall furthermore assume that p is constant along the interface. Visual inspection of Fig. 2.1 shows that distances between equipotential surfaces are smallest just ahead of surface bumps, and consequently ∇p is largest thereon. It follows then that humps advance faster than other pieces of the interface, that is, slight deformations -on all length scales- become amplified. Thus, flat interfaces are unstable.

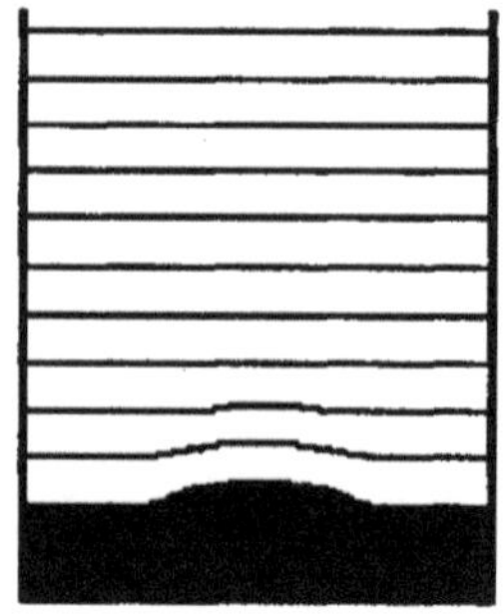

Figure 2.1: Equipotential lines. The invaded area (shown as black) is at a constant potential. It illustrates why interface humps advance faster.

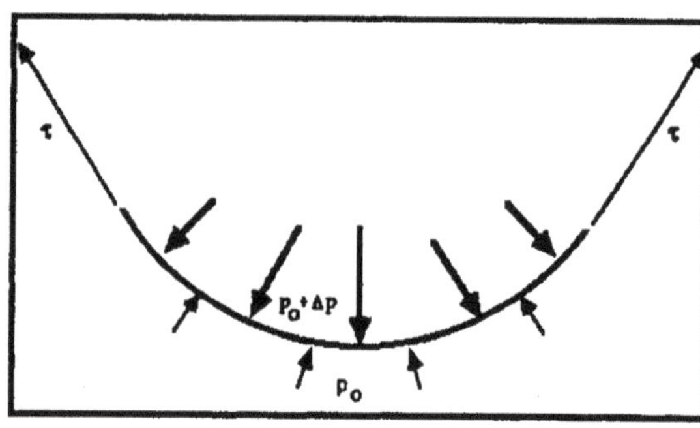

Figure 2.2: Pressure on the convex side is larger in order to balance surface tension.

The sort of instability just discussed must be at the root of fractal growth of the kind shown in Figs. 1.8 and 1.11.

The argument just given breaks down if there is a nonvanishing surface tension τ because there is then a pressure drop across the interface; consequently, interfaces that are not flat are no longer equipotential surfaces. To find how much the pressure drops then, consider some interface, as shown in Fig. 2.2, of constant radius of curvature R. As the interface is massless, forces acting on it have to balance. It is straightforward to show that there is then a pressure discontinuity Δp across the interface given by

$$\Delta p = \tau / R. \tag{2.1}$$

Consider now a sinusoidal deformation of a flat interface of amplitude A and wavevector k. It is fairly straightforward to show (see exercice 2 at the end of this lecture) that any vanishingly small deformation grows exponentially with time for any $\lambda > \lambda_c$, and

$$\lambda_c^2 = (2\pi)^2 \kappa / Ca, \tag{2.2}$$

where $Ca = U\mu/\tau$ is the so called capillary number, and U is the velocity of the viscous fluid downstream (notice that the fluid velocity is uniform far ahead of the interface) in a rectangular geometry. Furthermore (see exercice 2) deformations grow fastest for $\lambda^2 = 3\lambda_c^2$.

2.2. Similarity

When no solution can be found to the ruling equations of motion, *similarity* is a rewarding concept to turn to. Similarity methods have long been used in Mechanics[16] and in Hydrodynamics[17] to extract useful information from non-linear partial differential equations that cannot be solved in full. Trayectories or patterns corresponding to solutions of a given equation for different sets of parameters turn out to be (in some sense) geometrically similar; hence the name of the method. (*Scaling* is the word more often used in Statistical Physics.) The purpose of this section is three fold. It aims to show that fully developed viscous fingers have widths λ that follow the rule,

$$\lambda^2 \propto Ca^{-1}. \tag{2.3}$$

(Recall that Eq. (2.2) is valid only for incipient fingers.) In addition, similarity arguments help to understand why Laplacian growth is often fractal. The third purpose of this section is to lay the basis for scaling arguments given in lecture 3 with regard to flow in porous media.

It may be worthwhile to start with a simple example. Consider a simple anharmonic oscillator in one dimension. Let

$$\frac{d^2x}{dt^2} = wx^{2n+1}. \tag{2.4}$$

be its equation of motion. It follows by direct substitution that if $x_1(t)$ is a solution, then $x_2(t) = \lambda x_1(t\lambda^n)$ is also a solution. $x_2(t)$ and $x_1(t)$ are therefore similar in the sense exhibited in Fig. 2.3: x scales as λ while t scales as λ^{-n}.

Knowledge of such scaling rules are not merely of esthetic value. Useful relations, such as, $TA^n = constant$, between the amplitude A and period T of an oscillation follow inmediately from them (note that this relation reduces to the one for the harmonic oscillator for n=0).

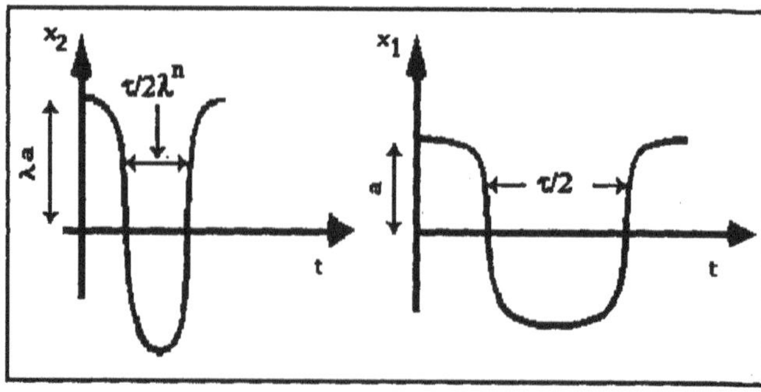

Figure 2.3: Two *similar* solutions of the non-linear equation (2.4).

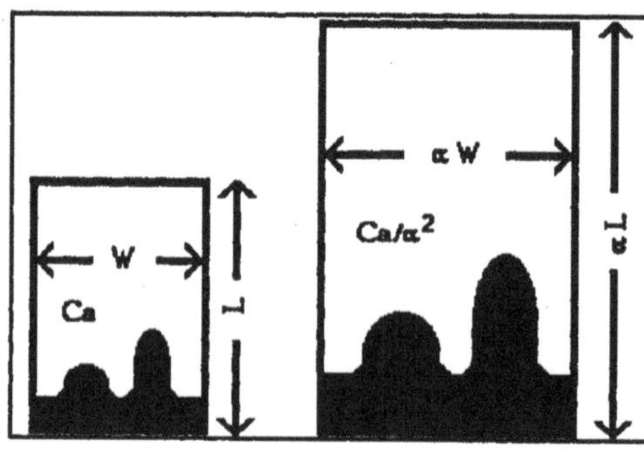

Figure 2.4: The *enlarged* evolution takes place physically if Ca scales as shown (see text).

Proceeding analogously (see exercice 3), a scaling rule is obtained for the time evolution of fluid-fluid interfaces from Eqs. (1.6), (1.7) and (2.1). Given a fluid-fluid interface development (as shown in Fig. 2.4), another time evolution on an enlarged (by a factor of α) scale can also take place if $Ca \longrightarrow Ca/\alpha^2$. That is, the scaling rule,

$$\lambda/L = f(L^2 Ca), \tag{2.5}$$

follows, where f is (since the equations have not been solve in full), an undetermined function, and λ is any distance (such as a suitably defined *finger width*) defined for the pattern.

Let $g(L^2Ca) = f(L^2Ca)/(LCa^{1/2})$; then

$$\lambda = Ca^{-1/2}g(L^2Ca). \tag{2.6}$$

Consider $L \to \infty$. One expects λ to stay finite (see Fig. 1.10), and, consequently, $g(\infty)$ to be finite. Eq. (2.3) follows.

Finally, consider similarity as exhibited in Fig. 2.4, in the $Ca \longrightarrow \infty$ limit (equivalently, in the $\tau \to 0$ limit), where finger widths vanish. Since both patterns would be evolving with the same value of Ca (infinitely large), the larger of the two patterns may just be a later development of the smaller one. Now, if there is no restructuring (i.e., if growth takes place mostly on the outermost regions of the pattern while its innermost parts remain frozen) then large and small parts of the pattern will be similar, since they correspond to late and early stages of growth -which are themselves similar, and are therefore fractals (as illustrated in Fig. 1.8).

2.3. Relation between DLA and Laplacian Growth

For the sake of concreteness, consider fluid-fluid displacement in a rectangular Hele-Shaw cell.[9] Let a non-viscous fluid be injected through the bottom be under pressure p_o as shown in Fig 2.5, and let the viscous fluid emerge through the top under no pressure ($p = 0$). As shown in section 1.5, $\nabla^2 p = 0$ within the viscous fluid. We shall *not* assume now that the pressure is continuous across the fluid-fluid interface; be it because the surface tension is significant or because flow takes place through a porous medium and non-negligible capillary effects exist, there is in general a pressure drop $\Delta p(s)$ at each point s along the interface. Let Δp_{max} be the maximum value of $\Delta p(s)$ along the interface and $p' = -p + p_o + \Delta p_{max}$. This definition is intended to lead to $p' \geq 0$ everywhere, which allows one to identify p' with a probability field.. We introduce a space lattice, in order to carry out simulations, and change notation accordingly ($\Delta p(s)$ becomes Δp_j, and so on).

As first noticed by Paterson[3], DLA models stochastically Laplacian growth. That is shown next. A modified (in order to handle $\Delta p \neq 0$) DLA algorithm is specified first, the connection with Laplacian growth is established afterwards. Consider N random walkers initially distributed over points at the top of the cell and on the interface as follows,

$$\frac{N(i \in T)}{N(j \in I)} = \frac{p_o + \Delta p_{max}}{\Delta p_{max} - \Delta p_i}, \tag{2.7}$$

where $N(i \in T)$ and $N(j \in I)$ are the number of walkers at site i on the top boundary and on site j on the interface, respectively. There are no random walkers whatsoever anywhere else initially. Note that the distribution of random walkers on the top boundary is uniform.

All N walkers take a first step at random at $t = 0$. Only 1/4 (on a square lattice) of the walkers initially at the top boundary take a first step downward; the rest (3/4 of them) are removed immediately. All survivors take additional random steps at times t+1, t+2, etc., until they either hit the top or the aggregate; they

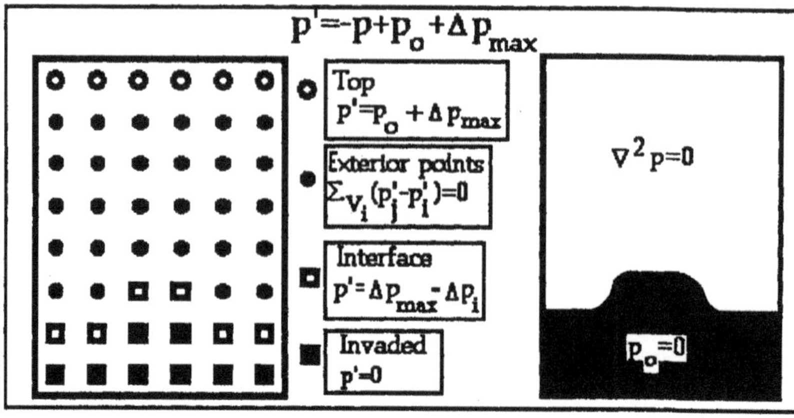

Figure 2.5: A non-viscous fluid (shown as black) has invaded the lower portion of a Hele-Shaw cell on the right hand side. The defending fluid (white) above is viscous. A discretized version of it is shown on the left hand side. Let $p' = -p + p_o + \Delta p_{max}$, then $p' = p_o + \Delta p_{max}$ on the top boundary; the discrete version of Laplace's equation (see exercice 4) holds on *exterior* (the region not yet invaded) points; p' is given on the interface $p' = \Delta p_{max} - \Delta p_i$; $p' = 0$ in the invaded region.

are then removed. The following equations are a restatement of these rules:

$$N(i, t+1) = \sum_j N(j,t)/4, \forall i \in \mathbf{E}, \tag{2.8}$$

$$N(i,t) = \{ \begin{matrix} N(i \in T)\delta_{t,0} \\ N(i \in I)\delta_{t,0} \end{matrix} \tag{2.9}$$

where $\mathbf{E}$ stands for exterior space (that is, space occupied by the viscous fluid), Σ_j stands for a sum over all points j which are nearest neighbors of i. If we subtract $N(i,t)$ from Eq. (2.8) and sum it over all t, from 0 to infinity, we obtain,

$$N(i,\infty) - N(i,0) = N \sum_j [n(j) - n(i)]/4, \tag{2.10}$$

where $n(i)[\equiv \Sigma_t N(i,t)/N]$ is the mean number of times that a random walker lands on site i before it either reaches the top boundary or the aggregate. On the other hand, all walkers are removed after a sufficiently long time, that is, $N(i,\infty) = 0$; furthermore, all exterior points are unoccupied at t=0, that is $N(i,0) = 0$. Therefore, Eq. (2.14) becomes

$$\sum_j [n(j) - n(i)] = 0. \tag{2.11}$$

This equation states that the mean number of times a random walker visits a lattice point satisfies the discrete Laplace equation on a lattice. Had our final aim been to obtain a numerical solution p_i to Laplace's equation on the exterior points (see Fig. 2.5), our task would have been accomplished by just setting $p_i \propto n_i$. Our final aim is however to find out how the interface moves. That is established next. Recall that $\mathbf{v} \propto \nabla p'$ (from Darcy's law) everywhere in the exterior space (including, of course, the neighborhood of the interface), and that $p_i \propto n_i$; it follows then that the probability $P_{i \to j}$ that the interface be advanced from site i to site j (that is, that the j-th site be aggregated to the growing cluster) fulfills $P_{i \to j} \propto (n_j - n_i)$. Finally, note that $(n_j - n_i)$ is proportional to the number of walkers arriving at i from j. After the corresponding set of sites is aggregated to the cluster, a new batch of N walkers is released, and so on. In order to save computer time, DLA is ordinarily carried out with $N = 1$; a random walker is released from a point either at the top of the cell or on the interface, with a probability given by Eq. (2.7); the last point visited by the walker before landing on the cluster is therefore aggregated; another walker is then released, and so on. DLA simulations carried out with $N \gg 1$ are, for obvious reasons, said to be done with *noise reduction*.

Finally, Kadanoff[4] has suggested a simple trick that does more than just save computer time. Consider a horizontal line (HL) at least one row beyond the aggregate, as in Fig. 2.6; now, random walkers that start walking from the top with a uniform probability distribution arrive at the HL with a uniform probability distribution. One may therefore launch them right off from the HL and save time. Furthermore, walks within the region between the HL and the top need not be executed because the outcome of such walks is easily computed beforehand as follows. A random walk that starts on a site just above the HL, as in Fig. 2.6, either reaches the top line first or it returns to the HL first. The probability that it reach the top and die is easily shown to be $1/l$, where l is the number of rows (4, in Fig. 2.6) above the HL. The probability that it return to a given site on the HL depends on the number of columns between the start off site and the arrival site (we assume periodic boundary conditions throughout) as well as on l. The dependence on l is inconvenient: it would have to be dealt with in simulations as the aggregate advances. To avoid it, we let the top of the cell recede to infinity; more precisely, let L be the length of the cell; we let $L \to \infty$ and $p_0 \to \infty$, such that $p_0/L \to < \nabla p >$ remains fixed. One therefore computes beforehand the probability for events of the type shown in Fig. 2.6a in a very long rectangular cell with lateral periodic boundary conditions; then, throughout the simulation, everytime a random walk visits a site inmediately above the HL, it is inmediately shifted to a site on the HL with the corresponding probability.

In this scheme, an infinitely long cell is simulated; that allows the definition of a pressure gradient $< \nabla p >$,

$$p_o/L \longrightarrow < \nabla p >, \tag{2.12}$$

as $L \longrightarrow \infty$, in agreement with the value taken by ∇p far beyond the aggregate, where it is uniform. It is therefore possible to have a non-ambiguous definition

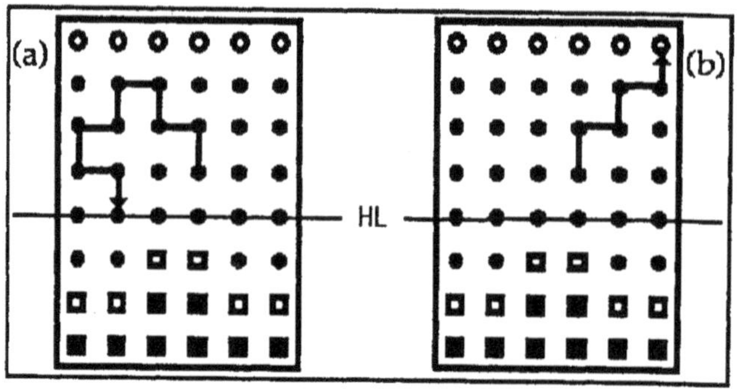

Figure 2.6: Random walks that start from a row above HL either end on HL, as in *a*, or at the top of the cell, as in *b*.

of the capillary number ($Ca = \kappa < \nabla p > /\tau$), or, equivalently, using Darcy's law, $Ca = U\mu/\tau$, where U is the *downstream* fluid velocity, which is, therefore, easily kept constant throughout a simulation.

We now need a rule for the probability to launch a random walker from the HL, instead of from the top. First, divide Eq. (2.7) by L/a (a is the nearest neighbor distance in a square lattice), and let $L \to \infty$; then $(p_o/L + \Delta p_{max}/L) \to < \nabla p >$, since Δp_{max} remains finite. Note next that $1/l$ is the probability that a random walk started at the top reach HL before it comes back to the top; therefore, we must set $N(i \in HL) = N(i \in T)a/L$ (since $L/(al) \to 1$ as $L \to \infty$). It follows from Eq. (2.12) therefore that,

$$\frac{N(i \in HL)}{N(j \in I)} = \frac{a < \nabla p >}{\Delta p_{max} - \Delta p_i}, \tag{2.13}$$

All the essential elements necessary to do DLA simulations are now at our disposal. The next lecture is about its application to various experimental conditions and the corresponding results obtained.

Exercices

1. Show that balance of forces implies Eq. (2.1) in d=2. Generalize it to three dimensions.

2. Equation (2.2) is usually obtained by the following linear instability analysis. Let $p_f(r)$ fulfill $\nabla^2 p_f = 0$ given a flat interface, and let $p_f \longrightarrow p_f + \delta p$ when the flat interface, say $y(x) = constant$, becomes slightly deformed initially, that is, when $y(x) \longrightarrow y(x) + \delta y(x)$. Let $\delta y(x) = A cos(qx)$ at t=0. Show that $\delta p = -\tau A q^2 cos(qx)$, to leading order in A, on the viscous fluid side of the interface, where τ is the surface tension. In

order to find how δp depends on y, let $\delta p(x,y) = -\tau Aq2cos(qx)f(qy)$, and substitute into Laplaces equation to obtain $f(qy)$. Finally, use Darcy's law [Eq. (1.6)] to obtain $A(t) = Aexp[-\omega(q)t]$, and $\omega(q)$. $A(t)$ increases with time for $\omega(q) < 0$. Equation (2.2) follows from $\omega(q) = 0$.

3. Follow the treatment, in section 2.2, for the anharmonic oscillator to obtain Eq. (2.3). (Hint: Equations (1.6), (1.7) and (2.1) rule the time evolution of a fluid-fluid interface; let x,t and Q stand for a spatial variable, time and curvature, respectively; define the unitless variables, $r = Ca^{1/2}x/a$, $s = (\tau Ca^{3/2}/a\mu)t$ and $q = QaCa^{-1/2}$, where $a^2 = \kappa$ (the permeability), and Ca (the capillarity) is defined below Eq. (2.2).)

4. Use

$$\frac{d^2f(x)}{dx^2} = \lim_{a\to 0}\frac{1}{a}\left[\frac{f(x+a)-f(x)}{a} - \frac{f(x)-f(x-a)}{a}\right]$$

to show that $\sum_j(p_j - p_i) = 0$, where $\sum_j$ stands for the sum over (j) all nearest neighbors of i, is the natural extension of Laplaces equation to a discrete space.

3. Surface Tension, Porous Media, and Poisson Growth

Growth patterns, as well as other numerical results, obtained with the DLA algorithm described in lecture II are shown and discussed in this lecture and compared with experiments. Surface tension effects are treated first; then, simulations and results for fluid-fluid displacement through porous media are discussed; finally, we treat fluid-fluid displacement of the sort that occurs when the gap in the Hele-Shaw cell varies with time.

3.1 Surface Tension Effects

Viscous fingers (such as the Saffman-Taylor finger) have non-vanishing widths because there is surface tension. One would hope to reproduce in DLA simulations the experimentally observed behavior. Surface tension comes into DLA through the term $\Delta p_{max} - \Delta p_i$ in Eq. (2.13), where Δp_i is the pressure discontinuity across the interface at i. It was shown in lecture II [see Eq. (2.1)] that $\Delta p_i = (\tau/R)_i$, where τ is the surface tension and $(1/R)_i$ is the curvature of the interface at i. The difficulty, of course, is in the definition of curvature at a lattice site.

Not only is the interface on a lattice a broken line (curvature is therefore undefined), but, in addition, Eq. (2.13) must be used to decide whether a random walk is to start from the interface or from the HL of the cell, in which case the interface changes by a finite amount, and so does the curvature *a fortiori*. That raises the question of which value of the curvature to use: the initial one, the final one, or perhaps some weighted average of the two. Vicsek[1] has proposed a method (a not too small neighborhood of point i is examined) that deals with the first difficulty but not with the latter one. We next give our own recipe[19].

Consider, for simplicity, an interface that can be represented by the single valued function $y(x)$, and let it move for $x_1 < x < x_2$, that is, let $y(x) \to y(x) + \eta(x)$ and $\eta(x_1) = \eta(x_2) = 0$. Let the length of the curve increase by δL and let the area swept be δA. It can be shown (see exercice 1 at the end of this lecture) that the curvature is given by

$$Q = \delta L / \delta A \tag{3.1}$$

if either $y(x)$ is is a circular arc, or if $\eta(x)$ is constant for $x_1 < x < x_2$.

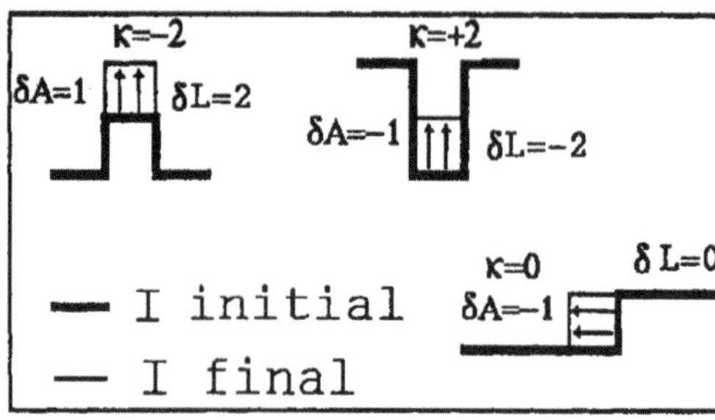

Figure 3.1: Three instances of interfaces on a square lattice illustrate how Eq. (3.1) is used. δL is in units of a (the nearest neighbor distance), δA is in units of a^2, and κ is in units of $1/a$.

Equation (3.1) can be easily applied to a moving interface on a lattice, and we adopt it for that reason. ($\delta A = 1, -1$ in units of a^2, for the aggregation or erosion of one particle, respectively, and, therefore, the expression $Q = -\delta L . \delta A$ may be used just as well[19]). Application of Eq. (3.1) is illustrated in Fig. 3.1. Note that there are only 3 possible values for Q on the square lattice: $-2/a, 0$ and $2/a$. The value to be used for Δp_{max} in Eq. (2.13) is clearly $2\tau/a$.

To apply Eq. (2.13), we therefore let $\Delta p_{max} = 2\tau/a$ and $\Delta p_i = 2m_i\tau/a$, where $m_i = -1, 0, 1$, as indicated in Fig. 3.1. Equation (2.13) then becomes

$$\frac{N(i \in HL)}{N(j \in I)} = \frac{a < \nabla p >}{(2\tau/a)(1 - m_i)} \tag{3.2}$$

Notice that there is only one parameter in this DLA simulation: $a^2 < \nabla p > /\tau$, which, using Darcy's law, becomes $(a^2/\kappa)Ca$ [recall that $Ca = (\mu U/\tau)$]. That is as should be: the outcome of the corresponding experiment also depends on Ca (and on the geometry of the Hele-Shaw cell, of course).

Given any small amount of noise or inhomogeneities, the Saffman-Taylor finger breaks up into ramified patterns at sufficiently large injection rates. Experimentally obtained patterns for $(\kappa/Ca)^{1/2} \approx 0.0016W$ are shown in Fig. 1.10; some

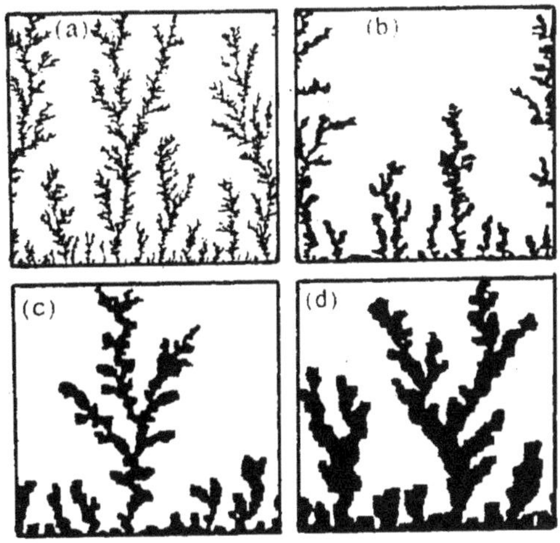

Figure 3.2: DLA patterns obtained on a square lattice for various values of r, where $r = 2Ca^{-1}\kappa/a^2$.

corresponding DLA patterns [19] obtained on square lattices are shown in Fig. 3.2. How acceptable are these results? What's to be used as a quantitative criterion for *acceptability*? It is found that finger widths increase as $Ca^{-1/2}$. That is in agreement with equation (3.2) which follows from Darcy's law and Laplace's equation. Thus, DLA simulations do yield viscous fingers with the correct dependence on Ca.

DLA simulations performed on triangular lattices[20] yield patterns that resemble the experimental ones more closely; moreover, patterns with reasonable resemblance to the Saffman-Taylor finger are obtained for smaller values of a^2Ca/κ. Fingers which look very much like the experimental ones have also been obtained[21] on square lattices, using Vicsek's curvature definition, with noise reduction, aggregating a boundary site j only after N arrivals at the aggregate coming from j (with $N \sim 10$).

3.2. Fluid-fluid Flow in Porous Media

The geometrical nature of porous media is not yet fully established. Fluid-fluid displacement laboratory experiments have been performed using packed glass beads[22]; extensive experiments in large nets of interconnected microchannels have been performed by Lenormand[23]. Figure 3.3 shows results of his. The growth pattern for the largest value of Ca is clearly DLA like, whereas an invasion percolation (see Fig. 1.5) like pattern is obtained for the smallest value of Ca. This is as expected: capillary forces dominate when flow is slow enough (i.e., Ca is small) which is precisely the condition under which IP applies; on the other hand, viscous forces dominate for high enough injection (i.e., large Ca) rates, and DLA like growth then takes place. Growth patterns are something in between for intermediate values of Ca.

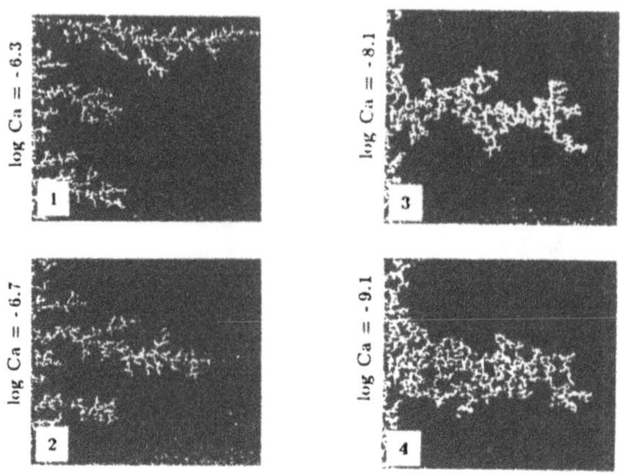

Figure 3.3: Experimental results of Lenormand for air displacing oil in a network of microchannels, whose widths vary randomly throughout space, for the values of Ca shown.

In order to simulate the observed experimental behavior, a model is needed for a microchannel network[24]. We do not attempt to describe details of motion within channels; we focus on large scale behavior. Accordingly, a small region of space, of linear dimension, say d, large enough to cover a few tubes and intersections, is represented in our model by a lattice site. We disregard surface tension effects across length scales larger than d; therefore, surface tension plays no role whatsoever in our calculation. Inhomogeneities (spatial variations of channel widths) have two effects: neither the permeability throughout the viscous fluid nor the capillary pressure on the interface are uniform. We disregard permeability variations. Capillary pressure effects come into DLA simulations through the term $(\Delta p_{max} - \Delta p_i)$ in Eq. (2.13). Note that whether the capillary pressure is positive or negative (that is, whether the invasion is wetting or non-wetting) does not matter: only variations in $(\Delta p_{max} - \Delta p_i)$ count.

Let Δp_{min} be the minimum value of Δp_i, $\Delta p \equiv \Delta p_{max} - \Delta p_{min}$, and $\Delta p'_i \equiv$

$\Delta p_i - \Delta p_{min}$; then equation (2.13) can be cast into,

$$\frac{N(i \in HL)}{N(j \in I)} = \frac{a < \nabla p >}{\Delta p(1 - x_i)} \tag{3.3}$$

where $x_i = \Delta p_i'/\Delta p$. It follows that there is only one parameter in this DLA simulation: $a < \nabla p > /\Delta p$, which takes a more familiar form for a rather disordered medium (for a completely ordered one, $\Delta p = 0$, of course); then $\Delta p \sim \tau/a$, whence, using Darcy's law and setting $\kappa \sim a^2$, it follows that $a < \nabla p > /\Delta p \sim Ca$.

Growth patterns thus obtained[24] are shown in Fig. 3.4 for various values of $\Delta p/(a < \nabla p >)$ (denoted by r in that figure). They resemble Lenormand's experimental patterns (in Fig. 3.3). Blow ups of portions of figures on the top row are shown on the lower one of Fig. 3.4.

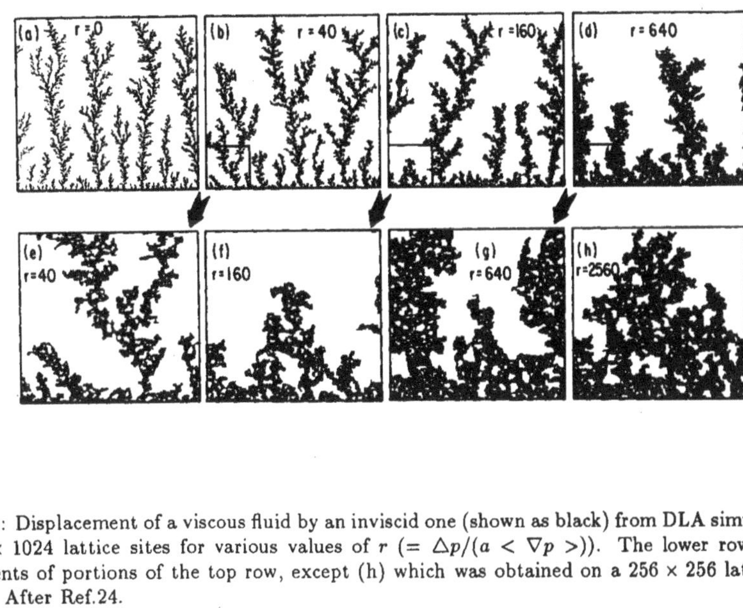

Figure 3.4: Displacement of a viscous fluid by an inviscid one (shown as black) from DLA simulations on 1024 × 1024 lattice sites for various values of r $(= \Delta p/(a < \nabla p >))$. The lower row shows enlargements of portions of the top row, except (h) which was obtained on a 256 × 256 lattice for $r = 2560$. After Ref.24.

Inspection shows that these enlargements look much like invasion percolation (IP) patterns (see Fig. 1.4). It seems therefore that these patterns resemble the ones obtained with *standard* DLA (that is, DLA in which all walks start on the top of the cell or on the HL line) on large length scales, but IP like on small length scales; furthermore, the crossover length between these two scales appears to increase as $\Delta p/a < \nabla p >$ increases.

These ideas are put on a more quantitative basis next. A crossover length L_c is first defined. First note that DLA patterns are ***open branched***, whereas ***trapping*** takes place in IP. *Branch* (and *tree trunk*) thickness provides a natural definition of L_c therefore. To determine it from a growth pattern, we put a cross at every point i on the pattern, and let both arms of the cross end at the ***skin of the tree.*** We let L_c be the average of the shortest of these two lines over the aggregate. Data points obtained[24] are shown in Fig. 3.5 for various values of $\Delta p/a < \nabla p >$ (denoted by r in the figure).

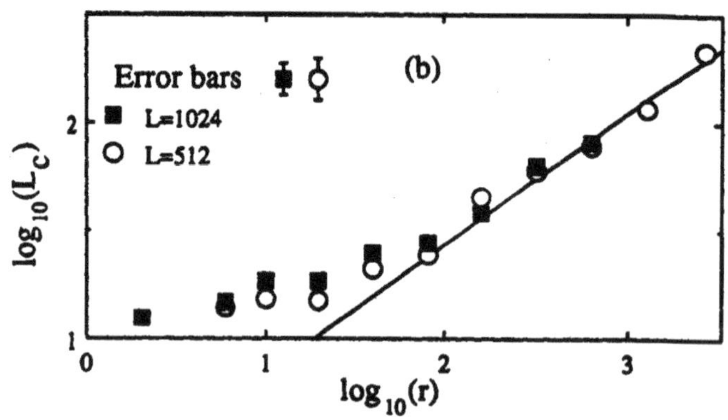

Figure 3.5: Plot of $log_{10}(L_c)$ vs. $log_{10}(r)$ for lattices of $L \times L$ sites for the values of L shown. The straight line is not a fit; it is a prediction (see text). After Ref.24.

There is an additional feature (in addition to the trapping versus open branch behavior) which helps identify the IP and standard DLA regimes in the patterns obtained. A log-log plot of the number of non-empty square boxes of side ε versus ε, gives [24]: (a) $D \simeq 1.70$ for $\varepsilon > L_c$, and (b) a value of D consistent with IP for $\varepsilon < L_c/2$ (see Ref. 24 for details).

There is an additional number which identifies the small length scale behavior obtained with IP: the interface fractal dimension for lengths significantly smaller than L_c turns out [Ref. 24] to be 1.3, in approximate agreement with its value for IP patterns.

Thus, this DLA simulation yields results in good agreement with experiments; but it does more than that, it unveils a scenario unseen through experiments thus far, mainly, the coexistence of two regimes in fluid-fluid displacement through porous media: one, IP like, on small length scales, and another one, standard DLA like, on large length scales.

We next try to understand the physics of the crossover length between these two regimes. Viscous forces lead to pressure drops along the flow direction that are proportional to the distance involved, whereas capillary forces are local forces. Growth is therefore expected to be Laplacian like on large length scales and IP like on small length scales. To be more specific, a scaling argument is now given. Consider flow as in Fig. 3.6; the image on the right is a blow up of the one on the left; clearly, everything moves α times faster on the right picture frame, and, since $v = \kappa \nabla p/\mu$ (recall that we disregard spatial variations in the permeability), it follows that $\kappa \nabla p/\mu$ also scales as α. Therefore, for all pressures, including capillary ones, $\kappa p/\mu$ scales[25] as α^2.

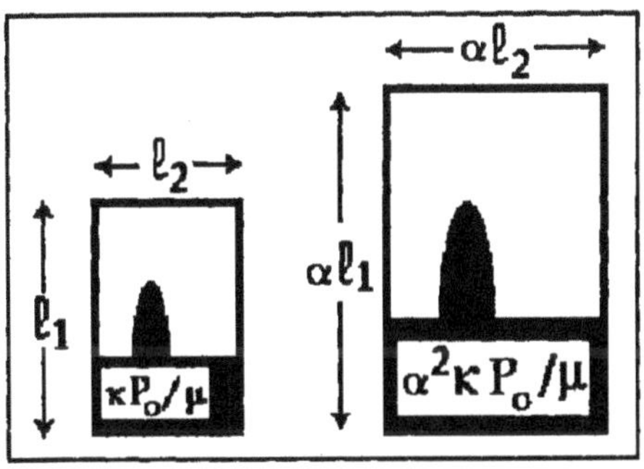

Figure 3.6: Flow in the right hand side cell is an enlargement of the one on the left.

Now, let $< \Delta p >_l$ be the mean capillary pressure over a piece of interface of length l (measured along a straight line). Clearly,

$$< \Delta p >_l \sim \Delta p / l^{D_s/2} \tag{3.4}$$

where D_s is the interface fractal dimension on short length scales, and l^{D_s} is the corresponding number of pores in length l.

On the other hand,

$$< \Delta p' >_l = \alpha^2 < \Delta p >_l \tag{3.5}$$

is required, where primed quantities are for the enlarged system. It follows then that

$$\frac{\Delta p'}{(\alpha l)^{D_s/2}} = \alpha^2 \frac{\Delta p}{l^{D_s/2}}, \tag{3.6}$$

that is,

$$\Delta p' = \alpha^{(2+D_s/2)}\Delta p. \tag{3.7}$$

Therefore r (given by $r = \Delta p/(a < \nabla p >)$) scales as $\alpha^{(1+D_s/2)}$. On the other hand r is the only parameter in the exercice; therefore, $L_c = f(r)$ and $L'_c = f(r')$, that is, $\alpha L_c = f(r\alpha^{(1+D_s/2)})$, which implies

$$L_c \sim r^{2/(2+D_s)} \tag{3.8}$$

The straight line in Fig. 3.5 represents this rule (setting $D_s = 1.3$, which is the value we find for it). It is reassuring to see that data points seem to approach the predicted slope as L_c becomes larger than about $10a$; lattice size effects produce deviations from the predicted behavior for smaller values of L_c. That is to be expected, since branch thickness cannot become less than a in a lattice, which necessarily leads to departures from Eq. (3.8) for r small enough.

Thus, DLA simulations lead one, in this case, to discover something that experiments had not unveiled: the spatial coexistence of IP and DLA like regimes, as well as the behavior of the corresponding crossover length. The reason why simulations turn out to be more powerful here is simple: runs were carried out on lattices with $\sim 10^6$ sites, whereas experiments were performed on networks of $\sim 10^5$ channels; it is the difference in spatial resolving power that makes the difference in this case.

3.3. Poisson Growth

A different Hele-Shaw cell type of experiment is considered in this section. Consider, for instance, a cell initially filled with some viscous fluid; if the gap distance is increased, the fluid will recede (assuming incompressibility), and will be invaded by whatever fluid it is inmersed in. We will assume that the latter fluid is an inviscid one. Viscous fingering takes place then[26]. It is explained below that Poisson's equation, rather than Laplace's, governs flow in this case. The DLA model must be modified accordingly to handle this. The patterns that develop do *not* belong to the class of *Laplacian fractals*.

We will only consider very small gap distances, even though they vary with time; consequently Darcy's law holds, just as for standard Hele-Shaw cell experimants. However, this exercice differs from the one we have studied here thus far because, even though the fluid is assumed to be incompressible, it does not appear to be so when looked at two dimensionally. This is best appreciated by examining Fig. 3.7.

Notice that, as the gap distance h increases, the amount of fluid in the column shown increases. Therefore the fluid looks compressible when viewed from the top (that is, two dimensionally); it follows that Laplace's equation for p (which follows from $\nabla . v = 0$ and Darcy's law) does not hold as a two dimensional equation.

More quantitatevly, from Gauss' theorem, $(\partial h/\partial t)\Delta x\Delta y = -\int\int v.dS$, and we define the divergence of v in d=2 as follows $\nabla_\perp . v = \lim_{\Delta x\Delta y\to 0}\left[\int\int v.dS/(h\Delta x\Delta y)\right]$. It

Figure 3.7: Top and front views of a Hele-Shaw cell whose plates are separating, the velocity field, and a column whose volume obviously increases with time, and, consequently, $\partial v_y/\partial y + \partial v_x/\partial x \neq 0$.

follows then that

$$h\nabla_\perp . v = -\partial h/\partial t \tag{3.9}$$

Substitution of Darcy's law into this equation gives,

$$\nabla_\perp^2 p = (\kappa/\mu)\partial h/\partial t \tag{3.10}$$

This is Poisson's equation, which, with Darcy's law, determines flow unambiguously. The corresponding time evolution of the two fluid interface has been called[26] *Poisson growth*. Retracing the steps followed in the second lecture that led to the DLA algorithm for Laplacian growth, one arrives at the DLA rule for Poisson growth. Assume, for concreteness sake, that the Hele-Shaw cell is of circular shape, and that its gap is open all the way round. Then, the DLA algorithm is as follows: start a random walk at a point x, y with uniform probability (assuming $h(x, y)$ is independent of x, y, that is, that the Hele-Shaw cell plates remain parallel) within the viscous fluid region, and stop walking on leaving the viscous fluid region (i.e, stop upon landing on the aggregate); the last site visited before landing on the aggregate becomes part of it.

Surface tension can be taken into account in the same manner as for Laplacian growth. A slightly modified algorithm has been used[26] to treat an experimental set up, in which the top plate was hinged at one end and lifted at the opposite one. Experimental and DLA results are shown in Fig. 3.8.

In contrast to Laplacian growth, Poisson growth leads to patterns that eventually fill all space available. That is because as long as there is any uninvaded region, random walks can start therein, and that leads therefore to their further invasion; clearly, such a process can only stop when all space has been invaded. Simulations do indeed show[26] that the fractal dimension D varies with time. Ini-

Figure 3.8: A digitized image from an experimer on a rectangular Hele-Shaw cell sealed at the top and sides (with open bottom) is shown on the left hand side. Different colors correspond to different stages of development. In contradistinction with standard DLA growth, there is growth deep within *fjords* here. The figure on the right shows the corresponding image produced by the DLA algorithm. After Ref. 26.

tially, patterns develop as standard DLA ones, and $D \approx 1.7$; but D increases with time until it finally equals the dimension of the embedding space.

Exercices

1. To obtain Eq. (3.1), start with the expression

$$L = \int_{x_1}^{x_2} [1 + (dy/dx)^2]^{1/2} dx$$

for the length of the initial piece of interface, let $y(x) \to y(x) + \eta(x)$, $[\eta(x_1) = 0 = \eta(x_2)]$ obtain the corresponding expression for δL; then, integrate by parts and collect terms to get

$$\delta L = -\int (d^2y/dx^2)/[1 + (dy/dx)^2]^{3/2} \eta(x) dx.$$

Note that the quantity multiplying η in the integrand is the curvature. Show that Eq. (3.1) follows if either η or the curvature κ are constant for $x_1 < x < x_2$, since the expression in curly brackets is κ and $\delta A = \int \eta dx$.

2. Experimental values of the Saffman-Taylor finger width extrapolate to $W/2$ as $Ca^{-1} \to 0$ (no data can be obtained for arbitrarily small Ca^{-1} because the finger breaks up into ramified patterns at sufficiently large injection rates); the theory for the W/2 limiting value can be found in Ref. 10. To get a feeling for what's involved, let flow be along the x axis, $y(x)$ be the finger interface, and v the finger tip velocity. Assume a steady state finger shape (that is, each point on the interface moves so that the end result is as though the finger moved rigidly forward); **show** that therefore (using Darcy's law) $| \nabla p | \propto v/[1+(dy/dx)^2]^{1/2}$, where ∇p is evaluated on the interface. On the other hand $\nabla^2 p = 0$ must be satisfied within the region occupied by the viscous fluid -subject to the boundary condition p=constant on the interface (since we are letting the surface tension vanish). Therefore, the problem of finding $y(x)$ (solved in Ref. 10, but not expected to be solved analytically by the reader within a reasonably short time) is equivalent to the following one: what is the shape of a (non-flat) conducting surface such that the charge density on it fulfills $\sigma \propto 1/[1+(dy/dx)^2]^{1/2}$ in two dimensions (the field must also satisfy the appropriate boundary condition on the cell boundaries).

Acknowledgments

Useful remarks by Dr. J. Rivero and by Prof. J. Stankiewicz are gratefully acknowledged. The very congenial hospitality enjoyed at Universidad de Granada while these lecture notes were written is deeply appreciated. The following grants contributed, in various ways, to the work reported here: SAB91-0273 of DGICYT (Dirección General de Investigaci/'on Científica y Técnica -of Spain), CONICIT (Consejo Nacional de Investigaciones Científicas y Tecnológicas -of Venezuela), FUNDAYACUCHO -of Venezuela, and CI1.0409.E of the Commission of the European Communities.

References

1. M. Eden, in *Proc. Fourth Berkeley Symp. on Math. Stat. and Prob.*, edited by F. Neymen, vol. IV (Univ. of California Press, Berkeley, 1961), p. 223.
2. T.A. Witten and L.M. Sander, *Phys. Rev. Lett.* **47** (1981) 1400.
3. L. Paterson, *Phys. Rev. Lett.* **52**, (1984) 1621
4. Other variants of the DLA algorithm exist. We are following here the scheme proposed by L.P. Kadanoff in *J. Stat. Phys.* **39** (1985) 267.
5. R. Chandler et al., *J. Fluid Mech.* **119** (1982) 249; D. Wilkinson and J. F. Willemsen, *J.*

Phys. A **16**, (1983) 3365.
6. L. Fururberg et al., *Phys. Rev. Lett.* **61** (1988) 2117.
7. B. Mandelbrot, *The Fractal Geometry of Nature* (Freeman, New York, 1982); see also, J. Feder, *Fractals* (Plenum Press, New York, 1988).
8. A. Bunde and S. Havlin, eds, *Fractals and Disordered Systems* (Springer-Verlag, Berlin, 1991).
9. P. G. Saffman and G. Taylor, *Proc. Roy. Soc. London* *A***245** (1958) 312.
10. R. Combescot et al., *Phys. Rev. Lett.* **56** (1986) 2036; B. Shraiman, *ibid* **56** (1986) 2028; D. C. Hong and J. Langer, *ibid* **56** (1986) 2032.
11. A. Arnéodo et al., *Phys. Rev. Lett.*, **63**, (1989) 984.
12. See, for instance, L. D. Landau and E.M Lifshitz, *Fluid Mechanics* (Pergamon Press, secon edition, Oxford, 1987) p. 46.
13. R. Brady and R.C. Ball, *Nature* **309** (1984) 225; see also, M. Matsushita et al., *Phys. Rev. Lett.* **52** (1984) 286; D. Grier et al., *ibid* **56** (1986) 1264; Y. Sawada, A. Dougherty, J.P. Gollub, ibid **56** (1986) 1260.
14. J.Stankiewicz, (unpublished).
15. L. Niemeyer, L. Pietronero, and H.J. Wiesmann, *Phys. Rev. Lett.* **52**, (1984) 1033.
16. V. I. Arnold, *Mecahnics* (Springer-Verlag, Berlin, 1978). pp. 50-52.
17. L. D. Landau and E.M Lifshitz, *Fluid Mechanics* (Pergamon Press, second edition, Oxford, 1987) pp 56-58.
18. T. Vicsek, *Phys. Rev. Lett.* **53** (1984) 2281; see also Kadanoff, in Ref. 4 in lecture I.
19. J.F.Fernández and J.M. Albarrán, *Phys. Rev. Lett.* **64** (1990) 2133.
20. J. Rivero and J. F. Fernández (to be published).
21. S. Liang, *Phys. Rev. A* **33** (1986) 2663
22. J. P. Stokes et al. *Phys. Rev. Lett.* **60** (1988) 1386 and references therein.
23. R. Lenormand, *J. Phys. C* **2** (1990) SA79 and references therein.
24. J.F. Fernández, R. Rangel, and J. Rivero, *Phys. Rev. Lett.* **67** (1991) 2958.
25. Up to this point, the picture on the right in Fig. 3.6 is just an enlargement of the one on the left. The question arises whether it can also represent an experiment. That obviously requires that the right injection pressure be applied, and that capillary pressures be just right. That is precisely what *scaling* the pressure does, as explained in section 2.2.
26. H. La Roche, J.F. Fernández, M. Octavio, A.G. Loeser, and C.J Lobb, *Phys. Rev. A* **44** (1991) 6185.

Glossary

D: fractal dimension (defined on section 1.4).
μ: viscosity (first introduced in section 1.5). Units of grams/(cm seg).
κ: permeability. (first introduced in section 1.5.) It depends on the geometry of the medium. Flux is proportional to it. Units of $length^2$.
τ: surface tension. (treated in section 2.1). Units of dyne/cm.
p: pressure.
Δp: pressure drop through an interface. (Introduced in section 1.5.)
U: fluid velocity far ahead of two fluid interface in a rectangular cell (introduced in section 2.1.).
Ca: capillary number. $Ca \equiv U\mu/\tau$. (defined in section 2.1.). Unitless.

INTEGRAL EQUATIONS FOR FLUIDS: CRITICAL PHENOMENA

LUIS F. RULL
Departamento de Fisica Atomica Molecular y Nuclear
Universidad de Sevilla, Aptdo.1065, 41080 Sevilla, Spain

1. Introduction and Summary

The aim of my lecture is to review the use of integral equation in the context of the statistical mechanics of fluids; numerical solutions of these equations are examined in detail. Integral equations have played an important role in the determination of structural and thermodynamic properties of macroscopic systems; they provide information on correlation functions. Integral equations may be obtained from diagramatic analysis of the pair distribution function. A more pedagogical description uses functional expansions of the appropriate functions. It is the latter approach which we have adopted here.

The so-called Yvon-Born-Green (YBG), Percus-Yevick (PY), and Hypernetted Chain (HNC) are the most familiar approximate integral equations. More recently, the Modified Hypernetted Chain (MHNC) Equation, which is an exact form of the HNC equation, has been developed using a *bridge function* which is obtained from exact simulation data. When hard-sphere simulation data are used to determine the parameters of the bridge function taking advantage of the universality of this function, the Referenced Hypernetted Chain (RHNC) Equation is obtained. The ideas of perturbation theory for fluids have been employed extensively in this approach. The approximate integral equations suffer from thermodynamic inconsistencies which do not occur within the RHNC equation.

Owing to the fact that integral equations can be solved analytically only for a very limited number of systems, numerical techniques has been used widespread. Numerical methods for integral equations will be discussed. We remark that a complete description of the use of integral equations to understand the structural and thermodynamic properties of fluids is not attempted here. Instead, I focus on the results obtained from integral equations for vapour-liquid coexistence in the critical region. The resulting critical constants and exponents are analysed in the light of some recent ideas about critical points.

2. Derivation of Integral Equations

A natural starting point is the reduced distribution function. For a system of N identical particles in a volume V at equilibrium at temperature T, it follows from the normalized canonical distribution function

$$F(x_1,...,x_N) = \frac{1}{N}\frac{1}{h^{3N}}\frac{\exp(-\beta H(x_1,...x_N))}{Q_N(V,T)} \quad . \tag{2.1}$$

Here, the Hamiltonian is

$$H(x_1,...,x_N) = T_N + U_N = \sum_{i=1}^{N}\frac{p_i^2}{2m_i} + U(\vec{r}_1,...,\vec{r}_N) \ ,$$

the canonical partition function is

$$Q_N(V,T) = \frac{1}{N}\frac{1}{h^{3N}}\int dx_1,...,dx_N \exp(-\beta H(x_1,...x_N)) \ , \tag{2.2}$$

k_B is the Boltzmann constant, h is the Planck constant, β=$1/k_BT$, and x_i=$(\vec{r}_i,\vec{p}_i)$ is the coordinate and momentum variable of particle i. The reduced distribution function is defined as

$$f_s(x_1,...,x_s) = \frac{N!}{(N-S)!}\int dx_{s+1},...,dx_N F(x_1,...,x_N) \quad s \leq N \ . \tag{2.3}$$

It can we written as

$$f_s(x_1,...,x_s) = \varphi_s(\vec{p}_1,...,\vec{p}_s)n_s(\vec{r}_1,...,\vec{r}_s) \ ,$$

where

$$\varphi_s(\vec{p}_1,...,\vec{p}_s) = \prod_{i=1}^{s}\varphi(p_i) \tag{2.4a}$$

and

$$n_s(\vec{r}_1,...,\vec{r}_s) = \frac{N!}{(N-S)!}\frac{1}{V^N Z_N}\int d\vec{r}_{s+1},...,d\vec{r}_N \exp(-\beta U(\vec{r}_1,...,\vec{r}_N) \ . \tag{2.4b}$$

The configurational partition function is

$$Z_N = \frac{1}{V^N}\int d\vec{r}_1,...,d\vec{r}_N \exp(-\beta U(\vec{r}_1,...,\vec{r}_N)) \ .$$

When considering a uniform fluid in equilibrium, there is a translational invariance of the distribution function, so that $n_s(\vec{r}_1,...,\vec{r}_s)$ only depends on s-1 positions. Furthermore, if the potential is pairwise, additive, and central one can write

$$H = \sum_{i=1}^{N}\frac{p_i^2}{2m_i} + \sum_{1\leq i}\sum_{<j\leq N} u(|\vec{r}_i - \vec{r}_j|)$$

corresponding to homogeneous and isotropic systems. Then,

$$n_1(\vec{r}_1) = \frac{N}{V} = \rho \tag{2.5a}$$

is the number density, and

$$f_2(x_1, x_2) = \varphi_1(p_1)\varphi_2(p_2)n_2(r_{12}) \tag{2.5b}$$

defines the pair distribution function:

$$g(r) = \frac{n_2(r)}{\rho^2} \ .$$

Frequently, one uses the distribution function

$$h(r) = g(r) - 1$$

which goes to zero when $r \to \infty$.

The thermodynamic functions can be expressed in terms of the afore mentioned distribution functions [1]. For example, the internal energy is

$$E(T, V) = N\left[\frac{3}{2} + 2\pi\rho\int_0^\infty drr^2u(r)g(r)\right] \ , \tag{2.6a}$$

the pressure is

$$P = \rho k_BT - \frac{4\pi}{6}\rho^2\int_0^\infty drr^3\frac{du(r)}{dr}g(r) \ , \tag{2.6b}$$

and the isothermal compressibility is

$$k_BT\left(\frac{\partial\rho}{\partial P}\right)_T = \rho k_BTK_T = 1 + \int_0^\infty 4\pi r^2h(r)dr \ . \tag{2.6c}$$

2.1. Derivation of the YBG Equation

A hierarchy of equations [2] can be obtained by differentiating the equation (2.4b) for $n_s(\vec{r}_1, ..., \vec{r}_s)$:

$$\frac{\partial}{\partial\vec{r}_1}n_s(\vec{r}_1, ..., \vec{r}_s) = \frac{N!}{(N-S)!}\frac{1}{V^NZ_N}\int d\vec{r}_{s+1}, ..., d\vec{r}_N\frac{\partial}{\partial\vec{r}_1}\exp(-\beta\sum_{i<j}u_{ij}) \ . \tag{2.7}$$

For a pairwise potential, $u(\mid \vec{r}_i - \vec{r}_j \mid)$, the latter can be written as

$$-k_BT\frac{\partial}{\partial\vec{r}_1}n_s(\vec{r}_1, ..., \vec{r}_s) = \sum_{i=2}^{s}\frac{\partial u_{1i}}{\partial\vec{r}_1}n_s(\vec{r}_1, ..., \vec{r}_s) + \int d\vec{r}_{s+1}\frac{\partial u_{1s}}{\partial\vec{r}_1}n_{s+1}(\vec{r}_1, ..., \vec{r}_{s+1}) \tag{2.8}$$

which provides an infinite set of equations in the thermodynamic limit that can be solved by approximate means. Dividing each side of Eq. (2.8) by $n_s(\vec{r}_1, ..., \vec{r}_s)$ we have

$$-k_BT\frac{\partial}{\partial\vec{r}_1}\ln n_s(\vec{r}_1, ..., \vec{r}_s) = \sum_{i=2}^{s}\frac{\partial u_{1i}}{\partial\vec{r}_1} + \int d\vec{r}_{s+1}\frac{\partial u_{1s+1}}{\partial\vec{r}_1}\frac{n_{s+1}(\vec{r}_1, ..., \vec{r}_{s+1})}{n_s(\vec{r}_1, ..., \vec{r}_s)} \ . \tag{2.9}$$

By defining

$$g^{(s)}(\vec{r}_1,...,\vec{r}_s) = \frac{n_s(\vec{r}_1,...,r_s)}{n^s} \equiv \exp(-\beta W_s(\vec{r}_1,...,\vec{r}_s)) \ , \tag{2.10}$$

$$\frac{\partial}{\partial\vec{r}_1}\ln n_s(\vec{r}_1,...,\vec{r}_s) = \frac{1}{n^s g^{(s)}}\frac{\partial n^s g^{(s)}}{\partial\vec{r}_1} = \frac{\partial \ln g^{(s)}}{\partial\vec{r}_1} = -\beta\frac{\partial W_s}{\partial\vec{r}_1} \ ,$$

we get:

$$\frac{\partial W_s}{\partial\vec{r}_1} = \sum_{i=2}^{s}\frac{\partial u_{1i}}{\partial\vec{r}_1} + \int d\vec{r}_{s+1}\frac{\partial u_{1s+1}}{\partial\vec{r}_1}\frac{n_{s+1}(\vec{r}_1,...,\vec{r}_{s+1})}{n_s(\vec{r}_1,...,\vec{r}_s)} \ . \tag{2.11}$$

From the above expression it is clear that if the s particles were isolated, the mean force acting on particle 1 would be given by the first sum of Eq. (2.11). But the particles are interacting with the medium that can be represented by the force acting on particle 1 due to particles $s+1$ averaged with the weighting function n_{s+1}/n_s which represents the conditional probability of finding particle $s+1$ at $\vec{r}_{s+1}$ with particles $1,...,s$ at $\vec{r}_1,...,\vec{r}_s$. If this potential of mean force is assumed to be pairwise additive for the system with s=3, i.e.,

$$W_3(\vec{r}_1,\vec{r}_2,\vec{r}_3) = W_2(\vec{r}_1,\vec{r}_2) + W_2(\vec{r}_1,\vec{r}_3) + W_2(\vec{r}_2,\vec{r}_3) \ , \tag{2.12}$$

then

$$\exp(-\beta W_3(\vec{r}_1,\vec{r}_2,\vec{r}_3)) = g^{(3)}(\vec{r}_1,\vec{r}_2,\vec{r}_3) = g^{(2)}(\vec{r}_1,\vec{r}_2)g^{(2)}(\vec{r}_1,\vec{r}_3)g^{(2)}(\vec{r}_2,\vec{r}_3) \ . \tag{2.13}$$

Substituting Eq. (2.13) into Eq. (2.11) we have for a system with s=2

$$-k_B T\frac{\partial}{\partial\vec{r}_1}\ln g(\vec{r}_1,\vec{r}_2) = \frac{\partial u_{12}}{\vec{r}_1} + \int d\vec{r}_3\frac{\partial u_{13}}{\partial\vec{r}_1}g(\vec{r}_1,\vec{r}_3)g(\vec{r}_2,\vec{r}_3) \tag{2.14}$$

Equation (2.12) is Kirkwood's approximation[3]. The integro-differential equation (2.14) is the Yvon-Born-Green (YBG) equation which, unlike expression (2.9), is a closed equation for the pair distribution function.

2.2. The PY and HNC Equations

The PY and HNC equations are closed approximate expressions for the radial distribution function. Two alternative routes can be used in their derivation. The first involves a well defined analysis of the diagrams of Eq. (2.9) some of which are ignored[4,5], whilst the second uses a functional Taylor expansion[6,7]. The latter route is simpler and more elemental but has the inconvenient feature that some of the functions used in the derivation are rather arbitrary. Its simplicity, however, allows for an elegant derivation.

Consider the system with an external field that depends only on the coordinates of the particles. The Hamiltonian is given by

$$H = \sum_{i=1}^{N}\frac{p_i^2}{2m} + \sum_{1\le i}\sum_{<j\le N} u(|\vec{r}_i - \vec{r}_j|) + \sum_{i=1}^{N}\varphi(\vec{r}_i) \ .$$

The partition function is

$$Q(N,V,T) = \frac{V^N}{N!}(2\pi m k_B T)^{\frac{3N}{2}} Z_N[\varphi] \,.$$

Assuming $Z_N(\varphi)$ is a functional of φ, it is easy to show that

$$\delta \ln Z_N = -\frac{1}{k_B T}\int d\vec{r}_1 \delta\varphi(\vec{r}_1) n_1(\vec{r}_1,\varphi) \,.$$

Also,

$$\frac{\delta \ln Z_N(\varphi)}{\delta\varphi} = -\beta n_1(\vec{r}_1,\varphi) \,.$$

Successive derivatives of $\ln Z[\varphi]$ can be written:

$$\frac{\delta^2 \ln Z_N[\varphi]}{\delta\varphi(\vec{r}_1)\delta\varphi(\vec{r}_2)} = \beta^2\{n_1(\vec{r}_1,\varphi)\delta(\vec{r}_1 - r_2) + n_2(\vec{r}_1,\vec{r}_2,\varphi) - n_1(\vec{r}_1,\varphi)n_1(\vec{r}_2,\varphi)\} \,.$$

Here, we have a relationship between the partition function and the radial distribution function. The direct correlation function, $c(r_1,r_2)$, is defined in such a way that

$$-\beta\frac{\delta\varphi(\vec{r}_1)}{\delta n_1(r_1,\varphi)} = \frac{1}{n_1(\vec{r}_1,\varphi)}\delta(\vec{r}_1 - \vec{r}_2) - c(\vec{r}_1 - \vec{r}_2) \,. \tag{2.15}$$

Using the equality

$$\delta(\vec{r}_1,\vec{r}_2) = \int d\vec{r}_3 \frac{\delta n_1(\vec{r}_1,\varphi)}{\delta\varphi(\vec{r}_3)}\frac{\delta\varphi(\vec{r}_3)}{\delta n(\vec{r}_2,\varphi)} \,, \tag{2.16}$$

one has for a homogeneous system with $(n_1(\vec{r}_1,0) = \rho)$:

$$h(r_{12}) = c(r_{12}) + \rho\int d\vec{r}_3 h(r_{13})c(r_{23}) \,. \tag{2.17}$$

This is the Ornstein-Zernike (OZ) equation which can be taken as a definition for the direct correlation function $c(\vec{r}_1,\vec{r}_2)$.

The Fourier transform of Eq. (2.17) is given by

$$\tilde{h}(k) = \tilde{c}(k) + \rho\tilde{c}(k)\tilde{h}(k) \tag{2.18a}$$

or

$$\tilde{h}(k) = \frac{\tilde{c}(k)}{1-\rho\tilde{c}(k)} \tag{2.18b}$$

which means that the isothermal compressibility, (2.6c), can be expressed in terms of the direct correlation function:

$$k_B T\rho K_T = 1 + \rho\tilde{h}(0) = \frac{1}{1-\rho\tilde{c}(0)} \,, \tag{2.19}$$

and

$$\frac{1}{k_B T}\left(\frac{\partial p}{\partial\rho}\right)_T = 1 - \rho\tilde{c}(0) = 1 - \rho\int d\vec{r} c(r) \,. \tag{2.20}$$

Now, consider a functional expansion of $A(\vec{r},\varphi)$ about $B(\vec{r},\varphi)$ which is also a functional of $\varphi(\vec{r})$:

$$A(\vec{r}_1,\varphi) = A(\vec{r}_1) + \int d\vec{r}_2 \left[B(\vec{r}_2,\varphi) - B(\vec{r}_2)\right] \left[\frac{\delta A(\vec{r}_1,\varphi)}{\delta B(\vec{r}_2,\varphi)}\right]_{\varphi=0}$$

$$+\frac{1}{2!}\int d\vec{r}_2 \int d\vec{r}_3 \left[B(\vec{r}_3,\varphi) - B(\vec{r}_3)\right]\left[B(\vec{r}_2,\varphi) - B(\vec{r}_2)\right] \left[\frac{\delta^2 A(\vec{r}_1,\varphi)}{\delta B(\vec{r}_3,\varphi)\delta B(\vec{r}_2,\varphi)}\right]_{\varphi=0} + \dots , \quad (2.21)$$

where

$$\frac{\delta A(\vec{r}_1,\varphi)}{\delta B(\vec{r}_2,\varphi)} = \int d\vec{r}_3 \frac{\delta A(\vec{r}_1,\varphi)}{\delta\varphi(\vec{r}_3)} \frac{\delta\varphi(\vec{r}_3)}{\delta B(\vec{r}_2,\varphi)} . \quad (2.22)$$

In order to obtain a closed expression for the radial distribution function, φ, B and A are arbitrarily fixed. This expression, together with the OZ equation provide us with a closed expression for $g(r)$.

2.2.1. The PY Equation

The external field φ is chosen to represent a particle which is identical to the other particles of the system and is placed at fixed position r_0. The reduced one particle distribution function is chosen for $B(r,\varphi)$, i.e.,

$$n_1(\vec{r}_1,\varphi) = N\frac{\int d\vec{r}_2,\dots,d\vec{r}_N \exp(-\beta\sum_{1\le i}\sum_{<j\le N} u(|\vec{r}_i - \vec{r}_j|))}{\int d\vec{r}_1,\dots,d\vec{r}_N \exp(-\beta\sum_{1\le i}\sum_{<j\le N} u(|\vec{r}_i - \vec{r}_j|))} = \frac{n_2(\vec{r}_0,\vec{r}_1)}{n_1(\vec{r}_0)} .$$

The following expression is chosen for $A(\vec{r}_1,\varphi)$:

$$A(\vec{r}_1,\varphi) = n_1(\vec{r}_1,\varphi)\exp(-\beta u_{01}) .$$

Substitution of these expressions into (2.21) gives

$$n_1(\vec{r}_1,\varphi) = \frac{n_2(\vec{r}_0,\vec{r}_1)}{n_1(\vec{r}_0)}\exp(\beta u_{01}) =$$

$$n_1(\vec{r}_1) + \int d\vec{r}_2 \left[\frac{n_2(\vec{r}_0,\vec{r}_2)}{n_1(\vec{r}_0)} - n_1(\vec{r}_2)\right] \int d\vec{r}_3 \left\{\frac{\delta[n_1(\vec{r}_1,\varphi)\exp(\beta u_{01})]}{\delta\varphi(\vec{r}_3)} \frac{\delta\varphi(\vec{r}_3)}{\delta[n_1(\vec{r}_2,\varphi)]}\right\} . \quad (2.23)$$

Using the definition of the direct correlation function and the functional derivative of $n_1(\vec{r}_1,\varphi)\exp(\beta u_{01})$ one can write

$$\frac{n_2(\vec{r}_{01})}{n}\exp(\beta u(r_{01})) = \rho + \int d\vec{r}_2 \left[\frac{n_2(r_{02})}{\rho} - \rho\right]\rho c(r_{12}) , \quad (2.24)$$

and, from the OZ equation (2.17) and from the equation for the pair distribution function one has:

$$c(r) = g(r)(1 - \exp(\beta u(r)) . \quad (2.25)$$

This is the Percus-Yevick (PY) equation.

2.2.2. The HNC Equation

The choices of φ and $B(\vec{r},\varphi)$ are the same for the HNC approximation as for the PY approximation. In this case, however,

$$A(\vec{r},\varphi) = \ln[n_1(\vec{r}_1,\varphi)\exp(\beta\varphi(\vec{r}_1))] = \ln\left[\frac{n_2(\vec{r}_0,\vec{r}_1)}{n_1(\vec{r}_0)}\exp(\beta u(r_{01}))\right] .$$

As for the PY equation, substitution into the series (2.21), and the use of the definitions of the direct correlation function and of the pair distribution function gives

$$c(r) = g(r) - \ln g(r) - \beta u(r) - 1 \ . \tag{2.26}$$

This is the so-called Hypernetted-Chain (HNC) equation.

2.3. Contents of the PY and HNC Equations

By considering the exact density expansion[8] of the radial pair distribution function, the following equation can be written

$$g(r_{12}) = \exp(-\beta u_{12}) + \rho\exp(-\beta u_{12})\int dr_3 f_{13}f_{23} + \rho^2\exp(-\beta u_{12})$$

$$\int d\vec{r}_3\int d\vec{r}_4\left\{f_{13}f_{24}f_{34} + f_{13}f_{14}f_{24}f_{34} + \frac{1}{2}f_{13}f_{14}f_{23}f_{24} + \frac{1}{2}f_{13}f_{14}f_{23}f_{24}f_{34}\right\} + \vartheta(\rho^3) \tag{2.27}$$

where $f = \exp(-\beta u_{ij}) - 1$ is the Mayer function. The above expression allows us to examine the content of the integral equations at each order in density. For the Percus-Yevick equation,

$$g(r_{12}) = \exp(-\beta u_{12})\left\{1 + \rho\int d\vec{r}_3(g(r_{13}) - 1)g(r_{23})(1 - \exp(-\beta u_{23})\right\} ;$$

one has up to zero order in density:

$$g(r_{12})\mid_0 = \exp(-\beta u_{12}) \ ,$$

up to first order in density:

$$g(r_{12})\mid_1 = \exp(-\beta u_{12})\left\{1 + \rho\int d\vec{r}_3(\exp(-\beta u_{13}) - 1)\exp(-\beta u_{23})(1 - \exp(\beta u_{23}))\right\}$$

$$= \exp(-\beta u_{12}) + \exp(-\beta u_{12})\rho\int d\vec{r}_3 f_{13}f_{23} \ ,$$

and up to second order in density:

$$g(r_{12})\mid_2 = \exp(-\beta u_{12})\left\{1 + \rho\int d\vec{r}_3(g(r_{13})\mid_1 - 1)g(r_{23})\mid_1(1 - \exp(-\beta u_{23}))\right\} .$$

Substituting for $g(r_{13})\mid_1$ and $g(r_{23})\mid_1$ we have

$$g(r_{12})\mid_2 = \exp(-\beta u_{12}) + \rho \exp(-\beta u_{12}) \int d\vec{r}_3 f_{13} f_{23}$$

$$+\rho^2 \exp(-\beta u_{12}) \int d\vec{r}_3 \int d\vec{r}_4 \{f_{14}f_{23}f_{34} + f_{13}f_{23}f_{24}f_{34} + f_{13}f_{14}f_{23}f_{24}\} \,. \tag{2.28}$$

Thus, the PY equation is exact in the limit of zero density; differences with the exact density expansion (2.27) occur in terms of order ρ^2.

In the case of the HNC equation,

$$\ln g(r_{12}) = -\beta u_{12} + \rho \int d\vec{r}_3 (g(r_{13}) - 1))[g(r_{23}) - 1 - \ln g(r_{23}) - \beta u_{23}] \,,$$

the problem is the occurrence of the logarithm of the pair distribution function instead of the function itself. This is overcome by writing

$$\ln g(r_{12}) = a + \rho b + \rho^2 c + \vartheta(\rho^3)$$

where

$$g(r_{12}) = e^a e^{\rho b} e^{\rho^2 c} \ldots$$

which can be written

$$g(r_{12}) = e^a(1 + \rho b + \frac{1}{2}\rho^2 b + \ldots)(1 + \rho^2 c + \frac{1}{2}\rho^4 c^2 + \ldots) = e^a + \rho e^a b + \rho^2(\frac{1}{2}b + c) + \vartheta(\rho^3) \,.$$

Following the same procedure as for the PY equation, we find to zero order in density:

$$\ln g(r_{12})\mid_0 = -\beta u_{12} \,,$$

to first order:

$$\ln g(r_{12})\mid_1 = -\beta u_{12} + \rho \int d\vec{r}_3 [\exp(-\beta u_{13}) - 1][\exp(-\beta u_{23}) - 1] \,,$$

and to second order:

$$\ln g(r_{12})\mid_2 = -\beta u_{12} + \rho \int d\vec{r}_3 [g(r_{12})\mid_1 - 1][g(r_{23})\mid_1 - 1 - \beta u_{23} - \ln g(r_{23})\mid_1] \,.$$

Substituting these expressions we can write the density expansion of $g(r_{12})$ as

$$g(r_{12})\mid_2 = \exp(-\beta u_{12}) + \rho \exp(-\beta u_{12}) \int d\vec{r}_3 f_{13} f_{23} + \rho^2 \exp(-\beta u_{12})$$

$$\int d\vec{r}_3 \int d\vec{r}_4 \{f_{12}f_{24}f_{34} + f_{13}f_{14}f_{24}f_{34} + f_{13}f_{34}f_{23}f_{24} + \frac{1}{2} f_{13}f_{14}f_{23}f_{24}\} \tag{2.29}$$

As can be seen from expression (2.29), the HNC approximation retains more of the diagrams of the full expansion (2.27) than the PY approximation. This does not necessarily imply that the HNC method provides a more accurate description of the system than the PY because, when all orders of density are considered, there may be a cancellation of errors to a certain degree.

2.4. The MHNC Equation

The density expansion (2.27) can be written

$$h(r) + 1 = \exp(-\beta u(r) + \theta(r) - E(r)) \ , \tag{2.30}$$

where

$$\theta(r) = h(r) - c(r) \ .$$

Here, $E(r)$ is called the bridge function which can be expressed, in the language of the graph theory, in terms of the sum of the elementary or bridge graphs[9]. Equation (2.30) reduces to the HNC equation (2.26) in the case of $E(r) = 0$. Furthermore, if one considers the "effective potential" $u(r) - \frac{1}{\beta}E(r)$, where the bridge function is known, then this would give a "modified" HNC equation (MHNC) which is exact.

The same argument can be used for the PY equation by introducing a function $D(r)$ which represents the set of diagrams freely connected at bridge points. The corresponding closure expression is

$$c(r) = \exp(-\beta u(r) - 1)g(r)\exp(\beta u(r)) + D(r) \tag{2.31}$$

This exact expression reduces to the PY equation (2.25) when $D(r) = 0$. Consequently, the expression is referred to as the modified PY equation (MPY).

There are, however, two reasons to favour the MHNC equation [9]. Firstly, $E(r)$ takes the form of an effective potential which can be given a physical interpretation. Secondly, $E(r)$ is a better behaved function than $D(r)$. In the case of hard spheres, for example, $E(r)$ is almost linear while $D(r)$ behaves exponentially.

Furthermore, $E(r)$ can be expressed as an infinite series of convolution integrals of $h(r)$; the universal nature of $g(r)$ for the hard sphere fluid, which is used in perturbation theories of more realistic systems, suggests that $E(r)$ is effectively the same for all systems. The hard spheres $E(r)$ can, thus, be substituted into the equation (2.30) to obtain an accurate closed expression for the radial distribution function. It is this form of the equation which is usually referred to as the modified hypernetted-chain (MHNC) equation.

The concept of thermodynamic consistency can be used in the choice of the bridge function $E(r)$. From equations (2.6b) and (2.20), and given that the integral equations are approximate relationships between $c(r)$ and $g(r)$ in terms of $u(r)$, we have the system equation of state from the pressure and the isothermal compressibility which will not be in general the same. This difference for all but low densities constitutes the basis of thermodynamic inconsistency.

Lado *et al*[10] have used the ideas of universality introduced in Ref. 9 for the bridge function of the MHNC equation, to derive an approximate expression for $E(r)$ in the case of a short range potential. Their approach is based on a functional of the free energy. As with perturbation theories, the potential of interaction is given in the form

$$u(r,\lambda_0,\lambda_1)=\lambda_0 u_0+\lambda_1\Delta u(r)$$

with

$$\Delta u(r)=u(r)-u_0(r)$$

where $u_0(r)$ is the reference short-range potential.

The free energy, A, for this system can be written in the form

$$\beta\frac{A}{N}=\frac{1}{2}\rho\int d\vec{r}\int_0^1 d\lambda_0 g(r,\lambda_0,0)u_0(r)+\frac{1}{2}\rho\int d\vec{r}\int_0^1 d\lambda_1 g(r,1,\lambda_1)\Delta u(r)\ , \tag{2.32}$$

where $g(r,\lambda_0,\lambda_1)$ corresponds to the pair distribution function of the system with the potential $u(r,\lambda_0,\lambda_1)$. The first term in the sum is the free energy of the reference system:

$$\beta\frac{A^0}{N}=\beta\frac{A_1^0}{N}+\beta\frac{A_2^0}{N}+\beta\frac{A_3^0}{N} \tag{2.33}$$

which can be calculated from the pair distribution function and the bridge function of the reference system. The second term in the sum is

$$\beta\frac{\Delta A}{N}=\beta\frac{\Delta A_1}{N}+\beta\frac{\Delta A_2}{N}+\beta\frac{\Delta A_3}{N}\ , \tag{2.34}$$

where

$$\beta\frac{\Delta A_j}{N}=\beta\frac{A_j}{N}+\beta\frac{A_j^0}{N}\quad j=1,2$$

and

$$\beta\frac{\Delta A_3}{N}=-\frac{1}{2}\rho\int d\vec{r}\int_0^1 d\lambda_1 E(r,1,\lambda_1)\frac{\partial g(r,1,\lambda_1)}{\partial\lambda_1}\ .$$

In essence, the system excess of free energy with the potential $u(r)$ is seen to be given by

$$\beta\frac{A}{N}=\beta\frac{A_1}{N}+\beta\frac{A_2}{N}+\beta\frac{A_3^0}{N}+\beta\frac{\Delta A_3}{N} \tag{2.35}$$

which is exact and has the following terms:

$$\beta\frac{A_1}{N}=-\frac{1}{2}\rho\int d\vec{r}\{\frac{1}{2}h^2(r)+h(r)-g(r)\ln[g(r)\exp(\beta u(r))]\} \tag{2.36a}$$

$$\beta\frac{A_2}{N}=-\frac{1}{2}\rho\int\frac{d\vec{k}}{2\pi^2}\{\ln[1+\rho\tilde{h}_0(k)]-\rho\tilde{h}_0(k)\} \tag{2.36b}$$

$$\beta\frac{A_3^0}{N}=-\frac{1}{2}\rho\int d\vec{r}\int_0^1 d\lambda_0 E(r,\lambda_0,0)\frac{\partial g(r,\lambda_0,0)}{\partial\lambda_0} \tag{2.36c}$$

$$\beta\frac{\Delta A_3}{N}=-\frac{1}{2}\rho\int d\vec{r}\int_0^1 d\lambda_1 E(r,1,\lambda_1)\frac{\partial g(r,1,\lambda_1)}{\partial\lambda_1}\ . \tag{2.36d}$$

Here, we have used

$$g(r)=g(r,1,1)$$

$$g_0(r) = g(r, 1, 0)$$

and the tilde denotes Fourier transforms. The first three terms can be calculated, and an approximate method can be used to determine the fourth

$$\beta\frac{\Delta A_3}{N} \simeq -\frac{1}{2}\rho \int d\vec{r} E(r,1,0)[g(r,1,1) - g(r,1,0)] \simeq -\frac{1}{2}\rho \int d\vec{r} E_0(r)[g(r) - g_0(r)] \,. \tag{2.37}$$

In this approximation one assumes that $E_0(r)$ is a short-range function, that $E(r)$ in Eq. (2.36d) only includes the first peak of $g(r)$, and that $E(r,1,\lambda)$ in this region is insensitive to the change of the potential from $u_0(r)$ to $u(r)$. All these approximations are collectively known as the reference hyppernetted chain equation (RHNC) approximation.

Up to now, the reference potential $u(r)$ has been arbitrary. We can impose a constraint on $u_0(r)$ in order to satisfy the fundamental condition of the canonical ensemble, i.e., that of minimal free energy A. Since A is a function of $g(r)$, $g_0(r)$ and $E_0(r)$ we can write

$$\beta\frac{\delta A}{N} = \frac{1}{2}\rho \int d\vec{r}\{c(r) - h(r) + \ln[g(r)\exp(\beta u(r))] - E_0\}\delta g(r) - \frac{1}{2}\rho \int d\vec{r}[g(r) - g_0(r)]\delta E_0(r) \tag{2.38}$$

where it follows that

$$c(r) = h(r) - \ln[g(r)\exp(-\beta u(r))] + E_0(r) \tag{2.39a}$$

$$\beta \int d\vec{r}[g(r) - g_0(r)]\delta E_0(r) = 0 \,. \tag{2.39b}$$

Equation (2.39a) is the RHNC relation, whilst condition (2.39b) is the constraint. If we now choose a reference potential

$$u_0(r) = u_0(r, \sigma, \epsilon) \tag{2.40}$$

which incorporates energy and length parameters which allow us to write

$$\rho \int d\vec{r}(g(r) - g_0(r))\sigma\frac{\partial E_0(r)}{\partial \sigma} = 0 \tag{2.41a}$$

$$\rho \int d\vec{r}(g(r) - g_0(r))\epsilon\frac{\partial E_0(r)}{\partial \epsilon} = 0 \tag{2.41b}$$

These two conditions allows the optimum values of ϵ and σ that minimizes the free energy to be determined. Without these, the RHNC would suffer from the same type of thermodynamic inconsistency as the original HNC equation[11]. In addition the thermodynamic relation

$$\left(\frac{\partial U}{\partial V}\right)_T = T\left(\frac{\partial P}{\partial T}\right)_V - P$$

proposed by Hiroike[12] as a strongest test of approximate integral equations is also satisfied.

3. Numerical Solutions

There are two possible solutions to integral equations, i.e., analytical and numerical solutions. In the former case, only the PY equation has been solved exactly for the two cases of hard spheres and sticky hard spheres. First, we shall see how the OZ equation (2.17) can be factorised to aid in the solution of integral equations for certain systems.

3.1. Factorisation of the OZ Equation

The use of the Wienner-Hopf[13] factorisation in the solution of the OZ equation was introduced by Wertheim[14] using the Laplace transform. In subsequent work, Baxter[15,16] used two new derivations and reformulations, the first also based on the Laplace transform and the second on the Fourier transform. Wertheim[17] has also derived the factorised form of the OZ equation using the combined methods of Fourier and Laplace, and has extended his results in various ways. We follow the second derivation of Baxter since it is the most transparent one[18].

The Fourier transform of the OZ equation (2.18b) can be written as

$$1 + \rho\tilde{h}(k) = \frac{1}{1 - \rho\tilde{c}(k)} \tag{3.1}$$

but here $1 + \rho\tilde{h}(k) = \tilde{s}(k)$ is the structure factor of the fluid that is real, positive, well defined and without singularities. Consequently $1 - \rho\tilde{c}(k)$ also has to be real positive and well defined. It is sensible to write $1 - \rho\tilde{c}(k)$ as the square of a function $q(k)$, i. e.,

$$1 - \rho\tilde{c}(k) = |\, q(k)\,|^2 = q(k)q^*(k) = q(k)q(-k) \tag{3.2}$$

where $q(k)$ is non zero and corresponds to an arbitrary function $q(r)$ in real space. Baxter[16,19], has shown that $q(k)$ has the following properties.

Firstly, $q(k) - 1$ has a Fourier transform $q(r)$ defined by

$$q(k) = 1 - 2\pi\rho \int_0^\infty dr \exp(ikr)q(r) \tag{3.3}$$

where $q(r)$ is a real function such that $q(k)$=$q(-k)$. Secondly, when $c(r)$ satisfies the condition

$$c(r) = 0 \quad r > R \tag{3.4}$$

$q(k)$ can be analytically obtained in such a way that

$$q(r) \neq 0 \quad r < R \tag{3.5a}$$

$$q(r) = 0 \quad r > R\,. \tag{3.5b}$$

Finally, $q(k)$ is analytical and is non-zero in the half complex k-plane.

Transforming equation (3.2) to real space we have

$$\int_r^\infty dr'r'c(r') = q(r) - 2\pi\rho \int_r^R dr'q(r')q(r'-r)\,. \tag{3.6}$$

If we use the left hand side of equation (3.1)

$$1 + \rho\tilde{h}(k) = | q(k) |^{-2} \tag{3.7}$$

and define $H(r) = \int_r^\infty dr'r'h(r')$, the real space transform is

$$H(r) = q(r) + 2\pi\rho \int_0^R dr'q(r')H(| r - r' |) \quad r \geq 0\,. \tag{3.8}$$

Differentiating with respect to r and using $q(r) = 0$ we have

$$rc(r) = -q'(r) + 2\pi\rho \int_r^R dr'q'(r')q(r-r') \quad 0 \leq r \leq R \tag{3.9}$$

$$rh(r) = -q'(r) + 2\pi\rho \int_0^R dr'q'(r')(r-r')h(| r - r' |) \quad r \geq 0\,. \tag{3.10}$$

The great advantage of these equations over the original OZ equation is that the one dimensional convolution type integrals in Eqs. (3.9) and (3.10) are over the reduced range (0,R), whereas in the original three dimensional equation, the integral is over the range (0,2R). The reduction of the range over which the integral has to be evaluated reduces the solution of the hard sphere PY problem for $c(r)$ with $r < R$ to a simple quadrature.

It is worth pointing out that the isothermal compressibility, and hence the equation of state, can be obtained from q by using

$$\rho k_B T K_T = q(k=0)^2$$

where $q(k=0)$ is real and is given by

$$q(k=0) = 1 - 2\pi\rho \int_0^\infty drq(r)\,. \tag{3.11}$$

The solution for the hard sphere system in the PY approximation is easy to obtain since in this case

$$h(r) = -1 \quad r < \sigma$$

$$c(r) = 0 \quad r > \sigma$$

$$c(r) = c_0 + c_1 \frac{r}{\sigma} \quad r < \sigma \tag{3.12}$$

where σ is the hard sphere diameter and the dimensionless constant are given by

$$c_0 = -\frac{(1+2\eta)^2}{(1-\eta)^4} \tag{3.13a}$$

$$c_1 = \frac{6\eta(1+0.5\eta)^2}{(1-\eta)^4} \tag{3.13b}$$

$$c_3 = \frac{1}{2}\eta c_0 \, . \tag{3.13c}$$

Here η is the "packing fraction".

The hard sphere equation of state can be written

$$\rho k_B T K_T = k_B T \left(\frac{\partial \rho}{\partial P}\right)_T = \frac{(1-\eta)^4}{(1+2\eta)^2} \, . \tag{3.14}$$

Integrating with respect to the density and taking $(\frac{P}{\rho k_B T})_{\rho \to 0} \to 1$ gives the compressibility factor

$$\frac{P}{\rho k_B T} = \frac{1+\eta+\eta^2}{(1-\eta)^3} \quad (\chi - route) \, . \tag{3.15}$$

The above equation was obtained from the isothermal compressibility equation (3.6c). The equation of state can also be obtained from the pressure equation (3.6b). For the hard sphere system we have

$$\frac{P}{\rho k_B T} = 1 + \frac{2}{3}\rho\sigma^3 g(\sigma^+) = \frac{1+2\eta+3\eta^2}{(1-\eta)^2} \quad (P - route) \tag{3.16}$$

Note that the two routes yield different equations of state. This is an example of the thermodynamic inconsistency and is a consequence of the PY equation being an approximate relation between $g(r)$ and $c(r)$. If we take the weighted average we obtain the phenomenological Carnaham and Starling[20] equation which gives an excellent agreement with simulation data.

By using the factorisation of the OZ equation, Baxter[21] obtained an analytical solution for the potential

$$u(r) = \infty \quad r < \sigma$$

$$u(r) = k_b T \lim_{a \to 0} \ln(12\tau \frac{a}{\sigma}) \quad \sigma < r \leq \sigma + a$$

$$u(r) = 0 \quad \sigma + a \leq r \tag{3.17}$$

where τ is a dimensionless temperature which is zero when T=0 and large for high temperatures. This potential has the form of a square-well potential in the limit of infinitesimal width and infinite depth. This model can be viewed as a hard sphere with surface adhesion. The temperature dependence of the potential makes it difficult to give it a physical justification.

3.2. Some Details of Methods

Apart from the hard sphere and sticky hard sphere systems mentioned in the preceding section, there are no general analytical solutions for the integral equations of fluids. A numerical solution of these equation provides the only way forward. Three numerical methods can be used.

The first one is an iterative method which uses the Fourier transform of the correlation function. In this case, the auxiliary function $\theta = h(r) - c(r)$ can be used which is smooth and which takes into account the "non-direct" part of the correlation function $h(r)$. The OZ equation can be then written as

$$\tilde{\theta}(k) = \rho \frac{\tilde{c}(k)^2}{1 - \rho\tilde{c}(k)} \tag{3.18}$$

and the integral equations as

$$c(r) = (\theta(r) + 1)(\exp(-\beta u(r)) - 1) \quad PY \tag{3.19a}$$

$$c(r) = \exp(\theta(r) - \beta u(r)) - \theta(r) - 1 \quad HNC \tag{3.19b}$$

$$c(r) = \exp(\theta(r) - \beta u(r) - E(r)) - \theta(r) - 1 \quad MHNC\ . \tag{3.19c}$$

The method of solution is performed by using the following iterative scheme

$$\theta(r) \rightarrow c(r) \rightarrow \tilde{c}(k) \rightarrow \tilde{\theta}(k) \rightarrow \theta^{out}(r)\ .$$

The advantage of this scheme is that the integrals involved in each transformation are performed in terms of the short range functions both in real and Fourier space. A number of convergence criteria have to be satisfied by these functions, and by the thermodynamic properties which are obtained from them, in order to achieve an acceptable solution.

The second numerical method for solving integral equations is to use the integral equations in a discrete form and to solve a set of non-linear equations by using the Newton-Raphson (NR) method[23]. Rapid convergence can be achieved if a good initial estimate is used. The main problem with the method is that it requires a considerable computational effort because the number of equations is usually of the order of 1000 or more.

Finally, the third numerical method that has showed promise is based on Baxter's form of the OZ equation[15,16]. Watts[22] solved Baxter equation using a modified Newton-Raphson method. Compared with the second scheme, the number of equations is reduced by a factor of about 1/2 but the corresponding matrix is tedious.

From the three numerical methods we can conclude that the direct iterative scheme is slow in converging and the NR method is not convenient when applied to a large number of equations. A combined iterative and NR method has been proposed by Gillam[24]. The function $\theta(r)$ is decomposed into a "coarse" part, written as a sum of roof functions $P_j(r)$, and a "fine" part $\Delta\theta(r)$, i.e.,

$$\theta(r) = \sum_{j=1}^{M} a_j P_j(r) + \Delta\theta(r)\ . \tag{3.20}$$

The coefficients a_j are evaluated by the NR method and $\Delta\theta(r)$ by direct iteration. M is normally not larger than 6 to 10. The function $\Delta\theta(r)$ takes values close to zero so that the iterative scheme is quickly convergent. Labik *et al*[25] have proposed a modification of Gillam's method that improves its efficiency. It is worthwhile examining the method in some detail. The OZ equation can be written as

$$\tilde{\Gamma}_j - \frac{\rho \tilde{C}_j^2}{k_j - \rho \tilde{C}_j} = 0 \tag{3.21}$$

where the transformed function are defined as

$$\tilde{\Gamma}_j = k_j \tilde{\theta}(k_j)$$

$$\tilde{C}_j = k_j \tilde{c}(k_j)$$

and the corresponding inverse

$$\Gamma_i = r_i \theta(r_i)$$

$$C_i = r_i c(r_i)$$

with $k_j = j\Delta k$, $r_i = i\Delta r$ $(i,j = 1,...,N)$ and $\Delta k \Delta r = \frac{\pi}{N}$. The method basically consists in solving the set of non-linear equations in terms of $\tilde{\Gamma}_j$ approximating them by a first order expansion about an initial $\tilde{\Gamma}_j^0$

$$\tilde{C}_j = \tilde{C}_j^0 + \sum_i \tilde{C}_{ji}(\tilde{\Gamma}_i - \tilde{\Gamma}_i^0) \tag{3.22}$$

where

$$C_{ji} = \frac{1}{N}[D(i-j) - D(i+j)] \tag{3.23a}$$

$$D(l) = \sum_i \left(\frac{\partial F}{\partial \Gamma_i}\right)_0 cos\left(\frac{\pi}{N} il\right) . \tag{3.23b}$$

Here, F is $C_i = F(\Gamma_i)$ which is the expression corresponding to equations (3.19a,b,c) and depends on the integral equation being used. Labik *et al* indicated that the equation only has to be solved for j=1... M (with M=30 if R=0.02 and N=512 where σ is a parameter which is characteristic of the range of the potential) and, then, it can be corrected by direct iteration. The NR technique used to solve Eq. (2.21) yields the following set of linear equations in $\Delta\tilde{\Gamma}_j = \tilde{\Gamma}_j^{n+1} - \tilde{\Gamma}_j^n$:

$$\sum_i^M J_{ji} \Delta \tilde{\Gamma}_i = \frac{\rho \tilde{C}_j^2}{k_j - \rho \tilde{C}_j} - \tilde{\Gamma}_j \quad j = 1,...,M , \tag{3.24}$$

where the elements of the Jacobian matrix are given by

$$J_{ji} = \delta_{ji} - \frac{\rho \tilde{C}_j}{k_j - \rho \tilde{C}_j}(2 + \frac{\rho \tilde{C}_j}{k_j - \rho \tilde{C}_j})\tilde{C}_{ji} \quad j = 1,...,M . \tag{3.25}$$

The set of equations (3.24) has to be solved for $\Delta\tilde{\Gamma}_j$. The iterations are then allowed to proceed until convergence is achieved. Lomba[26] has proposed a modification of

Broyles mixing iterative method[27] in order to improve the convergence of the direct iterations. Here, the news Γ_i^o and $\tilde{\Gamma}_j^o$ to be used in the equations, are obtained from

$$\Gamma_i^{o,n+1} = (1-\alpha)\Gamma_i^{o,n} + \alpha\Gamma_i \quad i = 1, ..., N \tag{3.26a}$$

$$\tilde{\Gamma}_j^{o,n+1} = (1-\alpha)\tilde{\Gamma}_j^{o,n} + \alpha\tilde{\Gamma}_j \quad j = 1, ..., N \tag{3.26b}$$

with a variable α determined by

$$\alpha = \alpha_i \quad if \quad \xi > \xi_m$$

$$\alpha = 1 - (1-\alpha_i)(\frac{\xi}{\xi_m})^2 \quad if \quad \xi < \xi_m$$

and $\xi = (\Delta r \sum (\Gamma_i - \Gamma_i^o)^2)^{\frac{1}{2}}$. He proposed setting α_i to 0.35 and ξ to 1 (when M=30, N=512 and Δr=0.02) in computations for the Lennard-Jones system. The latter iterative scheme works under conditions where the use of direct iterations ($\alpha = 1$) does not achieve convergence.

The efficiency of the method depends on the number of equations M used for the NR iterations. When M is too small, convergence is shown since the technique is then basically a direct iteration. When M is very large, the solution of the set of equations requires a considerable amount of computing time. As we have seen, the main difference of the method of Labik *et al.* to the Gillam is in the use of the trigonometric functions instead of roof functions.

In the case of the RHNC equation, its successive solution requires the incorporation of thermodynamic consistency in the optimisation procedure. This can be expressed in the form

$$4\pi\rho \int_0^\infty r^2[(g(r) - g_0(r,\sigma_0)]\frac{dE_0}{d\sigma}dr = 0 , \tag{3.27}$$

where $g(r)$ is the pair correlation function of the fluid and $g_0(r,\sigma_0)$ the pair correlation function of the reference hard-sphere fluid of atomic diameter σ_0. The choice of $E(r)$ is fundamental; Foiles *et al*[28] used an $E(r)$ with a highly negative region for $r \leq 1.2\sigma_0$. To solve the problem they proposed a crossover integral equation

$$c(r) = \exp\{-\beta u(r) + \theta(r) + (1 - l(r))E_0(r) + l(r)[\ln g(r) - g(r) + 1]\} - \theta(r) - 1 \tag{3.28}$$

where

$$l(r) = \frac{1}{2}[1 + \tanh(\frac{r-R}{W})] \tag{3.29}$$

R is the second point satisfying $g_0(r,\sigma_0) = 1$ and W is the distance from the first maximin to R.

Earlier, Malijevski and Labik[29] presented an empirical formula to express the bridge function for hard spheres in terms of the interparticle separation and the density. Their expression does not posses a region of highly negative values and is positive in all cases. This function was used by Lomba in his calculations for the Lennard-Jones potential, and excellent agreement with simulation data was obtained[30].

4. Critical Behaviour

4.1. Critical Exponents and Amplitudes

Fluids exhibit typically a phase transition between vapour and liquid phases. Assuming the existence of critical exponents as a basic hypothesis[33], it is convenient to define the reduced variables[34]

$$t = \frac{T - T_C}{T_C} \tag{4.1}$$

and

$$x = \frac{\rho - \rho_C}{\rho_C} . \tag{4.2}$$

The shape of the critical isotherm near the critical point is characterized by the exponent δ and the amplitudes W_- and W_+:

$$\left| \frac{p - p_c}{p_C} \right| \simeq W_\pm \mid x \mid^\delta \quad \rho - \rho_C \to 0^\pm \quad T = T_C . \tag{4.3}$$

The critical amplitude W_+ corresponds to the limiting behaviour of the isotherm to the right of the critical density ($\rho > \rho_C$) while W_- corresponds to the limit to the left of the critical density ($\rho < \rho_C$). For any consistent theory $\delta_+ = \delta_-$. At the critical point the isothermal compressibility K_T diverges to infinity. The way in which it diverges along the critical isochore for temperatures above T_C is described by the exponent γ and the amplitude C^+:

$$\rho k_B T K_T = \chi \simeq C^+ t^{-\gamma} \quad T - T_C \to 0^+ \quad \rho = \rho_c . \tag{4.4}$$

For $T < T_C$, χ is measured along the coexistence curve which defines a new critical exponent γ' and amplitudes $C'_\pm$:

$$\chi \simeq C'_\pm \mid t \mid^{\gamma'} \quad T - T_C \to 0^- \quad \rho = \rho_\pm(T) . \tag{4.5}$$

Here, the subscript (+) corresponds to the liquid branch of the curve, and the subscript (-) to the vapour branch. The coexistence curve close to the critical point is described by

$$x_\pm = \frac{\rho_\pm - \rho_C}{\rho_C} \simeq B_\pm \mid t \mid^\beta \quad T - T_C \to 0^- \tag{4.6}$$

where $\beta^+ = \beta^-$. To complete the description of the coexistence curve we can use the empirical law of rectilinear diameters which can be written

$$\frac{1}{2}(\rho_- + \rho_+) \simeq \rho_C(1 + a \mid t \mid) . \tag{4.7}$$

Another property which has a singularity close to the critical point is the heat capacity at constant volume, C_V. The critical exponents α and α' are introduced in order to characterise this behaviour

$$C_V \simeq D_+ t^{-\alpha} \quad T - T_C \to 0^+ \quad \rho = \rho_C \tag{4.8a}$$

$$C_V \simeq D_- t^{-\alpha'} \quad T - T_C \to 0^- \quad \rho = \rho_C \ . \tag{4.8b}$$

The exponents are not all independent of each other. The thermodynamic laws impose rigorous inequalities between combinations of these exponents [35–37].

$$\alpha' + 2\beta + \gamma' \geq 2 \tag{4.9a}$$

$$2 - \alpha' \leq \beta(\delta + 1) \tag{4.9b}$$

$$\beta(\delta - 1) \leq \gamma' \ . \tag{4.9c}$$

The scaling laws[31,38] convert these inequalities into equalities and since $\gamma = \gamma'$ and $\alpha = \alpha'$ only two of the macroscopic exponents are independent.

Ornstein and Zernike[39] suggested that the phenomenon of critical opalescence is due to the increasing range of correlations between the local fluctuations in density which become infinite at the critical point. As can be seen from expression (2.6c) for the isothermal compressibility, if the function diverges to infinity at the critical point, the spatial integral of the correlation function, $h(r)$, must also diverge. On the other hand, the integral has to tend to zero in the limit of large separations for obvious physical reasons. This means that at the critical point the asymptotic behaviour of $h(r)$ is of the form r^{-n}, where $0 \leq n \leq d$ and d is the dimensionality of the system.

The critical exponent η is often used to characterise the asymptotic behaviour of $h(r)$ at the critical point:

$$h(r) \propto \frac{\exp -(\frac{r}{\xi})}{r^{d-2-\eta}} \quad when \quad r \to \infty \quad T \to T_C \quad \rho = \rho_C \ . \tag{4.10}$$

The length ξ is a measure of the length of the correlations and must tend to infinity as the critical point is approached if the isothermal compressibility is also to do so. The exponents ν and ν' are associated with the length ξ

$$\xi \simeq f_+ \mid t \mid^{-\nu} \quad T - T_C \to 0^+ \quad \rho = \rho_C \tag{4.11a}$$

$$\xi \simeq f_- \mid t \mid^{-\nu'} \quad T - T_C \to 0^- \quad \rho = \rho_C \ . \tag{4.11b}$$

These definitions can be expressed in terms of the magnitude of the structure factor which is related to the correlation function, $h(r)$. Taking the Fourier transform:

$$S(k) \simeq S(0)((1 - k^2\xi^2)^{(1-\frac{\eta}{2})} \quad when \quad k \to 0, \quad T \to T_C \quad \rho = \rho_C \ . \tag{4.12}$$

The critical exponent ν describes how the correlation length ξ diverges. It can be determined from scattering measurements of S(k) via

$$\xi^2 = \lim_{k \to 0} k^{-2}\{(\frac{S(0)}{S(k)})^{-(1-\frac{\eta}{2})}\} \quad T \to T_C, \quad \rho = \rho_C \ . \tag{4.13}$$

The divergence of the structure factor can be described in terms of the exponent

$$S(k) \propto k^{-2+\eta} \quad k \to 0 \quad T = T_C \quad \rho = \rho_C \,. \tag{4.14}$$

We refer above to the critical point of simple fluids but other systems, e.g., ferromagnetic, paramagnetic, superconductivity, binary mixtures, etc., exhibit critical points which are formally equivalent, and critical exponents can be defined consequently[32,40]. It may appear surprising that critical exponents are almost the same for periodic crystalline or magnetic metal systems than for continuous systems such liquids or solutions. This is interpreted as an indication that the detailed nature of the intermolecular potential does not play an important role in describing the behaviour of the system near the critical point. In fact, studies of lattice models have lead to the idea of universality. According to this hypothesis, systems which have the same basic symmetry, such as the dimensionality of the lattice, spatial symmetry of the ordered phase and range of the interaction have the same critical exponents, *i.e.*, belong to the same universality class. The systems in a given class obey the same equation of state scaled with two adjustable parameters, which is confirmed by a number of experimental observations[41].

4.2. Classical Results

It is worthwhile report on the results that are obtained for the van der Waals equation of state using the Maxwell double tangent construction. Certain microscopic model systems obey this equation. The critical exponents obtained for this equation are $\delta = 3$, $W_- = W_+ = \frac{3}{2}$, $\gamma = 1$, $C_+ = C_- = \frac{4}{9}$, $\gamma' = 1$, $C'_+ = C'_- = \frac{2}{9}$, $B_- = B_+ = B = 2$, $\alpha = \alpha' = 0$. These values for the exponents are usually referred to as "classical". Other theories which describe critical phenomena such as the Weiss theory of ferromagnets or the generalised Landau theory correspond to the same values of critical exponents. These theories are collectively called classical or mean field theories.

The classical critical exponents can be described in terms of the OZ theory. As far as classical systems is concerned, the direct correlation function is assumed to remain of finite range in the neighbourhood of the critical point. This means that $\tilde{c}(k)$ does not have a singularity at the origin. Furthermore, one assumes that $\tilde{c}(k)$ can be expanded as a Taylor series about $k = 0$:

$$\rho\tilde{c}(k) = c_0(\rho, T) + c_2(\rho, T)k^2 + \vartheta(k^4) \tag{4.15}$$

$$c_0(\rho, T) = \rho\tilde{c}(0) = 1 - \frac{K_T^0}{K_T} \,, \tag{4.16a}$$

where K_T^0 is the compressibility of the ideal gas and

$$c_2(\rho, T) = -\frac{2\pi\rho}{3} \int c(r) r^4 dr = -R^2 \,. \tag{4.16b}$$

Here, R is the Debye persistence length which is finite even at the critical point.

The inverse structure factor is given by

$$\frac{1}{S(k)} = 1 - \rho\tilde{c}(k) = 1 - c_0(\rho, T)k^2 \,. \tag{4.17}$$

Experimental data for $\frac{1}{S(k)}$ versus k^2 have to be described by a straight line for different isotherms at the critical density. Here we can write

$$S(k) = 1 + \rho\tilde{h}(k) \simeq \frac{R^{-2}}{K^2 + k^2} \tag{4.18}$$

where

$$K^2 = R^{-2}\frac{K_T^0}{K_T} \,.$$

Using the inverse Fourier transform we obtain the asymptotic description of the correlation function

$$h(r) = \frac{1}{4\pi\rho R}\frac{\exp(-Kr)}{r} \,. \tag{4.19}$$

It is usual to identify K with the inverse correlation length. This approximation lead to values of $\eta = 0$ and $\nu = \frac{\gamma}{2}$, $\nu' = \frac{\gamma'}{2}$.

4.3. The YBG Equation

Exhaustive calculations have been undertaken for the square-well potential which is the simplest model that take into account both attractive and repulsive interactions. The square well interaction can be defined by

$$u(r) = \infty \quad r > \sigma$$

$$u(r) = -\epsilon \quad \sigma < r < \lambda\sigma$$

$$u(r) = 0 \quad r > \sigma$$

where ϵ and λ are parameters for the energy and range of the potential. The calculations were performed with the pressure equation and the compressibility equation[45-47]. Given that the YBG equation is not expressed in terms of the direct correlation function, $c(r)$, it is more convenient to use Eq. (2.6c) than Eq. (2.19). In the critical region $h(r)$ has larger range and the numerical evaluation of the integral of $h(r)$ can be significantly inaccurate if reliable values of $h(r)$ are not available for large distances. Because of this, $h(r)$ was only calculated over distances of up to $r = 50\sigma$ [47]. The resulting critical point values were $T_C \simeq 2.673\frac{\epsilon}{k_B}$ and $\rho_C \simeq 0.364\sigma^{-3}$. The corresponding values of the critical exponents were $\gamma \simeq 1.23, \delta \simeq 4.4, \beta \simeq 0.33$ and $\alpha \simeq 0.30$ (the exponent β was determined along the spinodal and not along the coexistence curve). These values are very close to the experimental values and to those of the three dimensional Ising model, which appeared to suggest that the YBG approach provides a good description of the critical region.

The values of the correlation function $h(r)$ remain positive for both intermediate and large distances. Analytical studies here show that $h(r)$ has to be less than zero for the critical point to exist in the case of dimensionality $d \leq 4$ and $\xi^{-1} \to 0$. When $h(r) \to 0$ the system does not exhibit a critical point in the sense that the isothermal compressibility remains uniform. This implies that one cannot talk of critical exponents and that the values obtained numerically are not significant.

In order to clarify the situation, Green *et al*[47] performed accurate numerical calculations in the neighbourhood of the critical point for values of $h(r)$ up to $r = 150\sigma$. These calculations indicate that the YBG equation for three-dimensional fluids does not describe a critical point and that: a) $h(r)$ is never negative; b) the inverse isothermal compressibility does not tend to zero along any isochore near the supposed critical point, and c) is never zero but takes a positive value of $\xi \simeq 12.5\sigma$. Finally, these results throw into question the suggestion made by Alexanian[51] that the exponential behaviour of $h(r)$ for $d \leq 4$ is compatible with the existence of a critical point for the YBG equation for positive values of $h(r)$ for large r.

In the case of the systems with $d > 4$, Jones *et al*[48] have reported numerical solutions of the YBG equation for a six dimensional square-well fluid. They obtained a critical point with $T_C \simeq 21.80170$ and $\rho_C \simeq 78.70\sigma^{-6}$ which was described by classical values of the corresponding critical exponents.

Summing up, the YBG equation exhibits values close to those obtained from renormalization group theory[38] for dimensionality in the beginning of $d = 4$.

4.4. The PY Equation

As mentioned above, the only analytical solution of the PY equation which describes a critical point is that for the sticky hard sphere fluid, in terms of the compressibility equation. Although the corresponding critical exponents take classical values, the critical amplitudes satisfy the non- classical asymmetry relations ($\frac{W_-}{W_+} \simeq 34$ and $\frac{\beta_+}{\beta_-} \simeq 2.2$) and, furthermore, the specific heat diverges logarithmically not only along the critical isochore but also along any isochore $\rho > \rho_C$ which crosses the critical isotherm. Fishman and Fisher[34] have shown that this anomaly originates from the non-classical character of the scaling functions used in the solution. The question remains as to whether this non-classical behaviour is only seen for the sticky hard sphere system with finite range or whether it is a general feature of the PY equation even though the critical exponents are classical.

Henderson and Murphy[52] have studied the PY equation close to the critical point for the Lennard-Jones potential

$$u(r) = 4\epsilon\left[\left(\frac{\sigma}{r}\right)^{12} - \left(\frac{\sigma}{r}\right)^{6}\right] \tag{4.20}$$

truncating at $r = 6\sigma$. This potential is the most commonly used for realistic potentials where the ϵ and σ parameters correspond to the energy and size of molecular interactions. The truncation allowed them to use Baxter's form of the OZ equation which was solved by using the algorithm of Watts[22]. The following critical constants were obtained using the compressibility equation of state:

$$T_C = 1.311\frac{\epsilon}{k_B}$$

$$\rho_C = 0.278\sigma^{-3} .$$

By examining the behaviour of the equation close to the critical point, the values $\lambda = 1$ and $\delta = 3$ were obtained. Furthermore, the specific heat obtained from the energy equation was finite at the critical point. They also obtained the classical $\beta = 1/2$ for the shape of the coexistence curve. Henderson and Murphy's data were used in Ref. 34 to estimate $\frac{W_-}{W_+} \simeq 1.6$ indicating an asymmetry similar to that found for the sticky hard sphere system.

Later, Brey *et al.*[53] solved the PY equation without any truncation by assuming that $c(r) = -\beta u(r)$ for $r > 5\sigma$. This allowed them to use an iterative scheme instead of Watts algorithm. These calculations confirmed the classical values found for the critical exponents and amplitudes. Using the same numerical scheme and the approximation for the direct correlation function, Velarde *et al*[54] solved the PY equation for a family of potentials described by

$$u_n(r) = \frac{\epsilon}{a(12-n)}\left[n\left(\frac{\sigma}{r}\right)^{12} - 6\left(\frac{\sigma}{r}\right)^{6}\right] \tag{4.21}$$

where $a = r_{min}^{-12}$, $r_{min} = 2^{\frac{1}{12-n}}$ which is the distance at which $u(r)$ attains its minimum.

A second set of potentials of the type

$$u_m(r) = \frac{6}{m}\left(1+\frac{m}{6}\right)^{1+\frac{6}{m}} \epsilon \left[\left(\frac{\sigma}{r}\right)^{m+6} - \left(\frac{\sigma}{r}\right)^{6}\right] \tag{4.22}$$

was also examined in order to determine the effect of the repulsive part of the potential on the critical exponents and amplitudes. Classical values of microscopic exponents were found both by using an iterative scheme [55] and by using Baxter's factorisation of the OZ equation[56].

4.5. The HNC Equation

The HNC integral equation exhibits more problems than the PY equation. The numerical results for the HNC are scarce and not very conclusive. Klein and Green[57] solved the HNC equation for a Lennard-Jones potential using an iterative procedure and determined the critical point from the spinodal which was in turn determined from the compressibility equation. They assumed that $c(r) = 0$ for $r > 12.5\sigma$ which neglected the long range behaviour of $c(r)$ for the HNC equation near the critical point. The same problem is seen with Watts's solution[22] for the truncated Lennard-Jones potential using Baxter's factorisation of the OZ equation[25]. This factorisation requires that $c(r) = 0$ above a certain cutoff distance, a condition which does not satisfy the HNC equation even the interaction potential is zero above that distance.

It is important to remark a peculiarity of Baxter's equation which yields two solutions in the PY and HNC approximations. The only physically acceptable solution is the one which corresponds to $c(r = 0) = -1$ in the limit of low densities and is an exact result. The presence of two solutions could be due to the fact the

OZ equation is linear in $h(r)$ whilst Baxter's equation is quadratic in $h(r)$. Watts has determined the critical point from the temperature at which there are two solutions for $h(r)$. The numerical accuracy of this point, specially the value of T_C , is very sensitive to the truncation distance of the PY equation, however, the isothermal compressibility does not diverge at the critical point but takes on a finite value. Consequently, one can not speak of a critical point in true thermodynamic sense of the word.

Guerrero *et al*[58] have solved the HNC equation using a iterative method for a LJ potential, truncated at $r = 6.4\sigma$. Assuming that $c(r) = 0$ for $r > 12.8\sigma$ they come to the same conclusion that K_T is finite at the "critical point", and that the corresponding exponent γ takes on a value much smaller than unity and which could be zero. This result does not correspond to classical or experimentally determined values.

Foiles and Ashcroft[59] have solved the HNC equation using an iterative scheme for the potential

$$u(r) = A\frac{\epsilon}{r}[\exp(-a(r-\sigma) - \exp(-b(r-\sigma))] . \tag{4.23}$$

Here, the constants are chosen to give a form similar to that of the LJ potential. Taking $c(r) = 0$ when $r > 20\sigma$ and examining the temperature dependence of the isothermal compressibility along the critical isochore they could not obtain the critical exponent. Furthermore, their results can not be described by logarithmic divergence of the isothermal compressibility at the critical point which would correspond to $\gamma = 0$.

Brey and Santos[60] used an iterative scheme to examine the behaviour for $c(r)$ near the critical point i.e.,

$$c(r) \sim \sin[h(r)]^2 . \tag{4.24}$$

The behaviour of $h(r)$ allowed them to write

$$h(r) = A\frac{\exp(-\frac{r}{\epsilon})}{r} \tag{4.25}$$

where A and ϵ were obtained by fitting these values to the numerically determined values for large values of r. The method does not converge for certain regions of the T-ρ plane which allowed to define a possible "critical point" as the limit at which the solutions seem to diverge. Nevertheless, a detailed study of the compressibility and correlation length led them to conclude that both of these properties tended to non-zero finite values and that the HNC equation does not exhibit a real critical point where ξ and K_T diverge . This is in agreement with earlier findings[61].

4.6. The RHNC Equation

Poll and Ashcroft[62] have solved this equation by using the hard sphere bridge function obtained from simulation data[63,64] and by using the bridge function proposed by Foiles *et al*[59], that crosses over from the short-range hard-sphere behaviour with a diameter obtained from thermodynamic consistency (Eq. (3.27)), to the long range mean-spherical behaviour (Eqs. (3.28, 3.29)). Using the potential (4.23) and an iterative fast Fourier algorithm, and, in order to ensure high accuracy, the calculations were performed on a grid consisting of N=4096 points and small variations in Δr did not have an appreciable effect on the results. They inferred that the RHNC approximation with the hard-sphere bridge function does lead to a singular point (ρ = $0252\sigma^3$ and $T_C = 1.395\frac{\epsilon}{k_B}$) but without the characteristic power law behaviour. In the case of the cross-over bridge function, the critical temperature was determined as $T_C = 1.255\frac{\epsilon}{k_B}$ and a power law described the limiting behaviour with a clear departure from $\gamma = 1$. By carefully examining their numerical results they reported a value of $\gamma = 2$ which is demonstrably non-classical. This behaviour is entirely consistent with numerical solutions of various perturbation equations with long-range mean-spherical behaviour.

Lomba[26] used the hard-sphere bridge function of Malijevski and Labik to solve the RHNC for the LJ potential. The chemical potential per particle can be determined from

$$\frac{\beta\mu}{N} = \frac{A}{Nk_BT} + \frac{P}{Nk_BT}$$

where A is the free energy given by the equation (2.35) and P is the pressure given by Eq. (2.6b). This expression allows one to determine the values of the densities on each side of the coexistence curve which satisfy the conditions of phase equilibria

$$\frac{\beta\mu(\rho_l, T)}{N} = \frac{\beta\mu(\rho_g, T)}{N} . \tag{4.26}$$

$$\beta P(\rho_l, T) = \beta P(\rho_g, T) .$$

In this way Lomba was able to obtain a vapour-liquid curve. A region of this coexistence curve, which includes the critical region, could not, however, be studied due to the lack of convergence of the solutions. He could not determine whether the problem was a numerical one or whether it was intrinsically linked to the RHNC closure. The behaviour in the critical region was estimated by simple extrapolation. The coexistence curve obtained by Lomba was in excellent agreement with Gibbs ensemble Monte Carlo data.

The later simulation study indicates that the square well system can be described by a universal critical exponent β for short ranges of the potential and there seems to be a transition to classical behaviour for the larger ranges[65].

Acknowledgements

I wish to acknowledge G. Jackson, E. de Miguel and E. Martin del Rio for useful comments and discussions and for a critical reading of the manuscript. Collaboration with the Centro Informatico Cientifico de Andalucia (CICA) is greatfully acknowledged. This work was supported by the DGICYT of Spain under Grant no.

PB-89-640.

References

1. R. Balescu, *Equilibrium and non-Equilibrium Statistical Mechanics*, (Wiley, New York, 1975). J. P. Hansen and I. McDonald *Theory of Simple Liquids* (Academic Press 2nd. Ed. 1990). J. A. Barker and D. Henderson. *Rev. Mod. Phys.* **48**, 587 (1976).
2. M. Born and M. S. Green *Proc. Roy. Soc. (London)* **A188**,10 (1946). J. Yvon *La Theorie Statistique des Fluid et l' Equation d'Etat, Act. Scient. et ind. 203*, (Hernann, Paris 1935).
3. J. G. Kirwood *J. Chem. Phys.* **3** 300 (1935).
4. J. M. J. van Leeuven, J. Groenveld and J. de Boer *Physica* **25**, 792 (1959).
5. T. Morite and K Hiroike *Prog. Theor. Phys.* **23**, 1003 (1960).
6. J. K. Percus and G. J. Yevick *Phys. Rev.* **1**, 1003 (1960).
   J. L. Lebowitz and J. K. Percus *J. Math. Phys.* **4**, 116 (1963)
7. R. Balescu, see Chapter 7 Ref. 1.
8. G. Stell *Physica* **29**, 517 (1963).
9. Y. Rosenfeld and N. W. Aschorft *Phys. Rev.* **A20**, 1028 (1979).
10. F. Lado, S. M. Foiles and N. W. Aschroft *Phys. Rev.* **25**, 792 (1959).
11. F. Lado *Phys. Rev.* **A8**, 2548 (1973).
    F. Lado *Phys. Letters* **89A**, 196 (1982).
12. K. Hiroike *J. Phys. Soc. Jpn.* **12**, 326 (1957).
13. B. Noble, *Methods based on the Wienner-Hopf technique for the solution of partial differential equations*, (Pergamon, New York, 1975).
14. M. S. Wertheim *Phys. Rev. Letters* **10**, 321 (1963). *J. Mat. Phys.* **5**, 463 (1964)
15. R. J. Baxter *Phys. Rev.* **154**, 170 (1967).
16. R. J. Baxter *Aust. J. Phys.* **21**, 563 (1968).
17. M. S. Wertheim *J. Chem. Phys.* **73**, 1398 (1980).*J. Chem. Phys.* **74**, 2466 (1981)
18. G. Gray and K. E. Gubbins, *Theory of Molecular Fluids* , (Clarendon, Oxford, 1984).
19. D. E. Sullivan and D. Levesque *J. Chem. Phys.* **72**, 1170 (1980).
20. N. F. Carnahan K. E. Starling *J. Chem. Phys.* **51**, 635 (1969). *J. Chem. Phys.* **53**, 600 (1970).
21. R. J. Baxter *J. Chem. Phys.* **49**, 2770 (1968).
22. R. O. Watts *J. Chem. Phys.* **48**, 50 (1968). *J. Chem. Phys.* **50**, 984 (1969).
23. W. R. Smith, D. Henderson and R. D. Murphy *J. Chem. Phys.* **61**, 291 (1974).
24. M. J. Gillam *Mol. Phys.* **38**, 1781 (1985).
25. S. Labik, A. Malijevski and P. Vonka *Mol. Phys.* **56**, 709 (1985).
26. E. Lomba *Mol. Phys.* **68**, 87 (1989).
27. A. A. Broyles *J. Chem. Phys.* **33**, 87 (1960).
28. S. M. Foiles, N. W. Aschroft and L. Reatto *J. Chem. Phys.* **80**, 4441 (1984).
29. A. Malijevski and S. Labik *Mol. Phys.* **60**, 663 (1987).
30. A. Paganiotiopoulos *Mol. Phys.* **61**, 813 (1987).
31. H. E. Stanley, *Introduction to Phase Transition and Critical Phenomena*, (Oxford University Press, Oxford, 1971).

32. A. Munster *Statistical Mechanics Vol. 1 and 2* (Springer Academic Press, Berlin New York, 1975).
33. R. B. Griffits *J. Chem. Phys.* **43**, 1958 (1965).
34. S. Fishmann and M. E. Fisher *Physica* **108A**, 1 (1981).
35. G. S. Rushbrooke *J. Chem. Phys.* **39**, 842 (1963).
36. D. A. Libermann *J. Chem. Phys.* **44**, 419 (1966).
38. M. E. Fisher *Rep. Progr. Phys.* **30**, 615 (1967).
37. M. E. Fisher *Rev. Mod. Phys.* **46**, 547 (1974).
39. L. S. Ornstein and F. Zenike *Proc. Acad. Sci. Amsterdam* **17**, 793 (1914).
40. P. A. Egeltaff and J. W. Ring *Physics of Simple Liquids* (Ed. H. N. V. Temperly, J. S. Rowlinson and G. S. Rushbrooke, North Holland Pub. Amsterdam 1968).
41. J. M. H. Levelt-Sengers *Progress in Liquid Physics* (Ed. C. Croxton,Willey, New York, 1978).
42. M. Kac , G. E. Uhlembeck and P. C. Hemmer *J. Mat. Phys* **4**, 216 (1963).
43. J. L. Lebowitz and O. Penrose *J. Mat. Phys* **6**, 1282 (1966).
44. I. B. Schrodt and K. D. Luks *J. Chem. Phys.* **57**, 200 (1972).
J. J. Kozak, I. B. Schrodt and K. D. Luks *J. Chem. Phys.* **57**, 206 (1972).
W. W. Lincon, Y. Tago and K. D. Luks *J. Chem. Phys.* **61**, 4129 (1972).
W. W. Lincon, J. J. Kozaks and K. D. Luks *J. Chem. Phys.* **62**, 1116 (1975).
W. W. Lincon, J. J. Kozaks and K. D. Luks *J. Chem. Phys.* **62**, 2171 (1975).
K. U. Co, J. J. Kozaks and K. D. Luks *J. Chem. Phys.* **64**, 2197 (1975).
45. K. A. Green and K. D. Luks *Phys. Rev. Letters* **42**, 985 (1979).
46. K. A. Green, K. D. Luks and E. Lee *Phys. Rev.* **A21**, 356 (1980).
K. A. Green, K. D. Luks and J. J. Kozak *Phys. Rev.* **A24**, 2093 (1981).
47. K. A. Green, K. D. Luks, G. L. Jones and E. Lee *Phys. Rev.***A25**, 1060 (1982).
48. G. L. Jones, E. Lee and J. J. Kozak *Phys. Rev. Letters* **48**, 447 (1982).
49. G. L. Jones, J. J. Kozak and E. Lee, S. Fishmann and M. E. Fisher *Phys. Rev. Letters* **46**, 795 (1981).
50. S. Fishmann and M. E. Fisher *Phys. Rev. Letters* **47**, 421 (1981).
51. M. Alexanian *Phys. Rev.* **A25**, 572 (1982).
52. D. Henderson and R. D. Murphy *Phys. Rev.* **A6**, 1224 (1972).
53. J. J. Brey A. Santos and L. F. Rull *Phys. Letters* **67A**, 383 (1978).
J. J. Brey A. Santos and L. F. Rull *Phys. Rev.* **A26** , 2993 (1982).
J. J. Brey A. Santos and F. Romero *Phys. Letters* **77**, 5058 (1982).
54. J. M. Velarde, P. Pinero, L. F. Rull and J. J. Morales *Mol. Phys.* **65**, 287 (1988).
55. J. J. Brey and A. Santos *J. Chem. Phys.* **79**, 452 (1983).
56. P. Piñero, L. F. Rull, J. J. Morales and J. M. Velarde *J. Comput. Phys.***88**, 490 (1990).
57. M. Klein and M. S. Green *J. Chem Phys.* **39**, 1367 (1963).
58. M. I. Guerrero, G. Saville and J. S. Rowlinson *Mol. Phys.* **29**, 1941 (1975).
59. S. M. Foiles and N. W. Ashcroft *Phys. Rev.* **A24**, 424 (1981).
60. J. J. Brey and A. Santos *Mol. Phys.* **57**, 149 (1986).
61. M. S. Green *J. Chem. Phys.* **33**, 1403 (1960).
62. P. D. Polls and N. W. Ashcroft *Phys. Rev.* **A32**, 1722 (1985).

63. L. Verlet and J. J. Weiss *Phys. Rev.* **A5**, 939 (1972).
64. D. Henderson and E. W. Grundke *J. Chem. Phys.* **63**, 601 (1975).
65. L. Vega, E. de Miguel, L. F. Rull, I. McLure and G. Jackson *J. Chem. Phys.* **96**, 2296(1992).

MODELLING OF MECHANICAL PROPERTIES OF ADVANCED MATERIALS

J. L. VALLES
Institute for Advanced Materials, Joint Research Centre
P.O.Box 2, NL-1755 ZG Petten, The Netherlands

ABSTRACT

New trends in the modelling of advanced materials for high temperature applications, involving the use of microstructural information, are presented for the non-specialized reader. Different physical mechanisms responsible for deformation, damage and fracture are reviewed briefly. Special attention is devoted to the modelling of two problems of high technological relevance, namely fatigue in nickel-based single-crystal superalloys and creep in ceramic matrix composites.

1. Introduction

1.1. Motivation

During the last decades, the condensed matter physics community has developed powerful experimental and theoretical techniques to study the atomic, electronic and magnetic properties of solids. Transistors or solid-state lasers are good examples of the tremendous impact that this field has had on technology. However, many other materials' issues like their mechanical properties have been mostly left as the domain of other academic disciplines[1].

Most properties of materials are determined by their collective microstructure, namely the size and orientation of grains, dislocation networks, inclusions, microvoids, etc. Consequently, in materials science the link between atomic structure and macroscopic properties is not direct, and it must be provided by an intermediate 'mesoscopic' scale. New experimental techniques and new theories are needed to explore this connection and this is a challenge not only for applied materials science but also for basic physics. In this context, ideas such as percolation, fractals, random fields or growth processes have been shown to have also relevance in materials physics when dealing with this intermediate scale [2], but new bridges with other materials-related disciplines such as metallurgy and mechanics have to be established.

On the other hand, engineering applications such as industrial design require the use of mechanical properties of the materials mainly from a macroscopic point of view. Since testing at all possible conditions of application is out of the question, and the complexity of the material microstructure can not be incorporated wholly in the calculation for a machine component, so-called constitutive laws have been developed. These phenomenological models incorporate rheological information and present adjustable parameters, which can be identified for each specific material through a series of mechanical tests. In many cases, however, a validation attempt fails for conditions or mechanical loading types different from the ones used to determine the model parameters.

Therefore, those models need to be improved. Clearly, the direction to follow is the inclusion of as much relevant microscopic and mesoscopic information as it is available and the feasibility of the industrial calculation allows. With the arrival of powerful computers the chances of having microscopically-informed models that can be used industrially are expected to increase notably.

This presentation is intended for young researchers in physics, chemistry or related fields. The main objective is to illustrate some of these recent attempts to incorporate microstructural information in the modelling of the mechanical properties of advanced materials. The first part will be a brief description of the physical mechanisms responsible for deformation and damage. A more complete and precise account on this subject can be found in textbooks [3–6]. The second part is devoted to the modelling of two specific problems, namely the study of fatigue in nickel-based single-crystal superalloys and of creep in ceramic matrix composites (CMC), two of the fields in which our group is active at present.

1.2. Advanced Materials for High Temperature Applications

The development of new materials with improved mechanical properties has been always at the forefront of technological progress. The fact that the strength-to-density ratio of modern structural materials is approximately 50 times higher than that of the cast iron of two centuries ago gives a clear idea of the dramatical changes in materials performance experienced throughout the industrial age [7].

Materials design itself has undergone radical changes. In the past, new ingredients or stages in the preparation of alloys were mostly decided on a trial and error basis. Present day techniques involving a profound knowledge of the microstructure and the intimate micromechanics of defects associated with deformation or damage processes allow the tailoring of advanced materials with an emphasis on the properties required for a specific application.

The conversion of heat energy to mechanical or electrical energy in engines or power plants has often been the application pushing for materials with improved performance. Since the efficiency of the conversion depends directly on the temperature at which the device can operate, materials that are strong at high temperatures are clearly preferred. In the case of the aerospace industry, materials capable to work at higher temperatures translate in a decrease in fuel consumption and polluting emissions, but also in lowering the cooling air usage, allowing to increase payload [8].

The components in a gas turbine are exposed to all kinds of extreme conditions: high temperatures, corrosion, high mechanical stresses caused by centrifugal forces, as well as cyclic variations in the stresses and temperatures susceptible to cause a fatigue failure. While the basic design of aircraft gas turbines has not changed much in the last 30 years, modern superalloys have raised the operating temperature by 300 C in the same period and can now operate at temperatures over 1100 C, and advanced ceramic matrix composites may push engine operating temperatures to 1400 C [9].

1.3 Modelling of Mechanical Properties

In industrial design, the prediction of the lifetime of a component is a very important issue involving several steps [10]. On the basis of the geometry and the loading conditions, and applying the constitutive model relating stresses and strains for that material and particular type of loading, the local stress-strain behavior is calculated first by means of a finite element analysis. The output of this analysis for the most critical locations of the component is then used in a lifetime model in order to predict the local failure life. The use of adequate constitutive and life models at the level of the material is therefore essential for a correct lifetime prediction of the component.

Experimental testing enters the scene of modelling at essentially two levels. The first level represents the measurement of the macroscopic response of the material to the loading in order for the model constants to be determined, coupled to a microstructural analysis of the deformation and damage processes if one attempts a mechanism-informed lifetime prediction. The second level encompasses the mechanical tests to validate the model predictions.

As mentioned above, the advantages of mechanism-informed lifetime prediction are evident. Phenomenological models are based on multiple experimental measurements and usually allow to interpolate the behaviour of the investigated material within the experimentally covered range of loading parameters. However, they are not suited for extrapolation purposes since they do not take into account the mechanisms which are governing the mechanical behaviour at the scale of the microstructure. For the same reason, phenomenological models cannot predict the mechanical response under loading types or sequences different from those covered experimentally. The incorporation of accurate descriptions of the microstructural events provides the most promising procedure to enhance the quality of a model.

2. Physical Mechanisms of Deformation and Fracture

2.1. Plasticity

During a tensile test a specimen is subjected to a continuously increasing uniaxial tensile force while simultaneous observations are made of the elongation of that specimen. The stress-strain curve constructed with those measurements presents initially an elastic region, characterized by the fact that the original dimensions of the solid are recovered when the load is removed. For most materials, as long as the load does not exceed the elastic limit the strain is found to be proportional to the stress. If the elastic limit has been exceeded the body will experience a permanent deformation when the load is removed. The body is then said to have undergone plastic deformation.

The general behavior of materials under load can be classified as ductile or brittle depending upon whether or not the material exhibits the ability to undergo plastic deformation. It is not an absolute property of a material and can vary depending on temperature or loading conditions. A completely brittle material would fracture at the elastic limit. Adequate ductility is an important engineering consideration because it allows the material to redistribute localized stresses. In brittle materials, localized stresses continue to build up when there is no local yielding and finally a crack forms and it spreads rapidly over the section. Ductile fracture, in contrast, occurs after extensive plastic deformation and is characterized by slower crack propagation.

An examination of a metal single crystal after deformation beyond its elastic limit shows step markings on its surface, which are called slipbands. The slipbands are caused by the slip or shear deformation of the atoms on specific cristallographic planes called slip planes. But large groups of atoms do not slide over each other simultaneously during plastic shear deformation, since the process requires too much energy. In order for large metal crystals to deform at their observed low shear strengths, a high density of crystalline imperfections known as dislocations must be present .

These dislocations are created in large numbers (approx. $10^6 cm/cm^3$) as the metal solidifies, and a highly deformed metal crystal may contain as high as $10^{12} cm/cm^3$ of dislocations. The motion of an edge dislocation, which can be pictured as an extra half plane of atoms, produces a unit step of slip under a low shear stress. A relatively small amount of stress is required for slip by this process since only a small group of atoms slips over each other at any instant. An analogous situation can be envisaged by the movement of a carpet with a ripple in it across a large floor. Dislocations in real crystals can be observed in the transmission electron microscope in thin metal foils and appear as lines due to the atomic disarrays which interfere with the transmission path of the electron beam of the microscope. They are also relevant in other areas of material science: for example, the so-called screw dislocations provide a mechanism for crystal growth at an interface between the crystal and a melt [11].

Dislocations produce atomic displacements on specific crystallographic slip planes and in specific crystallographic slip directions. The slip planes are usually the most densely packed planes, since a lower shear stress is then required for atomic displacement. Slip in the close-packed directions is also favoured since less energy is required to move the atoms from one position to the other if the atoms are closer together. A combination of a slip plane and a slip direction is called a slip system and each crystal structure has a number of characteristic slip systems. Slip begins within a single crystal when the shearing stress on the slip plane in the slip direction reaches a required level called the critical resolved shear stress. This value has a similar role as the yield stress of a polycrystalline metal or alloy which is found in a stress-strain tensile test curve. A second important plastic deformation mechanism which can occur in metals is twinning. In this process a part of the atomic lattice is deformed so that it forms a mirror image of the undeformed lattice next to it.

Almost all engineering metals and alloys are polycrystalline. In fact, single crystals are used mainly for research purposes and only in a few cases for engineering applications. The existence of grain boundaries strengthens metals and alloys because they act as barriers to dislocation movement, except at high temperatures where they become regions of weakness. For this reason, most metals are fabricated with a small grain size. During plastic deformation of metals, dislocations moving along on a particular slip plane cannot go directly from one grain into another in a straight line, because each grain has its own preferred slip planes which have different orientations from those of neighbouring grains. In this way, grain boundaries may act as a barrier to dislocation movement and cause dislocations to pile up.

With increasing plastic deformation new dislocations are created and must interact with those already existing. As the dislocations density increases with deformation it becomes more and more difficult for the dislocations to move through the existing forests of dislocations and thus the metal hardens, *i.e.* it becomes more difficult to deform plastically.

Dislocations frequently pile up at barriers, such as grain boundaries, second phase particles or dislocations which cannot glide, which are called sessile dislocations. The interaction leads to a high concentration of stress on the leading dislocation in the pile-up. This high stress either can initiate yielding on the other side of the barrier, or in other instances it can nucleate a crack at the barrier. Dislocations piled up against a barrier produce a back stress acting to oppose the motion of additional dislocations along the slip system.

Much of the strain energy involved in plastic deformation is stored in the form of dislocations. When a cold-worked metal is heated in the recovery temperature range, internal stresses are relieved and thermal energy is used to allow the dislocations to rearrange into lower energy configurations or annihilate. During recovery the strength is reduced slightly but the ductility is usually significantly increased.

2.2. Viscoplasticity

The successful use of materials at high temperatures involves a number of new problems. Certainly, high temperature will have an effect on diffusion-controlled processes, will result in a greater mobility of dislocations by the mechanism of climb and may activate additional slip systems. But moreover one has to consider in this case the deformation at grain boundaries, the effect on the microstructural stability of metals and alloys and the interaction with the environment.

An important characteristic of high-temperature strength is that it must always be considered with respect to some time scale. At elevated temperature the stress-strain curves become very dependent on strain rate, and this is a manifesta-

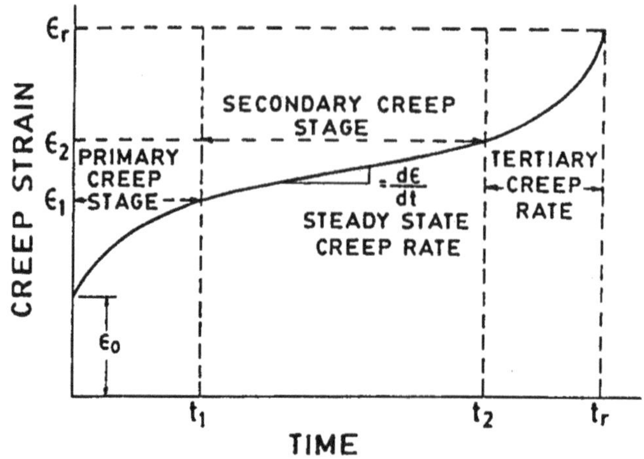

Figure 1: A typical creep curve showing the three stages.

tion of viscoplasticity.

When a metal or an alloy is under a constant stress it may undergo progressive plastic deformation over a period of time. This time-dependent strain is called creep. The creep of metals and alloys is very important for some types of engineering designs, particularily those operating at elevated temperatures. Let us consider creep of a pure polycrystalline metal at a high temperature and plot the strain-time curve. In the idealized creep curve (Fig. 1) there is first an instantaneous rapid elongation of the specimen. Following this, the specimen exhibits primary creep in which the strain rate decreases with time. In the second stage of creep the strain rate is essentially constant, and finally a tertiary stage may occur in which the creep rate rapidly increases with time up to the strain at fracture. Higher stresses and higher temperatures increase the creep rate. For many materials steady state creep during the secondary stage is well described by Norton's law

$$\frac{\partial \varepsilon}{\partial t} = A(T)\sigma^n \tag{1}$$

During primary creep the metal strain-hardens to support the applied load and the creep rate decreases with time. At high temperatures, during secondary creep recovery processes involving highly mobile dislocations counteract the strain hardening so that the metal continues to elongate at a steady rate. Finally, the creep rate accelerates in the tertiary stage of creep due to different processes like necking of the specimen, formation of voids along grain boundaries, microstructure degradation or environmental attack.

Viscoplastic solids can also be studied by means of relaxation tests, which are dual experiments to determine the stress response to a constant state of strain.

2.3. Fatigue

Metal components undergoing repetitive or cyclic stresses will fail at a stress much lower than the one required to cause fracture on a single application of load. These failures are called fatigue failures and are common in machine parts subject to repeated loading or vibration. It is often stated that today fatigue accounts for about 90% of all service failures due to mechanical causes. The most common fatigue test alternates tensile and compressive stresses of equal magnitude and provides curves in which the maximum stress is plotted versus the number of cycles N_f at which failure occurs.

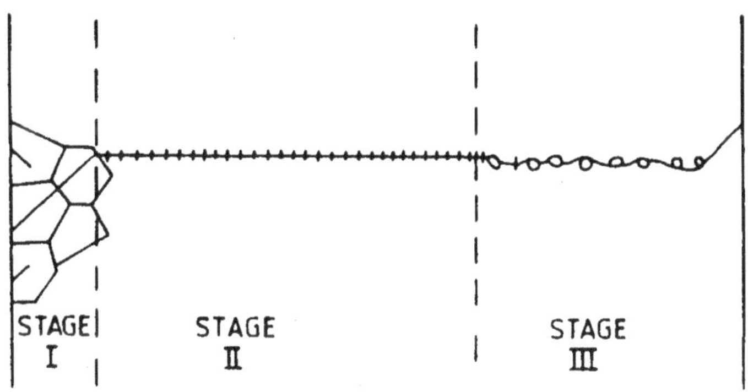

Figure 2: Schematic representation of the growth of a fatigue crack until final failure.

When a ductile homogeneous metal is subjected to cyclic stresses basic structural changes occur during the fatigue process. Crack initiation occurs because plastic deformation is not a completely reversible process. The stress alternation creates extrusions and intrusions on the surface as well as damage along persistent slipbands, and those surface irregularities and slipband damage nucleate cracks which propagate along slip planes subjected to high shear stresses. This is called stage I of fatigue crack growth (See Fig. 2). The rate of growth is very low ($10^{-10} m/cycle$) and in polycrystalline metals this stage only covers a few grain diameters. In stage II the crack changes direction to propagate in a plane perpendicular to the applied loads and advances at a relatively rapid rate ($10^{-6} m/cycle$) creating fatigue striations. Finally ultimate ductile failure occurs when the remaining section becomes so small that it can no longer support the load. A fatigue failure can usually be recognized from the appearance of the fracture surface, which shows a smooth region, fequently with a series of marks, due to the rubbing action as the crack propagated, and a rough region formed in a ductile manner when the load was too high for the remaining cross section.

The uniaxial cyclic deformation is usually characterized by the cyclic stress-strain curve. This curve shows transient behaviour until a stable limit cycle representative of the saturated state is reached. In the case of stress control, cyclic hardening or softening of the material is reflected by a reduction or an increase respectively in the axial strain amplitude. Under strain-controlled fatigue loading cyclic hardening or softening of the material causes an increase or decrease respectively in the axial stress amplitude. Many materials present also the Bauschinger effect, referring to the experimental result that after a certain amount of forward plastic deformation in tension or compression the material yields at a lower stress when the direction of loading is reversed than for continued forward deformation. The micromechanistic origins of this effect are related to the changes in dislocation substructure induced by reverse loading and in the ensuing changes in the internal stress systems. These phenomena are characterized in the unified viscoplastic models by means of the isotropic hardening and kinematic hardening variables, associated with the evolution of the elasticity domain in stress space.

The total fatigue life is defined as the sum of the number of cycles to initiate a fatigue crack and the number of cycles to propagate it subcritically to some final crack size. There are two types of approaches to fatigue design [12], which have also a more general scope. The total-life approaches involve the characterization of the total number of cycles to initiate a dominant crack and propagate it to failure, usually in terms of the stress range or the strain range. Since the crack initiation life constitutes a major component of the total life in smooth specimens, the classical stress-based and strain-based methods represent in many cases design against fatigue crack initiation. Under low stress fatigue situations, the material deforms primarily elastically, and the failure time has traditionally been characterized in terms of the stress range. However, when the stresses are high enough to cause appreciable plastic deformation prior to failure, the fatigue is characterized in terms of strain range. The defect-tolerant approaches assume that all components are inherently flawed and determine the initial size of the flaws. The useful fatigue life is then defined as the number of cycles to propagate the dominant crack from this initial size to some critical dimension and its prediction involves empirical crack growth laws based on fracture mechanics. Optimization of microstructural characteristics for improved resistance to both crack initiation and crack growth would require a trade off between the recommendations of these two types of approaches.

A typical characterization of fatigue life using the stress versus life and the strain versus life curves is the Coffin-Manson relationship

$$\frac{\Delta\varepsilon}{2} = \frac{\Delta\varepsilon_e}{2} + \frac{\Delta\varepsilon_p}{2} = \frac{\sigma'_f}{E}(2N_f)^b + \varepsilon'_f(2N_f)^c \tag{2}$$

which has found widespread application in industrial practice, and provides the number of cycles to failure N_f as a function of the strain range $\Delta\varepsilon$.

2.4. Constitutive Equations

Constitutive equations are expressions relating stresses and strains which are used to describe the deformation of a material. They are based on experimental observations, and are applied to characterize the deformation of materials under conditions which are not readily accessible by means of laboratory experiments and to perform the stress-strain analysis in components. A differential formulation of the constitutive equation is the most convenient approach when rate dependent material behavior is observed. In its most general formulation the stresses and the strains (and also the damage) are expressed in tensorial form in order to take account of multiaxiality effects. The instantaneous strain rate is a function of the stress sigma, the temperature T and the instantaneous internal structure of the material characterized by a set of internal variables S_i:

$$\frac{\partial\varepsilon}{\partial t} = \frac{\partial}{\partial t}\varepsilon(\sigma, T, S_i) \tag{3}$$

In the general case, the internal structure of a deforming material will change with time as the deformation proceeds. This leads to a second equation

$$\frac{\partial S_i}{\partial t} = \frac{\partial}{\partial t}S_i(\sigma, T, S_i) \tag{4}$$

which coupled to Eq. (3) forms the constitutive law for the deformation mechanism under consideration. The simplest constitutive equation is Hooke's law, which describes the behavior of a material in its linear elastic regime. The secondary or stationary stage of creep is well represented for many metals by Norton's law (Eq. (1)). In order to enable adequate descriptions of the mechanical behavior of structural materials at elevated temperatures viscoplastic models with more complex expressions are required.

2.5. Damage Mechanics

Mechanical loading of a material provokes changes at the scale of the microstructure which usually degrade the material in terms of its load carrying capacity. This degradation of the material is generally referred to as damage. The accumulation of damage eventually results in the failure of the material or component, either by fracture or by reaching a state of deformation which corresponds to a loss of function of the component. At the scale of the microstructure the concept of damage englobes a wide variety of categories and mechanisms of degradation which occur within specific limits of the loading parameters.

In creep deformation, for example, the major categories are damage by loss of internal section as the result of cavitation or of the growth of a single dominant crack, degradation of the microstructure or damage by environmental attack. In fatigue conditions the typical damage is localized, taking the form of a single crack or multiple cracking at slipbands or at grain boundaries. Mechanism maps are used to delineate the fields of operation of the different mechanisms. In the neighbourhood of the boundaries the damage mechanisms interact, which usually leads to an increase in the overall damage rates.

The total-life approach in lifetime prediction, which was mentioned above, essentially represents the attempt to predict failure from the evolution of a distribution of damage. In the framework of continuum mechanics the internal variable D, which was first proposed by Kachanov to describe the degradation of materials during tertiary creep, is introduced to quantify the damage in a phenomenological way [13]. This internal variable D is incorporated in the set of internal variables S_i appearing in the constitutive equations, and the evolution equation of D

$$\frac{\partial D}{\partial t} = \frac{\partial}{\partial t} D(\sigma, T, D) \tag{5}$$

is added to form a closed system of differential equations. The damage variable is taken to vary between $D = 0$ for the virgin material and $D = 1$ at failure. Integrating the equations between the appropriate limits yields the strain and the time to failure, and the shape of the deformation curve.

The easiest way to interprete the changes in the global mechanical properties in terms of D is by means of the concept of effective stress introduced by Rabotnov. In the case of isotropic damage one can consider that the presence of cavities and microcracks has reduced the effective area. The damage variable is taken as a measure of this reduction, $D = S_D/S$, and the effective area can then be expressed as $S' = S - S_D = S(1 - D)$. If an effective stress is defined as the one calculated using this effective area one obtains $\sigma' = \sigma/(1 - D)$. One assumes that this is the only effect that damage has in the deformation behavior. In this way, for example, for the case when the secondary creep rate is described by Norton's law the following equations have been suggested to describe tertiary creep

$$\frac{\partial \varepsilon}{\partial t} = \frac{\partial \varepsilon_0}{\partial t}\left[\frac{\sigma}{1 - D}\right]^n \quad \frac{\partial D}{\partial t} = \frac{\partial D_0}{\partial t}\left[\frac{\sigma}{1 - D}\right]^{n'} \tag{6}$$

where $\partial_t \varepsilon_0$ and $\partial_t D_0$ are temperature dependent amplitudes and n and n' are constant exponents.

2.6. Fracture Mechanics

The origin of the modern theories of fracture was the work of Griffith, who formulated criteria for the unstable extension of a crack in a brittle solid in terms of a balance between changes in mechanical and surface energies.

In damage tolerant applications the presence of a crack is accepted and the life is determined by the growth period of the crack up to a certain length. In that situation one identifies the crack length a as the relevant damage variable and applies fracture mechanics 'damage' equations of the form

$$\frac{\partial a}{\partial t} = \frac{\partial}{\partial t} a(\sigma, T, S_i) \tag{7}$$

which are integrated to calculate the time to failure. The essential feature in fracture mechanics involves the characterization, in the framework of continuum mechanics, of the local stress and deformation fields in the vicinity of the crack tip. The crack tip fields are uniquely characterized by the stress intensity factors K and the contour integral J in the case of linear elastic and non-linear elastic power-law hardening solids, respectively. In the case of creep crack growth in ductile alloys, the C^* integral is often the appropriate parameter to correlate creep crack growth rates.

Over the last decades the fracture mechanics research community has witnessed an increasing interest in the behavior of microstructurally and physically short cracks [14] which grow under medium to high fatigue loads. For applications where the growth of short fatigue cracks takes up a major fraction of the life of the component, the formulation of adequate short crack growth laws will expand the potential of the fracture mechanics approach in life time prediction at the expense of the total-life approach.

3. Modelling of Fatigue in Nickel-based Superalloys

3.1. Deformation Mechanisms in Superalloys

Turbine blades are the turbo-jet component that require a material with the highest strength and fatigue resistance when operating at high temperatures. Nowadays, they are increasingly made of single crystal nickel-based superalloys, which have the best combination of elevated temperature properties [15].

The reason of their strength lies in a special microstructure, which hinders the motion of the dislocations. Nickel-based superalloys consist of two phases, called γ and γ', which differ mainly in the degree of order [16]. The cubic precipitates of the γ' phase, FCC crystals of $Ni_3(Al,Ti)$ with a regular spatial distribution, are in coherence with the matrix of γ, which is an FCC disordered phase where the lattice sites are not occupied by a specific type of atom. The side of the cubes measures about $0.5\mu m$ and the narrow matrix channels are about $0.1\mu m$ wide.

As the material is strained, dislocations begin to move in the disordered γ matrix channels until their paths are blocked by the ordered precipitates. This is the reason why superalloys have a much higher strength than conventional alloys. If the internal stresses created by the pile-up at the boundary eventually force a dislocation through a γ' precipitate, this action will disturb the ordering sequence and imply a higher energy level. The passage of a second dislocation can restore the proper ordering and reduce the energy. The dislocations therefore need to travel in pairs in what is called the cutting of the γ' phase.

There is an optimum range of precipitate size. When the γ' particles are very small it is easy for dislocations to climb around them, resulting in reduced strength. If they are large, it becomes easier for dislocations to bow between them, also reducing strength. Superalloys are made by cooling the liquid mixture of their metallic ingredients. A solid γ phase is obtained first since its melting point is higher than that of the γ'. Further cooling of the solid alloy causes the precipitation of γ' cubes inside the γ matrix, whose final size can be controlled by changing the cooling rate.

Processing techniques permit nowadays an optimum usage of the microstructural information. The creep resistance of the nickel-based superalloys can increase drastically if they are prepared using directional solidification, so that the grain boundaries are oriented parallel to an applied uniaxial stress, as is the case of centrifugal forces in a turbine blade, avoiding damage generation in form of cavities along the grain boundaries.

Even better results are obtained using single crystals, where the grain boundaries have been eliminated. They can be grown with a procedure similar to directional solidification, but using a narrow bottleneck which allows the passage of only one of the growing crystals. The superiority of single crystals over polycrystalline superalloys is due to several reasons: the higher melting temperature allowing the γ' microstructure to be refined for increased creep strength, the absence of grain boundaries which act as failure initiation sites and their low modulus orientation providing better thermal fatigue resistance.

Plastic deformation in superalloy single crystals is mostly due to crystallographic slip. In FCC crystals there are three directions of easy slip in each of the four octahedral slip planes and two directions in each of the three cubic slip planes. Cross slip from the octahedral to the cubic planes can take place by thermal activation.

Creep curves of single crystal superalloys present marked differences with respect to those of polycrystals. In the initial stage deformation proceeds for extensive distances on single slip systems, resulting in high primary creep strains. The secondary creep rate is lower than for polycrystalline superalloys, as a result of a finer and more uniform γ'. Moreover, tertiary creep is quite extensive since the grain boundary failure initiation sites are absent. The creep properties of single crystal superalloys are anisotropic, and depending upon orientation large variations can occur, strongly affected by temperature and stress.

Cyclic straining both at room temperature and at elevated temperatures can also produce hardening, as a result of an increase in the dislocation density. In the intersection of slip planes dislocation tangles are formed, which hinder the motion of other dislocations in those planes. As cycling proceeds, eventually pairs of dislocations shear along the slip planes the γ' precipitates.

At high temperatures damage becomes more complex since creep and environmental effects become more important. Moreover, the γ' precipitates are unstable at elevated temperatures, and coarsening may have effects of a different sign: if after coarsening the precipitates are still sheared this may lead to strengthening, but if looping occurs because the particles are too large then the material will become weaker. Coatings can be applied to enhance the hot corrosion resistance but this may have an adverse effect on other mechanical properties like fatigue lifetime if, for example, they play a role in crack initiation.

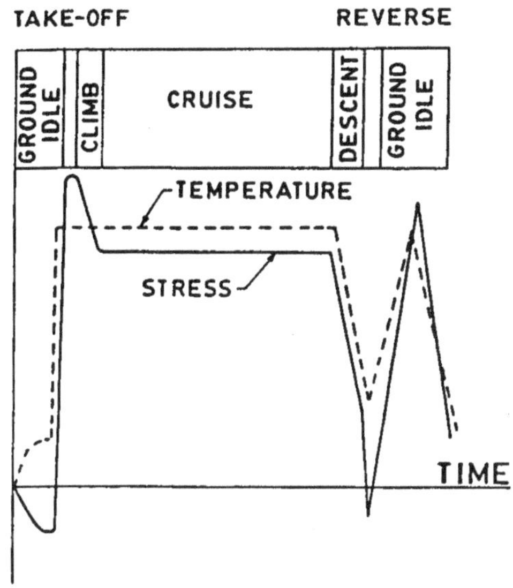

Figure 3: Typical stress and temperature cycles in a turbo-jet gas turbine.

3.2. Viscoplastic Models

In high temperature applications structural components are often subjected to complex loading patterns with changing loads and temperatures (see, for instance, Fig. 3). In order to describe the inelastic deformation behaviour under general loading conditions one applies usually as constitutive equations the so-called unified viscoplastic models, where a single inelastic strain rate equation is used to model both time independent and time dependent monotonic and cyclic material deformation. These models are phenomenological, and usually incorporate two internal variables R and X, reflecting isotropic and kinematic hardening and frequently associated with the size of the elasticity domain in stress space and with the position of its centre respectively. The main differences between the various viscoplastic models are the way in which the isotropic and kinematic hardening variables are chosen, how their evolution equations are formulated and the functional form of the inelastic strain rate [17].

One of the best known unified viscoplastic models, proposed by Chaboche, consists in its uniaxial form of the following Equations (4):

$$\frac{\partial \varepsilon_p}{\partial t} = \nu \left\langle \frac{|\sigma - X| - R - k}{K} \right\rangle^n \; ; \; \nu = sign(\sigma - X)$$

$$\frac{\partial X}{\partial t} = c\left(a\frac{\partial \varepsilon_p}{\partial t} - \left|\frac{\partial \varepsilon_p}{\partial t}\right| \right)$$

$$\frac{\partial R}{\partial t} = b(Q - R)\left|\frac{\partial \varepsilon_p}{\partial t}\right| \tag{8}$$

The equations for X and R can be integrated and one can invert the equation for the plastic strain rate to obtain:

$$\sigma = X(\varepsilon_p, X_0, \varepsilon_{p_0}) + \nu R(\varepsilon_p') + \nu k + \nu K(\varepsilon_p')\left|\frac{\partial \varepsilon_p}{\partial t}\right|^{1/n}$$

$$\frac{\partial \varepsilon_p'}{\partial t} = \left|\frac{\partial \varepsilon_p}{\partial t}\right| \tag{9}$$

In this case one needs to determine seven coefficients: in a fatigue test, for example, five can be obtained from the stabilized cycle, one from the initial behaviour and one from the transient cyclic behavior. These coefficients have been determined successfully for different alloys and at different temperatures. Extended versions of this model take into account also the elastic deformation.

3.3. Microstructural Input

When one tries to model the anisotropic cyclic behaviour of a superalloy single crystal, a macroscopic phenomenological model, in which each volume element is treated globally, is clearly insufficient. One can then follow a micro-macro approach, in which the constitutive equations of the macroscopic model are applied at the microscopic level, i. e. on the slip system scale. The approach followed by Cailletaud [18] in this direction is still clearly phenomenological, but it can be used in a finite element code and the integration of the slip system increases the predictive character of the constitutive model. As a microscale extension of Chaboche's model, the relevant variables describing stress and plastic strain for each slip system s of the single crystal are now the resolved stress tau s, found by taking the tensorial product of the macroscopic stress sigma by the orientation tensor ms, and the shear strain rate or *slip rate*, $\partial_t \gamma_v^s$, which is related to the macroscopic viscoplastic strain rate $\partial_t \varepsilon_v$ by $\partial_t \varepsilon_v = \sum m^s \partial_t \gamma_v^s$. The viscoplastic constitutive equations relating $\partial_t \gamma_v^s$ to τ^s are

$$\frac{\partial \gamma_v^s}{\partial t} = \left\langle \frac{|\tau^s - X^s| - r^s}{K} \right\rangle^n sign(\tau^s - X^s) \; ; \; \tau^s = m^s : \sigma$$

$$X^s = c^s \alpha^s \; ; \; \frac{\partial \alpha^s}{\partial t} = \varphi(v^s)\frac{\partial \gamma_v^s}{\partial t} - \left|\frac{\partial \gamma_v^s}{\partial t}\right| d^s \alpha^s$$

$$r^s = r_0^s + \sum_j H_{sj} b^j Q^j q^j \; ; \quad \frac{\partial q^s}{\partial t} = \left|\frac{\partial \gamma_v^s}{\partial t}\right| (1 - b^s q^s)$$

$$\varphi(v^s) = \phi + (1 - \phi)\exp[-\delta v^s]$$

$$v^s(t) = \int_0^t du \left|\frac{\partial \gamma_v^s(u)}{\partial t}\right| \tag{10}$$

and they include a kinematic hardening variable and an isotropic hardening variable for each slip system, denoted respectively by x^s and r^s for the slip system s. The interaction matrix H_{sj} reflects the interaction between the 18 different slip systems.

Once the material coefficients have been determined and a simple interaction matrix has been postulated, excluding interaction between the octahedral and cubic families, the numerical simulations with this micro-macro model present a very good agreement with the anisotropic behaviour of the experimental data from isothermal high-stress fatigue experiments.

A further improvement should come through an attempt to include more microscopic information. If one is also interested in modelling the behaviour during a thermomechanical fatigue test, where the specimen is subjected to simultaneous temperature and deformation cycling, an interpolation of the parameter values found in isothermal tests will not be sufficient, and wise modelling will probably require the incorporation of microscopic information regarding the temperature-related processes taking place in the superalloy.

3.4. Damage Equation for Lifetime Prediction

Having established a faithful constitutive model, a lifetime computation can be performed following the total life approach. At high temperatures one usually has to consider an interaction between the processes responsible for the fatigue and creep damage. A well known method [4] assumes that the material starts with no damage, and the fatigue and creep damages having additive macroscopic effects are then computed in a continuous evolution. For the creep damage one uses Kachanov's law and for fatigue damage a model that depends on the maximum and mean stresses. Both laws depend on the total damage:

$$dD = dD_c + dD_F$$

$$dD_c = \left(\frac{\sigma}{A}\right)^r (1 - D)^{-k} dt$$

$$dD_F = \left[1 - (1 - D)^{1+\beta}\right]^{\alpha} \left[\frac{\sigma_m - \bar{\sigma}}{M(\bar{\sigma})(1 - D)}\right]^{\beta} dN \tag{11}$$

The parameters A, r, k, and β are temperature-dependent, while α and M are functions of the stresses and the temperature. The values of all these parameters can be determined from experimental tests. Thermomechanical fatigue lifetimes can be computed also with this damage interaction model.

4. Modelling of Creep in Ceramic Matrix Composites

4.1. Distinctive Properties of Ceramic Matrix Composites

Monolithic ceramics show potential for application as high temperature structural materials. They are stronger than the nickel-based superalloys above 1000 *C*, have superior creep strength and oxidation resistance and are potentially cheaper. In addition their density is less than half that of the superalloys. Their problem, however, is their intrinsic flaw sensitivity, brittleness and consequent lack of reliability. The main source of the mechanic brittleness of ceramics derives from their crystalline structure which does not permit the existence or the movement of dislocations and hence local plastic flow under an applied stress. The only way to prevent catastrophic failure is to design ceramic microstructures with some tolerance to defects [19]. Fibre reinforced ceramic matrix composites have recently received a great deal of attention for use in high temperature structural applications. Strong ceramic fibres can indeed prevent catastrophic brittle failure in ceramics by providing various energy absorbing processes during a crack advance and thus provide some defect tolerance.

The incorporation of high strength continuous fibres in a metal or polymer matrix has been used for strengthening purposes. In the case of those ductile matrix composites, the high modulus fibres carry essentially all of the load and the matrix serves only to transmit the load to the fibres and to protect them from surface damage and from cracks arising from fibre breakage.

The aims and mechanisms of fibre reinforcement in CMC are very different from those in metal and polymer matrix composites, and this is reflected in the very different properties desired for the interface between matrix and the fibres [2]0. In the composites with metallic or polymeric matrix, the function of the interface is to assure maximum adhesion, and the contact between fibres and matrix at the interface is usually of a physico-chemical nature, by means of van der Waals forces or ionic chemical bonds. In the case of a CMC the introduction of the (ceramic) fibres is to tackle the brittleness of the ceramics. In fact, the deformation at fracture of the matrix is smaller than that of the fibres. When the deformation applied to the composite superates the fracture strain of the matrix, the matrix cracks and the load is transferred to the fibres. The role of the fibres is to facilitate the occurrence of new processes of dissipation of fracture energy, and these processes rely on particular interfacial characteristics. In order to achieve this, the interface must be the result of a compromise between a sufficient adhesion to assure the transfer of the load applied on the composite to the fibres, and an adhesion weak enough to allow these dissipative processes to take place. This is usually attained through a mechanical bonding due to the difference in the coefficients of thermal expansion of fibre and matrix (thermal mismatch) or in the Poisson coefficients. In the first case, the process of cooling down from the fabrication temperature to room temperature generates radial stresses which facilitate the load transfer through the interface. In the second case, it is reached through the application of a load. In other instances, the specific properties required from the interface can only be obtained by adding a new phase of a different material. It is very important to control the properties of the interface, since they are fundamental in determining the macroscopic behavior of the composite. Another role of the interface in CMC, since the matrix cracks earlier than the fibres, is to protect the fibres from environmental corrosion, especially at high temperatures.

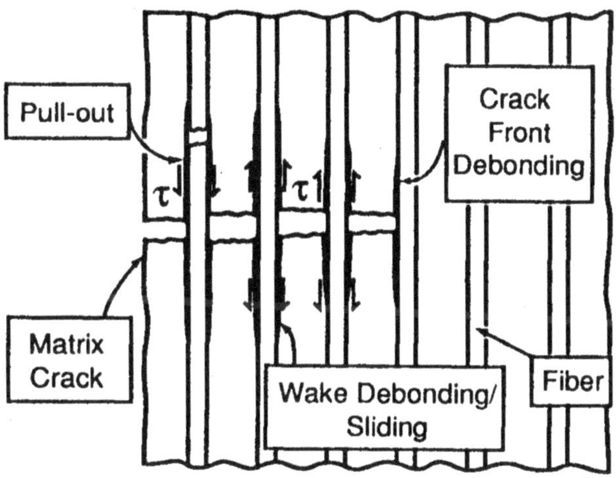

Figure 4: Processes dissipating fracture energy in a brittle matrix composite.

The processes dissipating fracture energy at the interface are well known and can be seen in Fig. 4 . When a matrix crack propagating perpendicular to the loading axis arrives to a fibre it is stopped momentarily. The fibre debonds then from the matrix and the crack can start to propagate in the direction of the fibre. Since Coulombian friction forces appear between the matrix and fibre surfaces still in contact, this prevents the crack from growing during a certain time. When the force transferred to the fibre attains a higher value than the fibre local strength, determined by the probability of finding a critical defect there, the fibre breaks but still takes part in the slowing down of the crack by simple friction between the surfaces of fibre and matrix. One talks then about the pull-out of the fibre.

The incorporation of continuous brittle fibres into a brittle ceramic matrix provides in this way the composite with a pseudo-plastic behaviour. In fact, the curve from a tensile test with fibre alignment parallel to the load direction presents several distinct regions. The first region reflects a purely elastic behaviour, until the first matrix cracking. In the second part, the number of matrix cracks increases and the behaviour is not linear. The third stage begins when the matrix cracking has reached saturation and it shows linearity because the load has been transferred completely to the fibres, which continue to deform elastically. In the final region, the fibres start to break, increasing thus the load tranferred to the neighbouring fibres and eventually provoking a controlled or catastrophic fracture of the composite. The strains of up to 0.7% attained in this way in a tensile test [21] are one order of magnitude larger than those found for matrix (or fibres) alone.

While fracture behavior in the CMC is now fairly well known, creep studies for these materials have only very recently started. The results available are also quite surprising [21]. While one would not expect any time-dependent deformation of relevance for the matrix, and the fibre ceramic would certainly break after the primary and secondary stages, the composite presents a creep curve showing a very large tertiary stage and featuring huge deformations of up to 5.6%.

4.2. Macromechanistic Models

Creep deformation at high temperatures in metallic matrix composites reinforced with continuous fibres can be succesfully described by the combined deformations of the constituent phases. The simple case of fibre and matrix individually deforming with steady state creep rates that are simple power law functions of stress was considered first, and one obtains a complex dependence of the composite creep rate on stress. Since in modern metallic matrix composites there is a very large difference in the creep rates of the matrix and reinforcing phases, a model where the fibres deform elastically constrained by the creep of the matrix gives also predictions consistent with the limited experimental data [22].

As a first approach, one could apply similar ideas to model the creep behaviour of ceramic matrix composites [23]. When the applied stress sigma is redistributed to maintain equal strains in the two phases the conditions of mechanical equilibrium read

$$\frac{\partial \varepsilon_f}{\partial t} = \frac{\partial \varepsilon}{\partial t} = \frac{1}{E_f}\frac{\partial \sigma_f}{\partial t} + \frac{\partial \varepsilon_{f_0}}{\partial t}\left[\frac{\sigma}{\sigma_{f_0}}\right]^{n_f}$$

$$\frac{\partial \varepsilon_m}{\partial t} = \frac{\partial \varepsilon}{\partial t} = \frac{1}{E_m}\frac{\partial \sigma_m}{\partial t} + \frac{\partial \varepsilon_{m_0}}{\partial t}\left[\frac{\sigma}{\sigma_{m_0}}\right]^{n_m}$$

$$\sigma = \sigma_f V_f + \sigma_m (1 - V_f) \tag{12}$$

where the subscripts f, m represent fibre and matrix respectively and V_f is the volume fraction of the fibres. Rearrangement of Eq. (12) leads to the following coupled differential equations, in which the creep rate of the composite is expressed as the sum of two power law terms, and depends on the parameter S_1, which is an internal stress variable increasing with strain

$$\frac{\partial \varepsilon}{\partial t} = a_1 \frac{\partial \varepsilon_f^*}{\partial t} S_1^{n_f} + a_2 \frac{\partial \varepsilon_m^*}{\partial t} (1 - S_1)^{n_m}$$

$$\frac{\partial S_1}{\partial t} = H \frac{\partial \varepsilon}{\partial t} - R S_1^{n_f} \tag{13}$$

Here, $\partial_t \varepsilon_m^*$ and $\partial_t \varepsilon_f^*$ are the creep rate with the full applied stress for matrix and fibre respectively, $S_1 = V_f \sigma_f / \sigma$, $H = V_f E_f / \sigma$, $R = H \partial_t \varepsilon_f^* / V_f^{n_f}$ and a_1, a_2 are constant with the applied stress.

This model (or constitutive law) predicts a creep rate that reaches a steady state. Power law exponents n_f and n_m determined from the creep curve corresponding to a calibration stress can be used to predict the behaviour at smaller stresses, where the creep life is expected to be much longer. However, the exponents resulting from the best fit are too large to be realistic. A more adequate formulation can be pursued through the incorporation of the micromechanical features of the composite.

4.3 Micromechanistic Models

The treatments based on purely combining the properties of the constituents disregard the basic idea behind a composite, which is that the improvement in the mechanical properties will come through the interaction between the constituents. The so-called micromechanistic models attempt to describe this interaction at the scale of a representative composite element, consisting of a fibre and the surrounding mantle of matrix.The best known of these approaches is the shear-lag model, originally developed by Cox [24], which is based on the hypotesis that a perfect bond exists between fibre and matrix. The outcome from these micromechanistic approaches are the stress and strain distributions over the different constituents, which are then used to explain a number of time dependent mechanical features, such as matrix or fibre cracking, debonding, sliding or composite strength.

One can also apply these ideas to improve the modelling of creep in a CMC [23]. We will have to take into account that the matrix is already cracked at the stress levels of interest when the creep process begins. We consider therefore a very long fibre embedded in a series of matrix cylinders of a finite length, as shown in Fig. 5 . The transfer of load will take place through shear stresses along the interface. Near a matrix crack we assume a region of slip where the shear stresses are purely due to friction, while in the rest of the interface there is non-slip mechanical contact. In both regions an equilibrium condition will relate the shear stress with the gradient of the axial stresses. We need to assume also compatibility in the radial or tangential deformations at the interface. Solving the differential equations corresponding to the two distinctive interface regions and connecting the solutions by imposing continuity in the shear stresses at the boundary between the regions, one can find the stress profiles along the fiber and the matrix block as a function of the external load, the magnitude of the thermal strains and the friction coefficient.

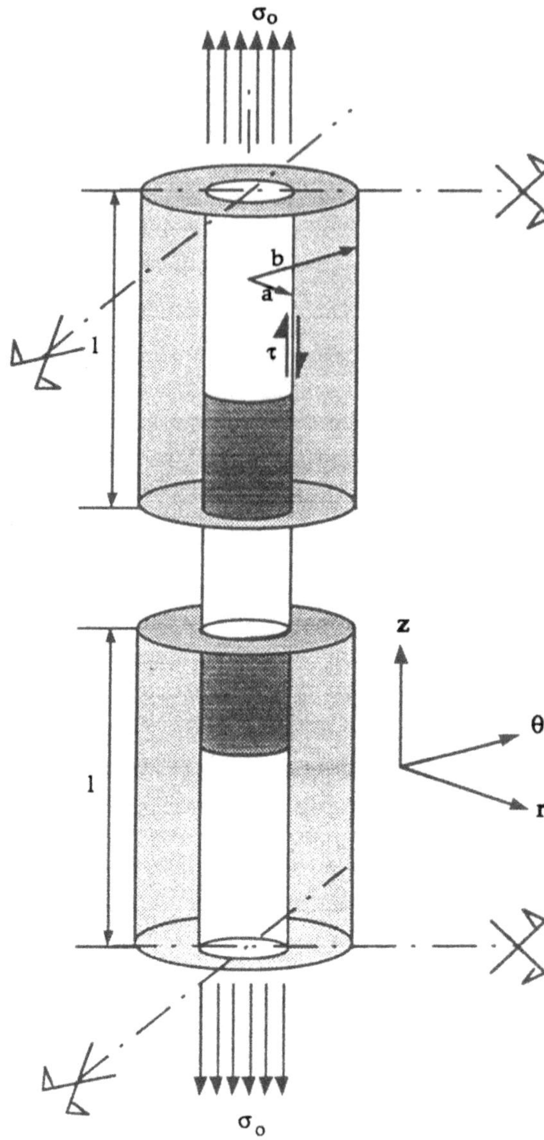

Figure 5: Micromecanistic model for a ceramic matrix composite after saturation of the matrix cracks.

A feature limiting the possibilities of application of this model is that it considers a purely mechanical bonding. When the thermal expansion coefficient of the fibres is larger than that of the matrix the required compressive radial stresses will only occur for temperatures higher than the manufacturing temperature. Since lower temperatures are also of interest, the condition of compressive stresses at the interface needs to be relaxed. The introduction of a coefficient of interfacial strength makes this feasible. The model will then allow three distinct interfacial regions, namely non-slip, slip and total detachment, but the three regions do not occur simultaneously. It is important to point out that the values of all the parameters in this model can be obtained from tensile experimental measurements, so that no curve fitting of the stress profiles is involved.

The stress profiles determined with this micromechanistic approach can be subsequently incorporated in the modelling of the creep curve. We assume that the blocks between matrix cracks are composed of fine transverse slices, which deform according to the rates resulting from solving the composite creep constitutive law in Eq. (13) for the stress value corresponding to the slice. In the region of exposed fibres between matrix blocks only the creep and elastic deformation of the fibres provide a contribution. The creep curves obtained with this finer approximation characterize well the primary and secondary stages in the experimental curves, and with the more reasonable exponents $n_f = 4$ and $n_m = 1$.

A satisfactory modellisation of the tertiary stage in the creep curves could be attempted also by including in the scheme the evolution of damage in the composite. In this case, it would probably require a quantitative description of the cooperative process of fibre breaking.

Acknowledgements

The author wishes to thank J. Bressers, M. Steen and J. Estevas-Guilmain for fruitful discussions and for stimulating his interest in the mechanical behavior of advanced materials. This work has been carried out within the framework of the specific research programme of the Institute for Advanced Materials of the Joint Research Centre, CEC.

References

1. S. T. Pantelides, *Physics Today*, September (1992) 67.
2. *Statistical Models for the Fracture of Disordered Media*, ed. H.J. Herrmann and S. Roux, (North-Holland, Amsterdam, 1990).
3. G.E. Dieter, *Mechanical Metallurgy* (McGraw-Hill, London, 1988).
4. J. Lemaitre, J.L. Chaboche, *Mechanics of Solid Materials* (Cambridge University Press, Cambridge, 1990).
5. W. F. Smith,*Principles of Materials Science and Engineering*, 2nd ed., (McGraw-Hill, 1990).
6. T. Mura, *Micromechanics of Defects in Solids* (M. Nijhoff Publishers, Dordrecht, 1987).
7. *Materials Science and Engineering for the 1990s* (National Academy Press, Washington

D.C., 1989).

8. J.R. Stringer, in *Heat-Resistant Materials*, ed. K. Natesan and D. J. Tillack (ASM, Metals Park, Ohio, 1992).
9. C. White, in *Research and Development of High Temperature Materials for Industry*, ed. E. Bullock et al., (Elsevier, London, 1989).
10. J. Bressers, M. Steen and J. L. Valles, in the Proceedings of the *International Symposium on Life and Performance of High Temperature Materials and Structures* (Tallinn, Estonia, 1992).
11. R. Thompson in *Current Trends in the Physics of Materials*, ed. G.F. Chiarotti, F. Fumi and M. P. Tosi, (North-Holland, Amsterdam, 1990).
12. S. Suresh, *Fatigue of Materials* (Cambridge University Press, Cambridge, 1991).
13. L. M. Kachanov, *Introduction to Continuum Damage Mechanics* (M. Nijhoff Publishers, Dordrecht, 1986).
14. K. J. Miller, in *Subcritical crack growth due to fatigue, stress corrosion and creep*, ed. L. H. Larsson, (Elsevier, London, 1982).
15. *Superalloys, Supercomposites and Superceramics*, ed. J. K. Tien and T. Caulfield, (Academic Press, Boston, 1989).
16. B. H. Kear, *Advanced Metals* (Scientific American, October, 1986, p. 136).
17. K. P. Walker, *Research and development programme for nonlinear structural modeling with advanced time-temperature dependent constitutive relationships*, NASA, Final Report CR-165533 (1981).
18. L. Meric, P. Poubanne and G. Cailletaud, *Journal of Engineering Materials and Technology* **113** (1991) 162.
19. H. Kent Bowen, *Advanced Ceramics*, Scientific American, October 1986, p. 146.
20. A.G. Evans in *High Performance Composites for the 1990's*, ed. S. K. Das, C. P. Ballard and F. Marikar, TMMMS (1991).
21. J. N. Adami, *Ph. D. Thesis*, Joint Research Centre, Petten (1992).
22. S. Goto and M. McLean, *Acta metall. mater.* **39** (1991) 153.
23. M. Steen and J. L. Valles, to be published.
24. H. L. Cox, *British Journal of Applied Physics* **16**, (1952) 72.

A PROBLEM OF COMPUTATIONAL ASTROPHYSICS: ORDER IN THE ORIENTATION OF GALAXIES

EDUARDO BATTANER
Departamento de Física Teórica y del Cosmos
Universidad de Granada, Campus de Fuentenueva
Granada, E-18071, España

ABSTRACT

The possible coherent orientation of two characteristic features of spiral galaxies is considered. The first one is the warping of their discs. A warping vector is defined for each individual galaxy and a statistical analysis is developed to study whether these vectors are randomly distributed in space or whether there exists a coherence alignment. The second one is the angular momentum of the galaxy. Two different methods are described, one of them based on the least squares principle and the other one is called method of "convergence of planes". Coherence for warping vectors is found for cells of characteristic size of about 25 Mpc. Angular momenta may have some degree of alignment, but this fact remains uncertain. The cosmological implications of these facts are commented.

1. Warps of Spiral Galaxies

It is an observational fact that the discs of spiral galaxies are often warped. Only the most external part of a galaxy presents this feature, which consists of a gradual separation of the disc from the mean plane at some azimuth and an antisymmetric separation at the opposite azimuth. As a result the galaxy resembles an integral sign when seen edge-on. The maximum separation is not larger than about 20 degrees.

Radio observations of the 21 cm spectral line of atomic oxygen permit the obtaining of measurements very far from the galactic centre. As warping is a peripherical phenomenon, these kinds of observations provide the best method of observing them. However, visible photometry as well as other wavelength ranges are also suitable.

Following a recent review by Binney[1], three current hypotheses may adequately explain warps:

a) Extragalactic magnetic fields are responsible for this deformation. This hypothesis has been analyzed by Battaner et al[2,3]. Therefore I will implicitly assume its validity, though the analysis that follows does not require this assumption.

b) Warps are normal modes of oscillations in discs, especially when it is considered that triaxial non axisymmetric bulges are frequent in spiral galaxies. One of the most representative papers considering such an explanation is that by

Sparke and Cassertano[4].

c) Infall of matter from the extragalactic medium or from the halo[1]. Other models also suppose the interaction with extragalactic matter. The hypothesis of tidal or gravitational interaction with a neighbour galaxy[5] has been extensively analyzed in the past.

Three observational measurements have been proposed by Battaner et al[6] that should be explained by the different theoretical models:

a) Colour dependence of the warp shape. Most theories in which gravitation is considered the force producing warps predict the same shape when gas, young stars (blue light) and old stars (red light) are observed. When the interstellar extinction is taken into account to interpret photometric observations, the expected order from higher to lower warping would be "gas-red-blue". The intergalactic magnetic field model however predicts the order "gas-blue-red". Observations have been carried out at La Palma, Spain, by means of the 2.5 m *Isaac Newton Telescope* with this specific purpose by Florido et al.[7,8]. Other observations reported in the literature may also be analyzed, particularly those of Sasaki[9] on the warped galaxy *NGC5907*. Of particular interest among these observations is that concerning the galaxy *NGC4013*, in which a dust line crossing the disc is more warped than the stellar background. As dust is a representative indicator of young objects and stellar formation, this photometry of this galaxy alone, seems to confirm the magnetic hypothesis. A different kind of observations could be carried out to verify the colour dependence of the warp: the study of our own galaxy, which shows a clear warp for young population components (atomic hydrogen, ionized hydrogen, molecular clouds, dust, OB associations, carbon monoxide...). Such an observation is now planned to be carried out with the Carlos Sanchez infrared telescope in Tenerife. Some attempts to find a warp in the old star population (Ichikawa and Sasaki[10], Guibert et al.[11]) have given negative results.

b) Bosma[12] in 21 cm, and Sanchez-Saavedra et al. [13] in the optical have made a statistical study on the frequency of warps, and obtained similar results: both concluding that the frequency is very large. About half of edge-on spirals are seen to be warped. Taking into account that some orientations do not permit their identification, this frequency may be estimated to be 83% or even larger. Of course, this is an important constraint for theoretical models.

c) Coherent alignments of warps. Most theoretical models predict a random distribution for the orientation of warps. On the other hand, the extragalactic field hypothesis would be consistent with a regular coherent alignment at least for some characteristic size cells. This is a typical favourable position for theoretical works: if this alignment is found, other theories are wrong; if it is not found, ours may be either wrong or right. To check this fact we have developed some computational techniques, which will be now discussed in some detail. They are general and may be applied to other problems, not purely astrophysical, and may have therefore be of interest in computational physics. On the other hand, if the theoretical ideas about magnetic induced warps were confirmed, then the alignment found could be used to determine the extragalactic magnetic field. This has obviously a high cosmological interest. Extragalactic magnetic fields could be determined at distances where warps could be identified. Therefore, the scale involved is 100 Mpc around the Milky Way.

2. The Individual Warp Vectors

The extragalactic magnetic field model has as a conclusion that the magnetic field produces warps if the field is neither parallel to the rotation axis nor perpendicular to it, i.e., contained in the galactic mean plane. Moreover, the greatest efficiency in producing warps is obtained when the magnetic field and the rotation axis form an angle of 45° (and therefore the magnetic field and the mean galactic plane also form an angle of 45°). This efficiency has a rather sharp maximum for this angle of 45°. Let us define the warping plane as the plane containing the greatest displacements from the mean galactic symmetry plane; the warping plane is perpendicular to the galactic mean plane. When the galaxy is seen edge-on, the warping plane coincides with the plane of the sky (i.e., the plane perpendicular to the line-of-sight). Let us define the warp vector b, as a unitary vector contained in the warping plane and forming an angle of 45° with the rotation axis. When the galaxy is edge-on the vector is in the plane of the sky and forms 45° with the mean galactic plane. If we assume the extragalactic magnetic field model to be correct, the warp vector would be exactly the same as the unitary vector along the magnetic field with greatest efficiency in producing warps. As the efficiency in producing warps has a rather sharp maximum at 45°, the warp vector is approximately parallel to magnetic field lines. We see that the above definition is incomplete, as the direction to which the vector points was not specified. Our model predicts that a magnetic field strength B and the opposite vector $-B$ produce the same warp. Therefore it is not possible to deduce the direction to which B points. An ambiguity remains. Similarly, when a vector b is deduced for an observed galaxy, $-b$ is also a solution. We will say that the "polarity" of b remains undetermined. The statistical techniques to be used must provide a mean value of b, and the individual polarities. More precisely, the polarity p of b may be defined as unity when the angle between north and b is less than 90°, and zero otherwise. We have for each galaxy b_i (or $-b_i$!) and must deduce a mean value $< b >$, and at the same time determine individual polarities p_i. But if $< b >$ is a solution, then $- < b >$ is also a solution. Therefore, the polarity of the solution also remains undetermined, and this indetermination is inherent to the method. If we arbitrarily assign a polarity to $< b >$, then individual polarities can be determined. We see that similar kinds of problems can be present in other research fields. Suppose now that we are observing an edge-on galaxy with coordinates α_0 and δ_0 (right ascension and declination). We know the direction of b, which is in the sky plane and has a position angle of β (the angle between b and north). The coordinates (α and δ) of the vector b, when its origin is transported to here, the Milky Way, are found by spherical trigonometric considerations:

$$\sin\delta = \cos\delta_0 \cos\beta \tag{1}$$

$$\sin(\alpha_0 - \alpha)\cos\delta = \sin\beta \tag{2}$$

Then we have all the coordinates (α_i, δ_i) of vectors b_i with their origin in the Milky Way. Of course $\alpha_i + 180^0$ and $-\delta_i$ are also "a priori" possible. We must now to describe the statistical methods that will provide $< b >$ and p_i.

3. Least Squares Method

It consists of carrying out the calculation for every possible value of the polarity matrix, for instance $(1,0,0,0,1,1,0...)$:

$$\mathbf{b} = \frac{\sum_i \mathbf{b_i}}{|\sum_i \mathbf{b_i}|} \tag{3}$$

and find out which of all combinations gives the minimum value of the function S:

$$S = \sum_i \left[cos^{-1}(\mathbf{b}.\mathbf{b}_i)\right]^2 \tag{4}$$

It is necessary to repeat the same calculation 2^n times, n being the number of galaxies. In practice, only 2^{n-1} calculations need be performed, because half of them must give the same result, i.e. the matrix (p_i) gives the same value as $(p_i + 1)$, in base 2. This fact is related to the inherent ambiguity of the polarity of the solution. The problem is of course that the number of individual computations increases very fast with n. When n is about 13 a personal computer requires several hours, even if the code has been written with good criteria of time economy. It can be seen how this method - a discrete numerical least squares - may be applied to other scientific fields.

Exercise 1

Consider the following list of galaxies adopted from Battaner et al.[3] (approximate values):

NGC	α	δ
2683	-51	56
3044	53	67
3245A	-51	58
3365	-103	68
3454	31	64
3510	-59	51
3600	-68	31
3628	60	52
3692	71	44
4206	-73	43
4565	12	63
5560	117	56

The angles are given in degrees. Write a computer program to calculate: i) The cartesian coordinates of $\mathbf{b}_i$ ($b_x = \sin\alpha\sin\delta, b_y = \cos\alpha\sin\delta, b_z = \cos\delta$). ii) $< \mathbf{b} >$. iii) The polarity matrix. iv) The standard deviation.

4. Convergence of Planes

Any plane is adopted initially and arbitrarily, for instance the equator. One of the resulting hemispheres (for instance the north one) is called the "upper" hemisphere. We set all individual polarities so that all b_i lie in the upper hemisphere. We calculate $< b >$ and adopt as the new plane the plane perpendicular to $< b >$. We iterate this procedure obtaining a series of $< b >$'s, which quickly converges, after no more than 4 or 5 iterations. The limiting value of this series is the adopted value of $< b >$. It is very interesting indeed to check how the result is the same for different initial choices of the initial plane, and the same as that obtained with the least squares method.

Exercise 2

Use the same list from *Exercise 1*, to write a computer program to obtain $< b >$ with the method of convergence of planes. Repeat the calculation for two initial planes other than the equator.

5. Interpretation of the Statistical Results

When about 40 edge-on warped galaxies were considered, a mean direction of the magnetic field in the Milky way neighbourhood given by $\alpha = 289^o$ and $\beta = -8^o$ was obtained. Let us remember again that the direction $\alpha = 109^o$, $\delta = 8^o$ is also possible.

However, the standard deviation was very high, of about 40°. Then a question arises: with respect to what was it high?. There is first a maximum value of the standard deviation, which corresponds to a theoretical continuous distribution, in which all b_i are uniformly distributed in the sphere. This may be exactly calculated, though the details of this analytical calculation are here omitted. This maximum value of σ, could be called 'rectangular' standard deviation σ_R, as in one dimensional problems a constant value of the probability density corresponds to a "rectangular" distribution function

$$\sigma_R{}^2 = \pi - 2 \tag{5}$$

Therefore, $\sigma_R = 61°$. There is another quantity related to the standard deviation, with a clear physical meaning. This is the degree of non uniformity, χ:

$$\chi = \frac{< \mathbf{R}_i{}^2 >}{\mathbf{U}^2} \tag{6}$$

The magnetic field may be written as $B_i = U + R_i$, where U is the uniform component and R the random component, so that $< B_i >= U$, because $< R_i >= 0$. Mean values are calculated in the space considered. It is possible to calculate that

$$\chi = 2 < \sin^2 \theta_i > \tag{7}$$

where θ_i is the angle between B and U, i.e., the angle between b_i and $< b >$. The use of squared quantities in Eq. (6) is due to the fact that the magnetic energy depends on the squared magnetic field strength. A rectangular distribution would correspond to $\chi_R = \infty$. However, the statistical methods used, both the convergence of planes and the least squares method, artificially put all warp vectors in the same hemisphere, so that $\chi_R \neq \infty$. With Eq. (7) it is obtained $\chi_R = 1.3$. Therefore, for large standard deviations, close to the rectangular distribution, the quantity χ no longer represents the degree of randomness of the magnetic field. On the other hand there is also a minimum value of σ, which corresponds to the scattering introduced by two systematic errors. First, we are assuming that the magnetic field lies in the plane of the sky. This effect can be analytically quantified as

$$\chi_P = \frac{1}{4} \tag{8}$$

Second, we are assuming that the magnetic field and the rotation axis form an angle of 45°. It can be exactly demonstrated that

$$\chi_{45°} = 2\left[\left(\frac{\pi}{4}\right)^2 - \frac{1}{2}\right] \tag{9}$$

Both values are so close, that we will use simply $\chi_{45°}$, though the quantity χ introduced by systematic errors would be $\sim 2\chi_{45°}$. With a reduced number of data in the samples it is possible to obtain $\chi < 2\chi_{45°}$. This would mean that the magnetic field is noticeably ordered, but it would not be possible to give a precise value to characterize this ordering. Adapted versions of the chi square distribution have also been developed (now the distribution function of angles should not be Gaussian!) and they are explained and discussed in a submitted paper[14].

6. Results for Warps

Though results for 10 groups of galaxies have been reported[3], we will present here results from those richer groups, which are on the other hand the most representative ones. The first group includes all edge-on warped NGC galaxies in the list of Sanchez-Saavedra et al[13], with a right ascension $6^h < \alpha_0 < 18^h$, a declination $\delta_0 < 45^o$ and a recession velocity lower than 2000 km.s^{-1}. It contains 12 galaxies and it was considered in the previous exercises. It will be called the Virgo group, as it contains the Virgo cluster. The second one has $6^h < \alpha_0 < 18^h$, $\delta > 45^o$ and a velocity lower than 3000 km.s^{-1}. It contains 14 galaxies, and will be called the Ursae group. (As they lie in the direction of Ursae major and Ursae minor). The third group will be the whole sample of 39 galaxies in the sample of Battaner et al.[3], including those belonging to the Virgo and the Ursae groups. Results are shown in the following table:

	Virgo Group	Ursae Group	All Galaxies
α	342	299	289
δ	54	10	8
l	106	50	43
b	-5	-10	-2
σ	31	32	40
χ	0.56	0.53	0.99
Size	25	30	150
Constellation	Lacerta	Aquila	Aquila

Where α and β are the equatorial coordinates of the direction of the mean magnetic field. l and b are the galactic coordinates of the same vector. The constellation around this direction is also written. σ is the standard deviation and χ the degree of non-uniformity (see text). All angles are in degrees and the characteristic sizes in Mpc. The direction of the magnetic field responsible for the warp of the Milky Way, corresponds to Hercules[2], very close to that of the Virgo group, and very close to the extragalactic magnetic field around the Andromeda nebula. We see that the value of χ for the Virgo and Ursae groups is slightly greater than $2\chi_{45^o}$, *i.e.* the field is very uniform, as the scattering of the points is mainly due to systematic errors inherent to the method of estimating the individual extragalactic magnetic field direction. Therefore, in cells of characteristic size of about 25 Mpc, the extragalactic magnetic field may be rather uniform. This conclusion is confirmed by the complete results[3]. However, when larger scales are considered, of the order of 150 Mpc, the magnetic field becomes tangled. For an even larger scale one must expect a completely random magnetic field, in agreement with the cosmological Principle of Isotropy.

7. Angular Momenta

The analysis of the possible warping coherence has been attempted by few researches, as the problem has recently arisen (at least under the magnetic field interpretation). However the analysis of the angular momentum coherence is very old. Galaxy formation is one of the less known cosmological topics and the existence of alignments could provide an important constraint. Reinhardt[15] discussed the possibility of alignments by magnetic fields. Oort[16] considered this possibility as a fossil eddy. Some models of galaxy formation predict such a coherence[17] and others a random distribution[18]. Therefore, noticeable attention has been paid to the observational detection of this effect. See the work of Garrido[19] for a review. A clear evidence of angular momenta coherence has not been found, although this possibility is not to be rejected. Therefore, we decided to use the statistical methods described in this work for a reexamination of the problem. We did not find a clear evidence of coherent alignments, in agreement with the hierarchical clustering theory of galaxy formation[18]. However in the Coma supercluster such an alignment was found in agreement with Hawley and Pebbles[20].

References

1. J. Binney, in *Dynamics of disc galaxies*, ed. Bjorn Sundelius, (Goteborgs University and Chalmers University of Technology, Sweden, 1991).
2. E. Battaner, E. Florido and M.L. Sanchez-Saavedra, *Astron. Astrophys.* **236** (1990) 1.
3. E. Battaner, J.L. Garrido, M.L. Sanchez-Saavedra and E. Florido, *Astron. Astrophys.* **251** (1991) 402.
4. L.S. Sparke and S. Cassertano, *Mon. Not. R. Astron. Soc.* **234** (1988) 873.
5. A. Toomre and J. Toomre, *Astrophys. J.* **178** (1972) 623.
6. E. Battaner, E. Florido and M.L. Sanchez-Saavedra, in *Warped disks and inclined rings around galaxies*, ed. S. Cassertano, P. Sackett and F. Briggs, (Cambridge Univ. Press ,1991).
7. E. Florido, M. Prieto, E. Battaner, E. Mediavilla and M.L. Sanchez Saavedra, *Astron. Astrophys.* **242** (1991) 301.
8. E. Florido, E. Battaner, A. Gros, M. Prieto and E. Mediavilla, *Astrophys. Space Sci.* **190** (1992) 293.
9. T. Sasaki, *Publ. Astron. Soc. Japan* **39** (1987) 849.
10. T. Ichikawa and T. Sasaki, *Proc. second asian-pacific regional meeting on Astronomy*, ed. B. Hidayat and M.W. Feast (Tira Pustaka Pub. House, Jakarta, 1984)
11. J. Guibert, J. Lequeux and F. Viallefond, *Astron. Astrophys.* **68** (1978) 1.

12. A. Bosma, *Warped disks and inclined rings around galaxies*, ed. S. Cassertano, P. Sacket and F. Briggs (Cambridge Univ. Press, 1991).
13. M.L. Sanchez-Saavedra, E. Battaner and E. Florido, *Mon. Not. R. astr. Soc.* **246** (1990) 458.
14. J.L. Garrido, E. Battaner, M.L. Sanchez-Saavedra and E. Florido, *submitted to Astron. Astrophys.* (1992).
15. M. Reinhardt, *Astrophys. Space Sci.* **10**,(1971) 363.
16. J.H. Oort, *11th Solvay Conference on the Structure and evolution of the Universe*, (1958) 163.
17. A. Dekel, *Astrophys. J.*, **298** (1985) 461.
18. S.F. Shandarin, A.G. Dorohkevich and Ya B. Zeldovich, *Soviet Phys.* **26** (1983) 46.
19. J.L. Garrido, J.L., *Ph.D.Thesis*, (Univ. Granada), in preparation.
20. D.L. Hawley and P.J.E. Peebles, *Astron. J.* **80** (1975) 477.

ON THE STRUCTURE OF QUASARS

ENRIQUE PÉREZ
Instituto de Astrofísica de Canarias
38200 La Laguna, Tenerife, Spain

ANDREW ROBINSON
Institute of Astronomy
Madingley Road, Cambridge CB30HA, UK

and

LUIS DE LA FUENTE
Instituto de Astrofísica de Canarias
38200 La Laguna, Tenerife, Spain

ABSTRACT

According to the reverberation model for broad emission line variability in quasars, the emission line luminosity at a given epoch maybe expressed formally as the convolution of the input ionizing continuum light curve with a transfer function, the form of which depends on the properties of the emitting material. In this lecture, we calculate transfer functions for several simple models of the broad line region in quasars characterized by various geometrical configurations and velocity fields. We dicuss how the structure of the transfer function is related to the model parameters and thereby show how several important properties of the emitting region, such as its size, geometry and velocity field, may be inferred. The velocity resolved transfer function is potentially a very powerful technique to dilucidate between competing models of quasars, such as those dominated by a supermassive black hole or those dominated by the formation of very massive stars.

1. Background

Quasars rank among the most spectacular and enigmatic objects studied by astronomers. In 1962 Schmidt and Green identified the emission lines in the spectra of two peculiar quasi-stellar objects which had been first discovered as strong radio sources. The fact is that the spectra are redshifted, and this can be explained by means of Doppler shifting if the objects are moving away from the observer. Two conflicting ideas were put forward to explain the nature of the red shift in these objects (of which soon many more were identified). In the first, the quasar is relatively nearby and it is assumed to have been ejected violently from the nucleus of a nearby galaxy. In the second, the velocity shift (which is *always* a red shift) is interpreted in terms of the general movement of expansion of the Universe (which Hubble had discovered in 1924); this implied that quasars are far away (as given by their redshift).

This controversy was the most important within the astronomical community during the mid 60's to mid 70's. Today, most astronomers believe that the Universe is expanding and that quasars are the bright nuclei of distant galaxies.

The first quasars discovered have a large apparent brightness which, at the distances implied by their redshift, translates into a huge *intrinsic luminosity*, often larger than the integrated luminosity of its host galaxy. It was soon noticed that quasars luminosity is variable with time, from where it can be deduced the size of the emitting volume; this volume works out to be quite small, which in turn implies that the *energy density* is also very large. This posed a serious problem, as in the 60's the only important cosmic source of energy was due to thermonuclear reactions, responsible for the generation of energy in the interior of stars. The theory of stellar evolution was not enough advanced at the time to produce stars sufficiently massive and hot to explain the luminosity and energy density observed in quasars.

In 1969 Lynden-Bell published the idea wherein gravity is the source of energy in quasars. In order to make it work, an intense gravitational field was needed, and it was proposed that this is generated by a supermassive ($\sim 10^8 M_\odot$ (*)) black hole. During the past 20 years this picture has been developed and the possible rôle played by stars in quasars was relegated to the oblivion.

From the mid 80's, several groups have started to consider again the rôle that stars may play in the quasar phenomenon. This is partly due to the significant development that the theory of stellar evolution has undergone during the ensuing years since quasars were discovered. A particular model stands as a real alternative to what has been the accepted wisdom during these two decades. Roberto Terlevich and co-workers argue that in the high metallicity, dense nuclear regions of a galaxy, massive stars ($\geq 100 M_\odot$) can form which, in their evolution through different phases, could explain most of the observed properties of quasars.

It is then important that we should understand the mechanism at work in quasars. Is there a supermassive black hole responsible for the conversion of gravitational into luminous energy or, on the contrary, is the nuclear fusion in the interior of very massive stars responsible for producing the observed luminosity? Our understanding of the Universe and its evolution will be different depending upon the answer we may give to this question. We should notice that although it is still not clear that such massive stars exist anywhere, at least we do know that stars exist! However, the existence of even stellar type black holes is still in the balance.

Because we cannot look directly into the interior of quasars and find what their structure might be, we have to develop an alternative procedure. Fortunately, it was observed early that the spectrum of quasars is variable with time; not only the overall luminosity level of the continuum and the intensity of the emission lines, but the profile of the emission lines is also time dependent. It is further observed that the emission lines (which can be as broad as more than $10^4 km\ s^{-1}$) vary with a time lag in response to continuum changes; this is a direct evidence that the gas surrounding the energy source is photoionized by the continuous spectrum. We can then use the variations in the ionizing continuum source as a search light with which

(*) $M_\odot$ is the symbol for the mass of the Sun

to probe into the structure of the surrounding ionized regions[1].

2. The Problem

The variability of the broad emission lines in quasars is widely interpreted in terms of what might be called the "point-source reverberation model" [1,2]. In this model, fluctuations in the ionizing luminosity of the central continuum source drive corresponding variations in the emission lines arising in the surrounding Broad Line Region (BLR[(*)]). It is assumed that the continuum source is point-like compared to the BLR and that the line intensity produced by any cloud in the BLR is linearly related to the local ionizing photon flux. Under the physical conditions prevalent in the BLR (*i.e.* high particle and radiation energy densities), the timescales ($\sim$ hours) on which the ionization and thermal structure of a given cloud adjust to changes in the local ionizing radiation field are certainly small compared to the light crossing time of the BLR ($\sim$ days). The response of the system will therefore be dominated by "reverberation effects" resulting from differences in photon flight paths between the continuum source, various points in the BLR and the observer (see figure 1a).

This is a problem of Information Theory and belongs to a general class very common in sciences, where we have an input function (our ionizing continuum source) which is modified by a transfer function (which embodies the properties we want to find out of our system), to result in an output (the observed quasar spectrum); $O = I * T$. The method explained by Dr. Cabrera elsewhere in this book, attempts to extract some information (e.g. the attractor dimension) of the studied system, T, or the input function, I, by analizing the behaviour of the time series alone, given by O; such method ignores one of the three variables. For example, from the analysis of stratigraphic data, O, it is possible to study the behaviour of the solar activity, I, without using any particular model for the Earth response, T. Other methods of solving this equation use two of the variables to find the properties of the third. In many general systems (as in economic models, population growth, etc.), we are interested in predicting the output from the input by modelling T with some adequate hypotheses. In our case, we know I and O (from observations) and want to find out about T; this is complicated by the fact that we have to invert the equation.

In this model, the relationship between the emission line flux density integrated over the BLR, $L(t, v^{||})$ (where $v^{||}$ is the line-of-sight emission velocity), and the continuum luminosity, $C(t)$, at time t, may be represented by[3]

$$L(t, v^{||}) = \int_0^t T(t - \tau, v^{||}) C(\tau) d\tau, \qquad (1)$$

(*) BLR is the accepted reference for the region of ionized gas responsible for the spectrum that we observe. Because this region is closest to the ionizing continuum source, we expect that the study of its properties will shed light on the nature of the quasar phenomenon.

where $T(t, v^{\parallel})$ is a transfer function which maps C into L. The form of the transfer function is determined by various properties of the BLR, such as the line emissivity distribution, geometrical configuration and velocity field. Thus, as pointed out by Blandford & McKee[3], given the line and continuum light curves, $L(t, v^{\parallel})$ and $C(t)$, the structure and dynamics of the BLR may, in principle, be "mapped" by inverting equation 1 for $T(t, v^{\parallel})$. These authors proposed to invert Eq. (1) by transforming into Fourier space and dividing by the transform of C, then taking the inverse transform; but, in practice, this method is very sensitive to the continuum light curve power spectrum. A much more successful attempt is that used by Keith Horne, who employs a technique based on the Maximum Entropy Method deconvolution.

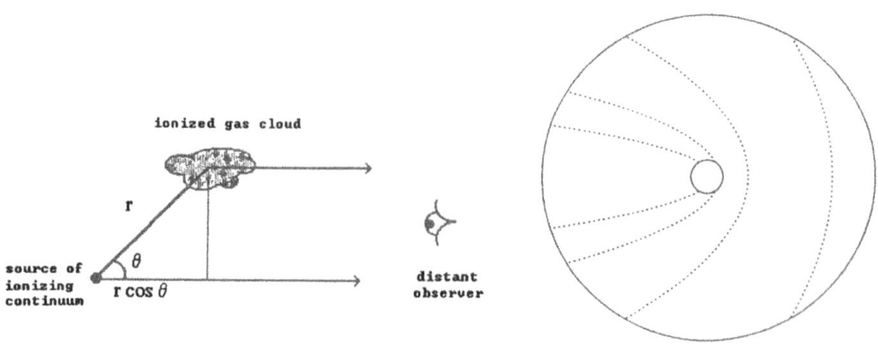

Figure 1: (a) The time lag, seen by a distant observer, between a continuum event and the emission line photons that this event induces when it ionizes a cloud of gas at a distance r, is given by the difference in the light path $r(1 - \cos\theta)/c$. (b) A BLR filled with gas clouds between an inner radius, r_0, and an outer radius, r_s. Four continuum pulses, which in the quasar reference frame have spherical symmetry, are seen by a distant observer as parabolic surfaces.

We present here a sample of transfer functions which correspond to simple models of the BLR characterized by various combinations of geometrical configuration and velocity field. These TFs are intended to give an insight into the reverberation signature produced by different classes of BLR model and thereby serve as a basis for the interpretation of reverberation maps recovered from observations. We show how the geometry, size and kinematics of the BLR can be obtained in a straightforward way from an analysis of the transfer functions.

3. The Method

The model TFs presented here are based on the conventional point-source reverberation picture, with the BLR described as an ensemble of discrete clouds distributed within some global geometrical configuration about the continuum source[6]. The TFs were generated using the computer code OBFRAM[4,5,6], which simulates the response of such systems to ionizing continuum fluctuations. In this study, the continuum light curve is taken to be a narrow top-hat function such that, effectively $C(t) \to \delta(t)$ and the line response is $L(t, v^{\|}) \equiv T(t, v^{\|})$ (equation 1).

The cloud system has inner and outer radii r_0 and r_s, respectively and is characterized by a radial volume emissivity distribution of the form $\epsilon(r) = \epsilon_0(r/r_0)^{\beta}$. All models presented here have inner and outer radii $r_0 = 10$ light-days (ld), $r_s = 100$ ld and $\beta = -3$. Three different geometrical configurations are considered: spherical, cylindrical and biconical. The models presented here each contain 25000 clouds.

The velocity field, $v(r)$, is assumed to have a radial dependence of the form

$$v(r) = v_0 \left(\frac{r}{r_0}\right)^p, \tag{2}$$

where v_0 is the modulus of the velocity field at r_0. The minimum velocity for each model (*i.e.* v_0 for $p > 0$) is always chosen to give a maximum velocity of $10^4 km\ s^1$.

At a given time t as measured by the distant observer, the continuum δ-pulse is received only by clouds distributed on the parabolic iso-delay surface,(*)

$$t = r(1 - \cos\theta), \tag{3}$$

where we take the speed of light $c = 1$. This surface may be regarded as the 'light front' corresponding to the propagation of the δ-continuum pulse through the BLR (figure 1b). Thus, the instantaneous line response at t is obtained by integrating the emission from these clouds (as a function of line-of-sight velocity) and the transfer function is then simply the variation of the line response with t for $0 \le t \le t_s$, where $t_s = 2r_s$, the BLR light crossing time.

3.1. Spherical Geometry

Let us consider the case in which the cloud system is embeded in a spherical geometry with a radial outflow velocity field. Transfer functions for radial velocity laws characterized by values of $p = 1$ and 2 are shown as grey-scale intensity plots in the $v^{\|} - t$ plane in Fig. 2. In this representation, the grey-scale image represents the evolution of the line profile as the δ-continuum pulse propagates through the BLR. At a given delay the grey levels define the instantaneous line profile. It is clear that

(*) It can be shown that spherically symmetric iso-delay surfaces in the source reference frame are viewed by a distant observer as parabolic.

the shape of the 2-DTF in the $v^{\|} - t$ plane depends strongly on the radial velocity dependence. We now consider the functional form of the outline of the TF in the $v^{\|} - t$ plane and discuss how it relates to the parameters defining the model BLR.

The velocity component projected along the line of sight, $v^{\|}$, is given by

$$v^{\|}(r) = -v(r)\cos\theta, \tag{4}$$

where θ is the angle between the radius vector and the line of sight. The minus sign indicates outflow, with negative projected velocities corresponding to blueshifts. Combining equations 3 and 4 we obtain

$$v^{\|}(r) = -v_0\left(\frac{r}{r_0}\right)^p + \frac{v_0}{r}\left(\frac{r}{r_0}\right)^p t. \tag{5}$$

for the relationship between projected velocity and time delay at a fixed radius. For an asymptotic thin shell of radius, r, therefore, the iso-delay surface maps onto a straight line in the $v^{\|} - t$ plane beginning at $(t, v^{\|}) = (0, -v_0(r/r_0)^p)$ and ending at $(2r, v_0(r/r_0)^p)$. The form of the 2-DTF for a thick shell can now be understood as a sequence of such lines corresponding to successive values of r between r_0 and r_s.

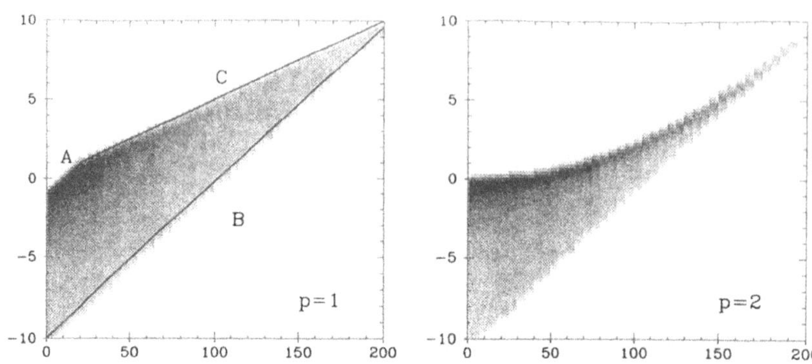

Figure 2: 2-D Transfer Functions for two different radial velocity fields in a spherical geometry, with p=1, 2. The horizontal axis represents time in days, while the vertical axis is velocity in units of 10^3 km/s. The greyscale represents the intensity of the line emission.

The outline of the 2-DTF is essentially defined by three lines. The two straight lines, marked **A** and **B** in Fig. 2, correspond to the inner and outer boundaries of the shell: $v^{\|} = v_0(1-t/r_0)$ and $v^{\|} = v_0(r_s/r_0)^p(1-t/r_s)$, respectively. The curved upper envelope, C, is the locus formed by the end-points of the lines associated with intermediate radii. It corresponds to the "leading edge" of the light front, where $t = 2r$, and its shape is therefore given by $v^{\|} = v_0(t/t_0)^p$. It follows that the precise form of the radial velocity law can, in principle, be obtained from the upper envelope, C, of the 2-DTF—which determines the evolution of the red wing of the line profile—for delays $t > t_0$.

3.2. Disk Geometry

In Figure 3 we present grey-scale maps of the velocity-resolved TFs for disk models in which the emitting material moves in circular Keplerian orbits ($p = -1/2$) confined to the disk plane. Since the 2-DTF reduces to a horizontal line at $v^{\|} = 0$ when the disk inclination $i = 0°$ we show only the $i = 45°$ and $90°$ cases.

In the case of a disk, the flight path of a line photon emitted in the plane of the disk is increased by a factor of $\sin i$. Thus, the projection of the iso-delay surface onto the disk plane is

$$t = r(1 - \cos\theta \sin i). \tag{6}$$

and the projected velocity is given by

$$v^{\|}(r) = v(r) \sin\theta \sin i. \tag{7}$$

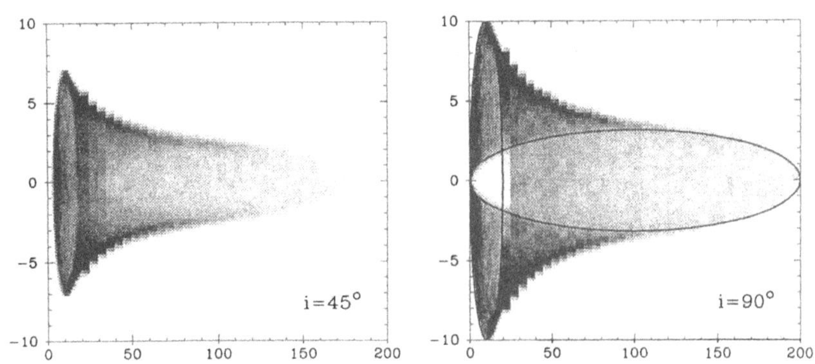

Figure 3: 2-D Transfer Functions for a Keplerian disk inclined at $i = 45°$ and $90°$. The solution for a face-on disk is trivial and is not shown. Axes as in Figure 2.

Combining equations 6 and 7, it may be seen that, for a fixed radius (*i.e.* a ring), the TF forms an ellipse in the $v^{\|} - t$ plane,

$$\frac{v^{\|2}}{(r/r_0)^{2p}\, v_0^2 \sin^2 i} + \frac{(t-r)^2}{(r \sin i)^2} = 1. \tag{8}$$

The ellipse is centred at $(t, v^{\|}) = (r, 0)$ and has semi-axes spanning $v_0(r/r_0)^p \sin i$ and $r \sin i$ in projected velocity and delay, respectively. Thus, as the radius of the ring increases, the velocity axis decreases in length (for $p = -1/2$) while the delay axis increases and the centre is displaced to larger delays. The series of ellipses corresponding to a sequence of incremental rings ranging in radius from r_0 to r_s produces the bell-shaped outline of the disk TF's in Fig. 3. The ellipses corresponding to r_0 and r_s are shown superimposed on the grey-scale plot for $i = 90°$.

The inclination of the disk affects the 2-DTF by reducing its dimensions along each axis. As the disk is tilted away from the edge-on position (*i.e.* to smaller values of i), the ellipse corresponding to each incremental ring has both semi-axes reduced by factors $\sin i$. For example, at $i = 45°$, the ellipse corresponding to r_0 spans $v^{\|}_{max}(r_0) = \pm 7\, 10^3 km\ s^{-1}$ in projected velocity and 7 days in time. Since the ellipse is still centred at $t = r_0$, the beginning of the response is shifted slightly along the delay axis; the smallest delay at which the TF has power being $t = r_0(1 - \sin i) = 3$ dy (Fig. 3). Similarly, the response associated with the outer edge of the disk, r_s, begins after a delay of $t = 30$ dy (rather than $t = 0$ for an edge-on disk) and ends at $t = r_s(1 + \sin i) = 170$ dy (instead of $t_s = 200$ dy). Note also that the zone of avoidance, which can be seen at $i = 90°$ as a prominent "clapper" in the bell-shaped TF, is filled in as the inclination is decreased and the TF is "compressed" by factors $\sin i$.

The maximum projected velocity, $v^{\|}_{max} = v(r) \sin i$, is attained at each radius after delay $t = r$, given by the centre of the corresponding ellipse. Thus, the outline of the 2-DTF for $r_0 < t < r_s$ is defined by the locus corresponding to the points $(t, v^{\|}) = (r, v^{\|}_{max})$, which is given by $v^{\|}_{max} = (2t/t_0)^p \sin i$, *i.e.* $v^{\|}_{max} \propto t^{-1/2}$ in this case.

3.3. Biconical Geometry

In the biconical configuration, the emitting clouds are confined to two opposing cones, aligned on a common axis passing through the continuum source. The axis is inclined at an angle i to the line-of-sight and both cones have the same opening 1/2-angle α.

As an illustration of the velocity-resolved transfer function produced by the biconical configuration, we consider the case of an accelerating radial outflow with $p = 1$. Examples of these TF can be found in reference [7]. The approaching and receding flows produce a distinctive bifurcated structure with the two arms merging for $i \geq \frac{\pi}{2} - \alpha$. The arms are narrower and their angular separation greater for the model in which the cone angle is small ($\alpha = 5°$). This structure can easily be understood if we regard the biconical TF as a special case of the corresponding spherically symmetric model in which allowed regions in the $v^{\|} - t$ plane are defined by the angles i and α.

For a radial flow confined to a cone, the projected velocity for an inclination i, is given by

$$v^{\|}(r) = -v(r)(\cos i \ \cos\theta + \sin i \ \sin\theta \ \sin\phi), \tag{9}$$

where the azimuthal angle, ϕ, and the zenith angle, θ, are measured with respect to the cone axis and have ranges $0 \leq \phi \leq 2\pi$ and $0 \leq \theta \leq \alpha$, respectively. The outline of the TF can be obtained by noting that for each cone, the limiting values of $v^{\|}$ occur when $\sin\phi = \pm 1$, *i.e.*

$$v^{\|}_{lim} = -v(r)\cos(i \pm \theta), \tag{10}$$

while the intersection of the iso-delay surface with the $\phi = (\frac{\pi}{2}, \frac{3\pi}{2})$ plane is

$$t = r(1 - \cos(i \pm \theta)). \tag{11}$$

The equations of the lines defining the TF outline in the $v^{\|} - t$ plane can now be obtained in a straightforward way from equations 10 and 11. For the approaching cone we have

$$v^{\|max}{}_{min} = -v_0 \left(\frac{2t}{t_0}\right)^p \frac{\cos(i \mp \alpha)}{(1 - \cos(i \mp \alpha))^p}. \tag{12}$$

Similarly, for the receding cone,

$$v^{\|max}{}_{min} = v_0 \left(\frac{2t}{t_0}\right)^p \frac{\cos(i \mp \alpha)}{(1 + \cos(i \mp \alpha))^p}. \tag{13}$$

4. Discussion

Interpreted according to the point-source reverberation model as represented by equation 1, the variability of the broad emission line intensities and profiles is given by the convolution of the ionizing continuum light curve $C(t)$ with transfer functions similar to those presented here. The evolution of the line profile will clearly be strongly influenced by the shape of the velocity-resolved transfer function. Indeed, simulations employing "realistic" light curves suggest that in some situations it may be possible to discern the basic structure of the 2-DTF directly from a 'trailed' sequence of spectra and thus distinguish between various broad classes of kinematical or geometrical model [6,7,8].

The calculations in this paper are based on a very simple version of the standard model of the BLR, involving an ensemble of isotropically emittinglinear reprocessors distributed-around a point source of ionizing continuum. These calculations illustrate the basic structure of the transfer functions corresponding to a variety of geometrical and kinematical models. However, it is possible that a number of effects, not taken into account in this work, may result in significant modifications. Thus, the clouds may emit anisotropically, General and Special Relativistic effects may be important, scattering processes may create an effectively extended continuum source or increase photon flight times within the BLR. Furthermore, few emission lines are, in reality, well approximated by linear reprocessing and therefore to calculate accurate TFs it will be necessary to compute the line emissivities explicitly. These issues will be addressed in future work.

A perhaps even more important task is to calculate theoretical transfer functions for 'non-standard' scenarios, such as the wind/shock model of Perry & Dyson[9] or the supernova model proposed by Terlevich *et al.* [10]. In the latter model, Terlevich *et al.* dispense entirely with the conventional notion that AGN are powered by accretion onto supermassive black holes and propose instead that they can be explained in terms of the evolution of a cluster of massive stars. The BLR is formed by supernovæ remnants when their kinetic energy is rapidly thermalized by dense interstellar gas. An analogue of the velocity-resolved transfer function can, in principle, be computed for such a system and it might be expected that it will have distinctive features which may be directly compared with transfer functions for point-source models such as those presented here. Potentially, therefore, the 2-DTF is the key to discriminating decisively between starburst and black hole models for AGN.

Acknowledgements

We acknowledge the software provided by the Starlink Project which is funded by the UK SERC. AR is grateful to the Royal Society for financial support.

References

1. B.M. Peterson, *Publs. astr. Soc. Pacif.* **100** (1988) 18.
2. M.V. Penston, in *Variability of active galactic nuclei*, ed. H.R. Miller & P.J. Wiita (Cambridge University Press, 1991) 343.
3. R. Blandford and C. F. Mckee, *Astrophys. J.* **255** (1982) 419.
4. A. Robinson and E. Pérez, *Mon. Not. R. astr. Soc.***244** (1990) 138.
5. A. Robinson, E. Pérez and L. Binette *Mon. Not. R. astr. Soc.***246** (1990) 349.
6. E. Pérez, A. Robinson and L. de la Fuente, *Mon. Not. R. astr. Soc.***255** (1992) 502.
7. E. Pérez, A. Robinson and L. de la Fuente, *Mon. Not. R. astr. Soc.***256** (1992) 103.
8. W.F. Welsh and K. Horne, *Astrophys. J.* **379** (1992) 586.
9. J. J. Perry and J.E. Dyson, *Mon. Not. R. astr. Soc.***213** (1985) 665.
10. R. J. Terlevich, G. Tenorio-Tagle, J. Franco and J. Melnick, *Mon. Not. R. astr. Soc.***255** (1992) 713.

Contributions

SIMULATION OF NONLINEAR DIFFUSION PROCESSES

J. M. LOPEZ and M. A. RODRIGUEZ
Departamento de Física Moderna, Universidad de Cantabria,
Av. de los Castros s/n, 39005- Santander, Spain

ABSTRACT

We consider models of nonlinear diffusion used in the study of flow through porous media. We compare the results obtained from scaling and renormalization group theories with random walk simulations. We conclude that the commonly used Barenblatt equation is dominated by inherent fluctuations.

1. Introduction

Models with nonlinear diffusion are commonly used to study flow in porous media[1]. Among others, the Barenblatt model has recently received a lot of attention because it is the object of the first nonstatistical problem solved by renormalization group techniques[1]. In one dimension the Barenblatt equation is written:

$$\frac{\partial \Phi}{\partial t} = D\Phi_{xx} + \epsilon\theta\left(\Phi_{xx}\right)\Phi_{xx} \tag{1}$$

where θ is a Heaviside function:

$$\theta(x) = \begin{cases} 0, & \text{if } x > 0; \\ 1, & \text{if } x < 0 \end{cases}$$

and Φ_{xx} is the second spatial derivative.

Three facts characterize this Barenblatt equation: First, it is not conservative and as a consequence it is not a purely diffusive model. Second, the nonlinear term has the form of a step function which is typical of systems exhibiting avalanche dynamics and self-criticality[2]. Finally the nonlinearity refers to the curvature of the propagating front.

In this paper we study the consequences of the above mentioned characteristics as concerns fluctuations. It is well known that this model is stable by instantaneous perturbations but nothing is known about noisy perturbations. Since the nonlinear coefficient depends strongly on the sign of the curvature, fluctuations, even of very low intensity, can dramatically affect the dynamics of the propagating front. Noise effects are more related to the correlation length than to the intensity of fluctuations. In what follows we analyse this problem simulating conservative and nonconservative models with the use of generalized random walks. Fluctuations appear as inherent to the statistical process with relative intensity proportional to $(\sqrt{N})^{-1}$ where N is the number of walkers.

It is worth remarking that simulation of nonlinear propagating fronts by means of random walkers is not a usual technique. Hence our work is focused on two points, on the one hand we study the stability of nonlinear diffusive models with respect to fluctuations and on the other hand we use these models as a benchmark for the application of random walk techniques in nonlinear diffusion problems.

2. The Models

We have studied the Barenblat model as well its conservative versions with random-trap (RT) and random-barrier (RB) diffusivities[3]. The corresponding equations are written in the form:

$$\frac{\partial \Phi}{\partial t} = \frac{\partial^2}{\partial x^2}\left(D + \epsilon\theta\left(\Phi_{xx}\right)\right)\Phi \qquad for\ RT \tag{2.a}$$

$$\frac{\partial \Phi}{\partial t} = \frac{\partial}{\partial x}\left(D + \epsilon\theta\left(\Phi_{xx}\right)\right)\frac{\partial}{\partial x}\Phi \qquad for\ RB. \tag{2.b}$$

They are genuine nonhomogeneous diffusion equations, modelling trapping or reflection with obstacles, respectively. As a consecuence $\Phi(x,t)$ can be thought of as being a probability density and the process can be simulated by means of random walks with probability dependent coefficients. The Barenblatt equation has not a direct physical interpretation but if we write Eq. (1) as:

$$\frac{\partial \Phi}{\partial t} = \frac{\partial}{\partial x}\left(D + \epsilon\theta\left(\Phi_{xx}\right)\right)\frac{\partial}{\partial x}\Phi - \epsilon\frac{\partial}{\partial x}\theta_x\left(\Phi_{xx}\right)\Phi + \epsilon\theta_{xx}\left(\Phi_{xx}\right)\Phi \tag{3}$$

it can be seen as a random barrier model with a density dependent drift and a term of creation-annihilation of particles. Simulation with random walkers is then more evident.

3. Scaling and Renormalization Group Analysis

In conservative cases (RT and RB) the scaling form $\Phi(x,t) \sim \ell^{-1}(t)\, f(x/\ell(t))$ with $\Phi_{xx}(x,t) = 0$ in $x = \pm\ell(t)$, holds. In order to show it we consider Eq. (2) in integral form with $\Phi(x,0) = \delta(x)$ as initial condition:

$$\Phi(x,t) = G(x,t) + \epsilon\int_0^t ds \int_{-\ell(s)}^{+\ell(s)} dy\ \frac{\partial^2 G(x-y,t-s)}{\partial y^2}\ \Phi(y,s) \qquad for\ RT \tag{4.a}$$

$$\Phi(x,t) = G(x,t) - \epsilon\int_0^t ds \int_{-\ell(s)}^{+\ell(s)} dy\ \frac{\partial G(x-y,t-s)}{\partial y}\ \frac{\partial \Phi(y,s)}{\partial y} \qquad for\ RB \tag{4.b}$$

where $G(x,t) \sim t^{-1/2} exp\left(-x^2/4Dt\right)$ is the Green function of the homogeneous diffusion equation. From these equations and the condition $\int_{-\ell(t)}^{\ell(t)} \Phi(x,t)dx = cte.$ imposed by the scaling form we deduce the long time evolution of $\ell(t)$ given by $\ell(t) \sim \sqrt{t}$. Substituting the scaling function with $\ell(t) \sim \sqrt{t}$ in Eq. (4) and operating we obtain:

$$f(z) = exp(\frac{-z^2}{4D}) + \epsilon\int_0^1 \frac{d\eta}{\eta}\int_{-1}^{+1} d\xi\ f(\xi)\ \frac{\partial^2 \tilde{G}(\xi,z,\eta)}{\partial \xi^2} \qquad for\ RT$$

$$f(z) = exp(\frac{-z^2}{4D}) + \epsilon \int_0^1 \frac{d\eta}{\eta} \int_{-1}^{+1} d\xi \; f(\xi) \; \frac{\partial^2 \tilde{G}(\xi, z, \eta)}{\partial \xi^2} \; -$$

$$- \epsilon \int_0^1 d\eta \left(f(\xi) \; \frac{\partial \tilde{G}(\xi, z, \eta)}{\partial \xi} \right)_{\xi=-1}^{\xi=+1} \qquad for \;\; RB$$

with $\tilde{G}$ given by

$$\tilde{G}(\xi, z, \eta) = (4\pi D(1-\eta))^{\frac{-1}{2}} exp\left(\frac{\left(z - \xi\sqrt{\eta}\right)^2}{4D(1-\eta)} \right).$$

These equations are well defined and guarantee the existence of a scaling function. Althought they can not be solved, the scaling gives the needed information about the system evolution. The diffusion coefficient can be obtained through the calculation of $< x^2 >$ from Eq. (4). We multiply by x^2, integrate on both sides of the equations and use the scaling function, obtaining for long times $< x^2 >= 2D_{eff}t$, with:

$$D_{eff} = D + \epsilon A \qquad for \;\; RT \tag{5.a}$$

$$D_{eff} = D + \epsilon(A - 2B) \qquad for \;\; RB \tag{5.b}$$

where the constants A and B are related to the scaling function: $A = \int_{-1}^{+1} f(\omega)\, d\omega$ and $B = f(1)$.

The main conclusions are that for long times the behaviour is diffusive with a renormalized diffusion coefficient depending linearly on ϵ and the discontinuity point in the diffusion coefficient $\ell(t)$ behaves like a diffusion length $\ell(t) \sim \sqrt{t}$. It is worth mentioning that these results are not perturbatives.

As we said in the introduction the Barenblatt equation is the first nonstatistical problem solved by renormalization group techniques. The nonconservative part is the origin of the anomalous dimension α found by this technique[1], as can be easily seen by applying the same scaling that in the conservative cases. Indeed, operating like before (RT and RB cases) we obtain from Eq. (1) an integral equation for the scaling function $f(z)$ with well defined terms proceeding from the diffusive and drift parts and a divergent term given by

$$\int_0^1 \frac{d\eta}{\eta} \left(\tilde{G}(\xi, z, \eta) \; f\prime(\xi) \right)_{\xi=-1}^{\xi=+1}$$

proceeding from the nonconservative term. This divergence is avoided by using a different scaling given by

$$\Phi\,(x,t) \sim t^{-\left(\frac{1}{2}+\alpha\right)} f\left(\frac{x}{\sqrt{t}}\right) \tag{6}$$

with $\alpha \neq 0$. The value of α has been found in Ref(1) with classical renormalization techniques; it is $\alpha = (2\pi e)^{-1/2}\epsilon + O\left(\epsilon^2\right)$.

4. Simulations

We have performed simulations on a 1-D network with N initial random walkers, taking Δ the space between sites and h the evolution time step. In fact we simulate a master equation that in the continuous limit $\Delta, h \to 0$ yields exactly the correct equations, Eq. (1) or (2).

The random-trap and random-barrier models are defined from general one-step Markovian master equations in a chain:

$$\frac{\partial P_n(t)}{\partial t} = \omega_{n+1}^- P_{n+1}(t) + \omega_{n-1}^+ P_{n-1}(t) - (\omega_n^- + \omega_n^+) P_n(t)$$

ω_n^+ (ω_n^-) is the transition probability per unit time for going from site n to site $n+1$ (site $n-1$). The RB and RT models differ on the symmetry requeriments imposed on the transition probabilities $\omega_n^\pm$. In the RB case one models a situation in which:

$$\omega_n^- = \omega_{n-1}^+ \equiv \omega_n$$

and in the RT model

$$\omega_n^+ = \omega_n^- \equiv \omega_n.$$

Then the master equations can be written as:

$$\frac{\partial P_n}{\partial t} = (1 - \hat{E}^-)(\hat{E}^+ - 1)\omega_n P_n \qquad for\ RT \tag{7.a}$$

$$\frac{\partial P_n}{\partial t} = (1 - \hat{E}^-)\omega_n(\hat{E}^+ - 1)P_n \qquad for\ RB \tag{7.b}$$

where we have introduced shifting operators $\hat{E}^\pm$ defined as $\hat{E}^\pm f_n = f_{n\pm1}$.

For the Barenblatt model this master equation is written:

$$\frac{\partial P_n}{\partial t} = (1 - \hat{E}^-)\omega_n(\hat{E}^+ - 1)P_n + \hat{H}\beta_n P_n + \hat{I} M_n P_n \tag{8}$$

where $\hat{H}$ and $\hat{I}$ are operators defined as:

$$\hat{H} \equiv (\hat{E}^+ - 1)(1 - \hat{\delta}) - (1 - \hat{E}^-)\hat{\delta}$$

$$\hat{I} \equiv 2\hat{\delta} - 1$$

$\hat{\delta}$ being a new operator given by

$$\hat{\delta} f_n = \begin{cases} 1, & \text{if } f_n > 0; \\ 0, & \text{if } f_n < 0 \end{cases}.$$

$P_n(t)$ is the probability of being at site n at the time t and ω_n, β_n, and M_n are transition probabilities.

The continuous limit can be easily performed by considering:

$$\Phi(x,t) \sim \frac{1}{\Delta} P_{\frac{x}{\Delta}}(t)$$

$$\hat{E}^\pm \sim exp(\pm\Delta\frac{\partial}{\partial x}),$$

substituting P and $\hat{E}^{\pm}$ in Eq. (7) and (8) and taking $\Delta \longrightarrow 0$ with

$$\Delta^2 \omega_{\frac{x}{\Delta}} \longrightarrow D(x)$$

$$\Delta \beta_{\frac{x}{\Delta}} \longrightarrow v(x)$$

$$M_{\frac{x}{\Delta}} \longrightarrow M(x).$$

Then the coefficients of the master equations (Eq. (7) and (8)) reproducing continuous equations (Eq. (2) and (3)) are well defined. In order to simulate the master equation (Eq. (8)) we must generalize the usual method, in which transition probabilities are constant, for a case where they are probability dependent. It is possible by considering from the very begining the simultaneous evolution of N random walkers with a random-barrier diffusive behavior, a drift and a probability of walker production or annihilation. The transition probabilities are dependent of the number of walkers at the corresponding site:

$$\omega_{\frac{x}{\Delta}} = \frac{D(x)}{\Delta^2} = \frac{D + \epsilon\theta(\Phi_{xx})}{\Delta^2}$$

$$\beta_{\frac{x}{\Delta}} = \frac{V(x)}{\Delta} = \frac{\epsilon\theta_x(\Phi_{xx})}{\Delta}$$

$$M_{\frac{x}{\Delta}} = M(x) = \epsilon\theta_{xx}(\Phi_{xx}).$$

Fluctuations arise due to the finite number of walkers. A measure of the intensity of these inherent fluctuations is $(\sqrt{N})^{-1}$, with N the number of walkers. Obviously these fluctuations are spatially uncorrelated. We can add some kind of correlation by smoothing the front in the course of the simulation. We present in tables (1) and (2) results of simulations for RB and RT models varying the number of walkers and the smoothing length. This corresponds to varying noise intensity and correlation lengths. One smoothing corresponds to averaging between nearest-neighbours $P_i \longrightarrow (P_{i-1} + P_i + P_{i+1})/3$. We have smoothed 1,2,3,8,16 successive times.

Smoothing length	0	1	2	8	16
1000 walkers	$1.936+5.141\epsilon$	$1.996+1.337\epsilon$	$1.950+2.668\epsilon$	$1.987+1.360\epsilon$	$2.013+1.176\epsilon$
5000 walkers	$1.949+2.343\epsilon$	$1.992+1.265\epsilon$	$1.988+1.491\epsilon$	$2.006+1.056\epsilon$	$2.016+0.899\epsilon$
10000 walkers	$1.959+2.197\epsilon$	$1.992+1.136\epsilon$	$1.987+1.446\epsilon$	$1.995+0.896\epsilon$	$2.000+0.842\epsilon$

Table 1: Random-barrier: linear fit of $2D_{eff}$ versus ϵ with several smoothing lengths and different number of walkers. ϵ varies in the interval (0,0.5) and $D = 1$.

Smoothing lenght	0	1	2	8	16
1000 walkers	$2.003+1.216\epsilon$	$2.027+0.624\epsilon$	$2.014+1.186\epsilon$	$1.992+1.370\epsilon$	$2.007+1.448\epsilon$
5000 walkers	$2.005+0.958\epsilon$	$2.015+0.771\epsilon$	$2.003+1.121\epsilon$	$2.009+1.166\epsilon$	$2.018+1.298\epsilon$
10000 walkers	$1.996+1.171\epsilon$	$2.000+0.793\epsilon$	$1.998+1.117\epsilon$	$2.000+1.191\epsilon$	$2.008+1.244\epsilon$

Table 2: Idem Table 1 for Random-trap.

In all cases we obtain the global behavior predicted by the scaling theory, Eq. (5). The process is diffusive with a diffusion coefficient linear in ϵ. Note that this result is not perturbative, ϵ is not always small. Then we can conclude that in conservative situations fluctuations of different intensities and correlation lengths are not relevant as concerning the global behavior. They only change diffusion coefficients and the linearity in ϵ is conserved.

Comparing both tables, we observe that the RB model is more sensitive to fluctuations than the RT one. In principle is not an expected result since the RB model seems to be more regular (detailed balance holds). However in this model symmetry is more easily broken by fluctuations since it involves two differents sites. For the same reason the smoothing with an even or odd number of sites gives very different results.

On the other hand we present in Figure 1 and 2 the evolution of $\int_{-\infty}^{+\infty} \Phi(x,t)dx$ for the Barenblatt equation. In this case we observe a great sensitivity to fluctuations. Instead of obtaining $t^{-\alpha}$, as the scaling predicts (see Eq. (7)), we obtain exponential decay. This decay depends strongly on the initial number of walkers. Also we have observed a high sensitivity to symmetry changes in the fact that odd or even smoothing gives linear growth or exponential decay respectively.

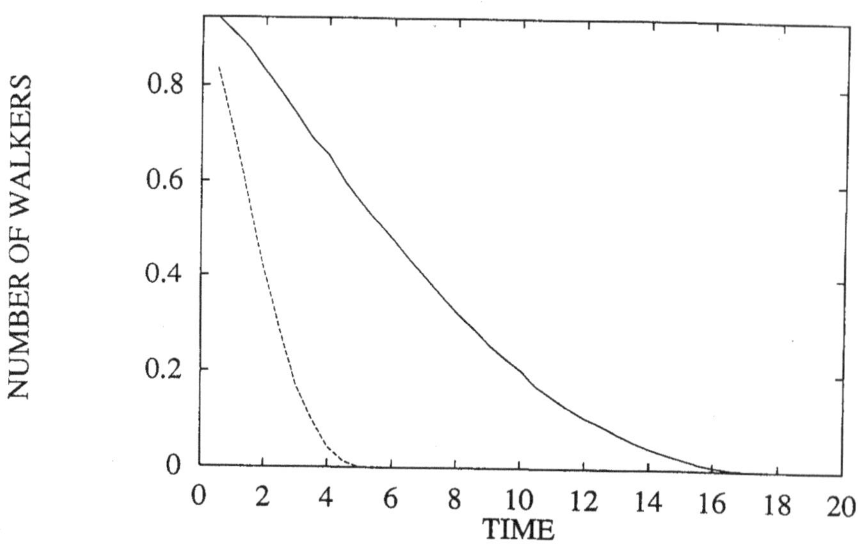

Figure 1: Evolution of $\int_{-\infty}^{+\infty} \Phi(x,t)dx$ with time, for $\epsilon = 0.01$ (solid line) and $\epsilon = 0.05$ (dotted line), without smoothing and with 10000 initial walkers.

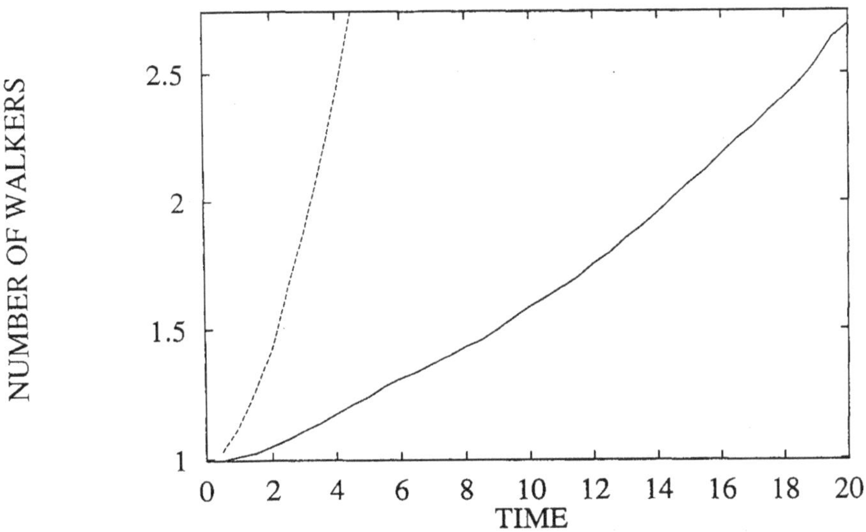

Figure 2: The same as Fig. 1 with nearest-neighbour smoothing one time and with 10000 initials walkers. $\epsilon = 0.01$ (solid line) and $\epsilon = 0.05$ (dotted line).

In conclusion, fluctuations are not relevant in conservative cases and simulations agree with the scaling theory, obtaining a diffusion coefficient linear in ϵ. Contrary to this, simulations of the Barenblatt equation denote a strong sensitivity to fluctuations. The limit of low fluctuations does not exist and fluctuations dominate the nonconservative part of model.

Acknowledgements

We are grateful to L. Pesquera for a careful reading of this contribution and suggestion for improvements. Also we acknowledge financial support from the Dirección General de Investigación Científica y Técnica Project No. PS90-0098.

References

1. N. Goldfeld, O. Martin, Y. Oono, and F. Liu, *Phys.Rev.Lett.* **64** (1990) 1361; L. Chen, N. Goldenfeld, and Y. Oono, *Phys.Rev.A* **44** (1991) 6544.
2. P. Bántay and I.M. Jánosi, *Phys.Rev.Lett.* **68** (1992) 2058.
3. J.W. Haus and K.W. Kehr, *Phys.Rep.* **150** (1987) 263; N.G. Van Kampen, *Stochastic Processes in physics and chemistry* (NH PL 1981), Chap. 10.

4. C. Van Den Broeck, in *Stochastic processes applied to physics* , ed. L. Pesquera and M. A. Rodriguez (World Scientific, Singapore 1985); C.A. Whitney, *Random processes in physical systems* (Wiley Interscience, 1990), p. 192.

MULTIFRACTAL ANALYSIS OF THE GROWTH PROBABILITY DISTRIBUTION IN ZINC ELECTRODEPOSITION

J. MACH, P.P. TRIGUEROS, F. MAS and F. SAGUES
Deptartamento de Química-Física, Universidad de Barcelona
c/Martí i Franquès 1, E-08028 Barcelona, Spain

ABSTRACT

The scaling properties of the growth probability distribution (GPD) for fractal-like electrodeposits grown two-dimensionally are analyzed. There are two ways to compute the GPD, assuming that the growing dynamics is Laplacian controlled, and from the evaluation of the normal velocity of the growing electrodeposit surface. Both methods are compared using the multifractal analysis.

1. Introduction

Scientists working in various fields of physics, chemistry, biology, etc. have reported that many of the structures observed in their experiments possess a special kind of geometrical complexity. This awareness is largely due to the activity of Benoit B. Mandelbrot[1-3] who drew attention to the particular geometrical properties of some objects such as the shore of continents, the branches of trees, or the surface of clouds. Mandelbrot coined the name *fractal* for these complex shapes to express the fact that they can be characterized by a *non-integer* (fractal) dimensionality. [(+)]

An important field where fractals have been observed is that of far-from-equilibrium growth phenomena which are common in many fields of science and technology. Examples of such processes include dendritic solidification in an undercooled medium[4], viscous fingering when a viscous fluid is injected into a more viscous one[5], dielectric breakdown[6], and electrodeposition of ions onto a electrode[7-12].

We report here on the process of the quasi two-dimensional electrodeposition of zinc. Electrodeposition constitutes a fascinating example of growth phenomena under non-equilibrium conditions[13]. As compared to other pattern growing phenomena, electrodeposition offers the unique opportunity to show very different morphologies, from largely disordered fractal to apparently well-ordered anisotropic dendritic structures. Here, we concentrate on the open-fractal morphology which has been commonly invoked as a paradigmatic experimental realization of a popular simulation routine of non-equilibrium growth processes known as diffusion-limited-aggregation (DLA)[14]. However, whether or not electrodeposition is generically a diffusion-limited process is, to our understanding, an open question. Indeed, the

(+) Cf. the paper by J.F.Fernandez in this volume, and references therein (editors note)

resulting effective field governing the motion of metal ions is strictly the electrochemical potential, in terms of which one establishes the well-known Nernst-Planck equation. Under quasi-static approximation, such a relationship adopts a Laplace form commonly invoked as the macroscopic realization of the DLA simulation code. Therefore, it is reasonable to try to check further the Laplacian nature of real DLA-like electrodeposited patterns in terms of the scaling properties of their growth probability distribution (GPD).

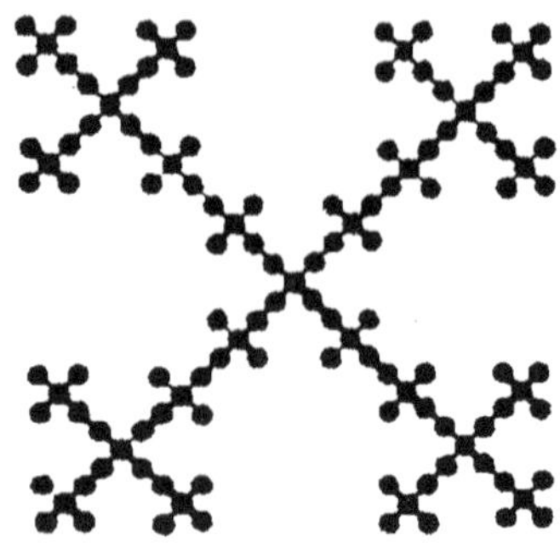

Figure 1: Example of a deterministic fractal embedded into two dimensions. The dimension of this fractal is $D = 1.465$ (Vicsek 1989).

2. What is a Fractal?

Fractals are crinkly objects that defy conventional measures, such as length and area, and are most often characterized by a fractional dimension[15]. One of the common features of fractal objects is that they are self-similar (scale invariant). This means that if we first cut out a part of them, and then we blow this piece up, the resulting object (in a statistical sense) will look the same as the original one. The self-similar structure is perhaps the main reason for the striking beauty of so many fractals. Self-similarity also implies a scale-invariant property. A set is strictly self-similar if it can be expressed as a union of sets, each of which is a reduced copy of (is geometrically similar to) the full set. See figure 1.

Another typical property of fractals is related to their volume with respect to their linear size. To demonstrate this we first need to introduce a few notions. We call *embedding dimension* the Euclidean dimension d of the space in which the fractal can be embedded in. Obviously, the volume of a fractal, $V(l)$, can be measured by covering it with d-dimensional box of edge l. This approach is called box-counting. Then the expression

$$V(l) = N(l)l^d \tag{1}$$

gives an estimate of the volume, where $N(l)$ is the number of boxes needed to cover the object completely and l is much smaller than the linear size L of the whole structure. For the ordinary objects $V(l)$ quickly attains a constant value, while for fractals typically $V(l) \to 0$ as $l \to 0$.

Considering that the fractal has a real volume ($\neq 0$) and in order to maintain the identity (1), the exponent d must be a non-integer. This dimension is called fractal dimension, D. In order to calculate this dimension one can assume *V(l)=constant*, and from the expression (1) one obtains

$$N(l) \sim l^{-D} \tag{2}$$

Infact, this expression should be understood in terms of a normalized unit of the box size: $\varepsilon = l/L$

3. Multifractal Analysis

This section introduces a formalism for the description of the situation when a singular distribution is defined on a support which is fractal. The structure of fractals plays an essential role in the physical processes they are involved in and, as a result, one needs an infinite number of dimension-type exponents to characterize these distributions.

The term *multifractality* expresses the fact that points corresponding to a given type singularity typically form a fractal subset of the total support whose dimension depends on the type of singularity[16].

In more complex situation, when a non-uniform fractal measure (i.e., multifractal) with a distribution having infinitely many singularities is considered, we are led to assume the general form

$$p(\varepsilon) \sim \varepsilon^{\alpha} \tag{3}$$

where $p(\varepsilon)$ is the normalised measure contained in a box of size ε, and α corresponds to the strength of the local singularity of the measure and is also called Hölder exponent. There are usually many boxes with the same α. In general, the number of such boxes scales with ε as

$$N_{\alpha}(\varepsilon) \sim \varepsilon^{-f(\alpha)} \tag{4}$$

where, in view of (2), $f(\alpha)$ is the fractal dimension of the subset of boxes characterized by the exponent α. The exponent α can take any value from the interval $[\alpha_{\infty}, \alpha_{-\infty}]$. When the size ε of the boxes vanishes the number of the boxes increases; this implies that $f(\alpha) > 0$. If we remember the expression $N(\varepsilon) \sim \varepsilon^{-D}$ (2) it is obvious that $N_{\alpha}(\varepsilon) \leq N(\varepsilon)$. So,

$$max_{\alpha} f(\alpha) = D \tag{5}$$

The plot of $f(\alpha)$ in a $f - \alpha$ plane typically adopts a convex shape, where the maximum is the fractal dimension of the measure support. In calculating the $f(\alpha)$ spectrum of an ordinary fractal one obtains a single point on the $f - \alpha$ plane.

An important quantity which can be determined from the weights p_i is the sum over all boxes of the qth power of box weights[17-20]

$$Z_q(\varepsilon) = \sum_{i=1}^{N(\varepsilon)} p_i^q(\varepsilon) \tag{6}$$

for $q \in R$. It is easy to see for $q = 0$

$$Z_0(\varepsilon) = N(\varepsilon) \sim \varepsilon^{-D} \tag{7}$$

and for $q = 1$

$$Z_1(\varepsilon) = 1 \tag{8}$$

In general, $Z_q(\varepsilon)$ depends on q in a non-trivial way

$$Z_q(\varepsilon) \sim \varepsilon^{(q-1)D_q} \tag{9}$$

where D_q is called generalized dimension. For an ordinary fractal, $D_q = \alpha = f(\alpha) = D$ for any q.

Working with these definitions we can obtain the following relationships:

$$\tau_q \equiv (q-1)D_q \tag{10}$$

$$\frac{\partial \tau_q}{\partial q} = \alpha \tag{11}$$

$$\frac{\partial f(\alpha)}{\partial \alpha} = q \tag{12}$$

A more detailed discussion on the meaning of these expressions can be found in Vicsek[16] and in Halsey[21].

In order to compute the scaling properties we use the following expressions:

$$Z_q(\varepsilon) = \sum_{i=1}^{N(\varepsilon)} p_i^q(\varepsilon) \sim \varepsilon^{\tau_q} \tag{13}$$

$$D_q \equiv \frac{\tau}{q-1} \tag{14}$$

$$D_1 \equiv lim_{\varepsilon \to 0} \left(\sum_{i=1}^{N(\varepsilon)} \frac{p_i(\varepsilon) log p_i(\varepsilon)}{log \varepsilon} \right) \tag{15}$$

$$\alpha_q = lim_{\varepsilon \to 0} \left(\sum_{i=1}^{N(\varepsilon)} \frac{\hat{p}_i(\varepsilon) log p_i(\varepsilon)}{log \varepsilon} \right) \tag{16}$$

$$f(\alpha) = lim_{\varepsilon \to 0} \left(\sum_{i=1}^{N(\varepsilon)} \frac{\hat{p}_i(\varepsilon) log \hat{p}_i(\varepsilon)}{log \varepsilon} \right) \tag{17}$$

where

$$\hat{p}_i(q,\varepsilon) = \frac{p_i^q(\varepsilon)}{\sum_{j=1}^{N(\varepsilon)} p_j^q(\varepsilon)} \tag{18}$$

This direct procedure for calculating $f(\alpha)$ and α_q is from Chhabra and Jensen[22] and Ball and Spivack[23]. Typically, the two fractal measures calculated on a aggregation pattern are the mass distribution measure and the growth probability distribution measure (GPD).

4. Growth Probability Distribution in Zinc Electrodeposition

The formalism introduced in the previous section is applied here to two-dimensional electrodeposits growing under a parallel geometry. We want to determine the GPD of the growing structure, which is the measure on the fractal set. There are two ways to compute this GPD, which are based on different ways of calculating the normal flux of particles at the growing surface.

The first method assumes that the growing dynamics is Laplacian controlled[24]. In this case, GPD is proportional to the normal gradient of this Laplacian field

$$p_i \propto |\vec{\nabla}_n \phi|_i$$

where $\nabla^2\phi = 0$. Periodic boundary conditions are imposed on the lateral directions[25].

A completely experimental procedure consists of obtaining the GPD from the evaluation of the normal velocity of the growing electrodeposit surface

$$p_i \propto |\mathbf{V}_n|_i$$

We need two successive images of the growth process to evaluate this velocity[4]

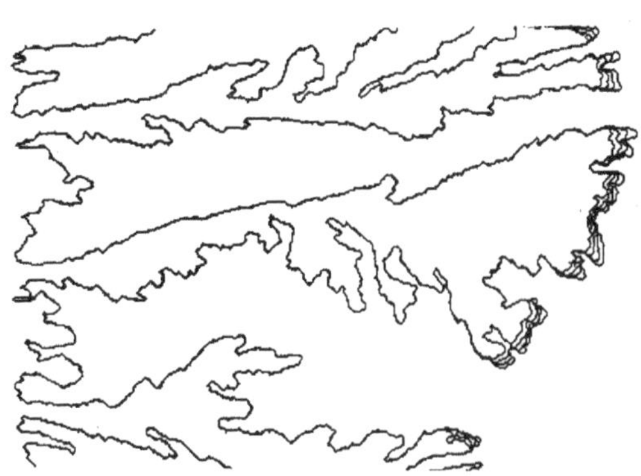

Figure 2: Open fractal pattern observed at $[Zn^{2+}] = 0.4M$ and $\Delta V = 6V$. Perimeters obtained at 163.21, 173.54, 183.52 and 194.74 sec.

5. Experimental Procedure

Two-dimensionally grown electrodeposits are obtained experimentally with a simple arrangement consisting of a pair of parallel zinc electrodes $70\mu m$ thick, separated by a distance of $3cm$ and sandwiched between two glass plates which encloses an aqueous $0.4M$ solution of $ZnSO_4$[11]. The sample is electrolyzed ($\Delta V = 6V$) during several minutes until typical metallic trees develop on the cathode up to a length of several millimetres. Images are digitized with a resolution of 512×512 pixels and binarized to define, relative to an intensity threshold, the cluster boundaries. See figure 2.

The numerical estimation of the normal gradient of the Laplacian field is carried out by solving the discretized Laplace equation, with periodic boundary conditions in a square lattice[6]. This iterative method was carried out to a relative convergence of 10^{-5} at all perimeter sites. The normal gradient of this field is calculated from the values of the Laplacian field in the perimeter sites[25]. The growth probabilities are finally normalized to unity.

The experimental way to evaluate the GPD consists in calculating the normal velocity on the perimeter. This is computed as the length of a segment of a line perpendicular to the curve between the perimeters of two successive digitized images[4]. In order to have more growing sites, this procedure is here repeated for six successive perimeters taking always the first one as fixed, and averaging over the results obtained in each realization. Finally, the normal velocities are normalized to unity.

The scaling properties of the GPD are analyzed in relation to its moments Z_q, here calculated in terms of the growth probabilities accumulated within square boxes of lattice constant ε composing a grid which covers the cluster boundary[26]. This is a box-counting fixed-size algorithm, which is a good method of calculating the scaling properties for $q > 0$. For $q < 0$ the way to obtain good results is by computing the scaling properties with a box-counting fixed-mass algorithm[27–29].

6. Results and Discussion

In Table 1, we show the values of D_q for different q when the measure is the mass distribution on the fractal set, i.e., the zinc electrodeposit. The fact that one gets approximately equal values of D_q is a clear signature of autosimilarity and permits the conclusion that the analyzed pattern is a fractal set. In the practical computation of the above mentioned measures one is compelled to use a limited range of about two decades on a pixel-based scale in order to circumvent both lower and upper cut-offs directly originated in the image acquisition techniques. The selected intermediate range of length scales where (14) is applied to obtain D_q corresponds to $2^{-6.5} < \varepsilon < 2^{-2.5}$. The regression techniques which may be used finally to obtain a numerical estimate for the generalized dimension are based on

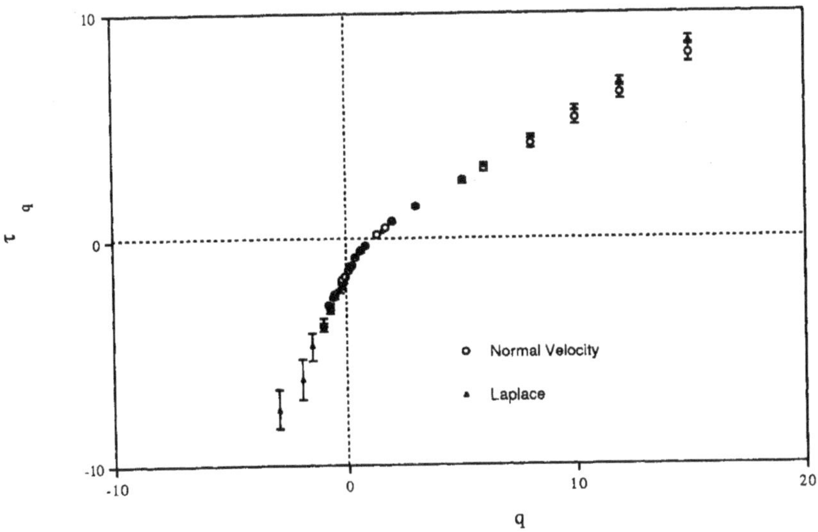

Figure 3: Plot of τ_q vs. q for $q > -2$. Results for the two methods explained in the text, normal velocity and Laplace method, are displayed.

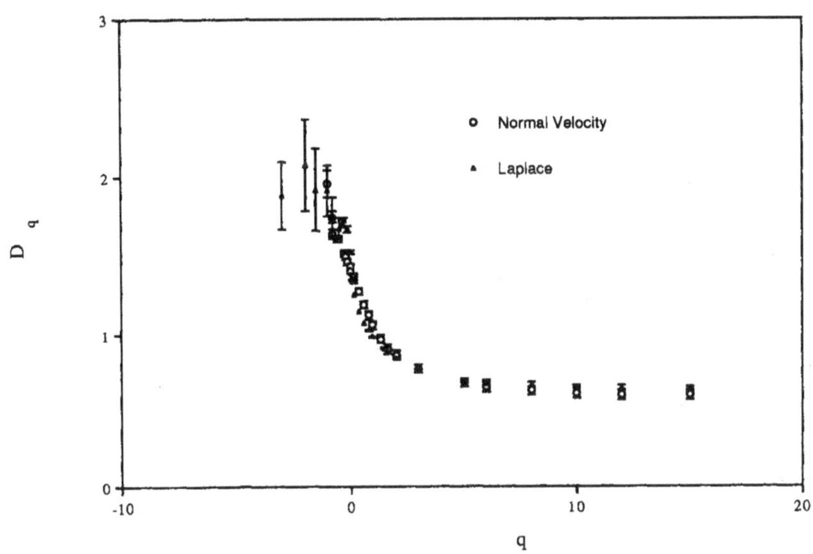

Figure 4: Plot of D_q vs. q for $q > -2$. Results for the two methods explained in the text, normal velocity and Laplace method, are displayed.

q	D_q
0	1.53±0.01
1	1.51±0.01
2	1.51±0.01
3	1.51±0.01
6	1.49±0.01

Table 1: Some values of D_q for different q. The fractal measure is the mass distribution of the open-fractal electrodeposit.

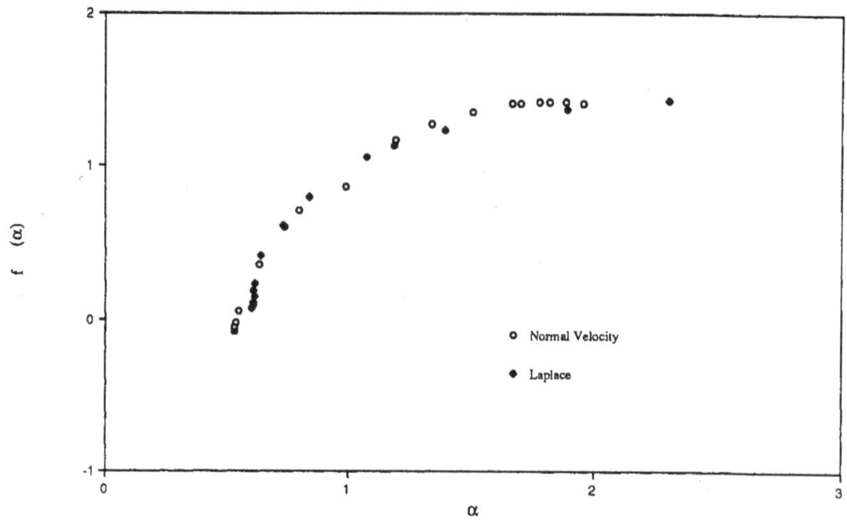

Figure 5: Plot of $f(\alpha)$ vs. α for $q > -1$. Results for the two methods explained in the text, normal velocity and Laplace method, are displayed. For this particular range, error bars are small enough to be shown in the picture

least-squares statistically weighted linear fits as adapted to *log-log* plots. This technique is also applied when the measure is the growth probability.

Focusing now on what is the central aim of this presentation, the following results refer to the scaling indices of the GPD in fractal electrodeposition. In Figures 3, 4 and 5 we show D_q, τ_q versus q and $f(\alpha)$ versus α for the two methods above mentioned. It can be seen that there is a good agreement between the experimental (normal velocity) and the theoretical (normal gradient of the Laplacian field). Typical error bars for $q < 0$ are clearly observed in all the plots due to previously mentioned limitations of the box-counting fixed-size algorithm. In Figure 5 the spectrum of singularities $\alpha(q)$ and associated fractal dimensions $f(\alpha)$ are plotted; the right branch not shown would corresponds to $q < -1$ values. These results enable us to conclude on the relevance of a Laplacian growth dynamics in controlling fractal growth in electrodeposition, within the length scale at which this morphology has been analyzed in this contrinbution.

Remarkably enough, the fractal nature of a given pattern does not necessarily coincide with that of this growth mechanism: the open-fractal electrodeposits analyzed here show that a multifractal growth behaviour can give rise to a well-defined self-similar structure.

Acknowledgements

We are grateful to the "CESCA" for providing us with computation facilities. One of us (J. Mach) benefited from a F.P.I. grant from the Ministerio de Educación y Ciencia (Spain). Financial support from the Dirección General de Investigación Científica y Tecnológica (DGICYT) under project PB87-0014 is also acknowledged. We also thank J. Claret for their help in the image processing techniques.

References

1. B. B. Mandelbrot, *Fractals: Form, Chance and Dimension*, Freeman, San Francisco, (1977).
2. B. B. Mandelbrot, *The Fractal Geometry of Nature*, Freeman, San Francisco, (1982).
3. B. B. Mandelbrot, *Fractals and Multifractals: Noise, Turbulence and Galaxies*, Freeman, San Francisco, (1982).
4. S. Ohta and H. Honjo, *Phys. Rev. Lett.*, **60** (1988) 611.
5. J. Nittmann, H. E. Stanley, E. Touboul and G. Daccord, *Phys. Rev. Lett.*, **58** (1987) 619.
6. L. Niemeyer, L. Pietronero and H. J. Wiesmann, *Phys. Rev. Lett.*, **52** (1984) 1033.
7. R. M. Bradi and R. C. Ball, *Nature*, **309** (1984) 225.
8. D. Grier, E. Ben-Jacob, R. Clarke and L. M. Sander, *Phys. Rev. Lett.*, **56** (1986) 1264.
9. Y. Sawada, A. Dougherty and J. P. Gollub, *Phys. Rev. Lett.*, **56** (1986) 1260.
10. F. Sagués and J. M. Costa, *J. Chem. Educ.*, **66** (1989) 502.
11. F. Sagués, F. Mas, M. Vilarrassa and J. M. Costa, *J. Electroanal. Chem.*, **278** (1990) 351.
12. P. P. Trigueros, J. Claret, F. Mas and F. Sagués, *J. Electroanal. Chem.*, **312** (1991) 219.
13. L. M. sander in *The Physics of Structure Formation*, ed. W. Güttinger and G. Dangelmayr (Eds.), Springer, Berlin, (1987).
14. T. A. Witten and L. M. Sander, *Phys. Rev. Lett.*, **47** (1981) 1400.
15. J. Theiler, *J. Opt. Soc. Am.*, A, Vol. 7, No.6, (1990) 1055.
16. T. Vicsek, *Fractal Growth Phenomena*, World Scientific, (1989).
17. A. Rényi, *Probability Theory*, North Holland, Amsterdam, (1970).
18. H. G. E. Henstchel and I. Procaccia, *Physica*, **D8** (1983) 435.
19. P. Grassberger, *Phys. Lett.*, **A97** (1983) 227.
20. P. Grassberger and I. Procaccia, *Physica*, **D13** (1984) 383.

21. T. C. Halsey, M. H. Jensen, L. P. Kadanoff, I. Procaccia and B. J. Shrainon, *Phys. Rev.*, **A33** (1986) 1141.
22. A. Chhabra and R. V. Jensen, *Phys. Rev. Lett.*, **62** (1989) 1327.
23. R. C. Ball and D. R. Spivack, *J. Phys.*, **A23** (1990) 5295.
24. R. F. Voss, *Phys. Rev.*, **B30** (1984) 334.
25. F. Mas and F. Sagués, *Europhys. Lett.*, **17** (1992) 541.
26. J. Mach, F. Mas and F. Sagués, *in preparation*, (1992).
27. R. Badii and A. Politi, *Phys. Rev. Lett.*, **52** (1984) 1661.
28. R. Badii and A. Politi, *J. Stat. Phys.*, **40** (1985) 785.
29. R. Badii and G. Broggi, *Phys. Lett.*, **A131** (1988) 339.

NUMERICAL SIMULATION OF STOCHASTICAL MODELS FOR EXTERNALLY–DRIVEN LASERS

M. S. TORRE, H. F. RANEA SANDOVAL
Instituto de Física "Arroyo Seco"
Facultad de Ciencias Exáctas, UNCPBA
Pinto 399, (7000) Tandil(BA). Argentina

and

R. C. BUCETA
Departamento de Física
Facultad de Ciencias Exáctas y Naturales, UNMdP
Funes 3350, (7600) Mar del Plata. Argentina

ABSTRACT

A model which describes the dynamics of an externally–driven laser in addition to white and colored noise, is presented. The system is known to have a Hopf bifurcation at the phase locking point. The resonances become apparent before the bifurcation point is reached and are a result of gain saturation and phase excursion phenomena. Well above this point, the spectrum is the same as a frequency locked laser when noise is present. Results for periodic (i.e. time dependent) driving forces are also presented.

1. Introduction

The response of a non linear system to an external coherent injection, and to noise, has been the subject of numerous publications since the phenomena of frequency locking and line width narrowing in lasers was first reported in external cavity seeding[1–7].

Homogeneously broadened lasers, at the vicinity of the threshold, can be modeled by the *Hopf Stochastic Normal Form* for the electric field in the cavity[8,9]. We have shown that it is possible to give, for an arbitrary time-independent driving force, a *Stochastic Normal Form Equation* which qualitatively displays the spectral power density main features.

Several examples were inspected by a numerical simulation procedures. The model was analyzed for stochastic white and color (Ornstein–Uhlenbeck) noise, the dominant spectral contribution show results that agree well with the theory.

Our approach allows to extend the analysis to driving injections amplitudes not necessarily weak, which is responsible for the appearance of the harmonics and the spectral shift for the fundamental frequency, and to time–dependent driving forces, case in which the system is not homogeneous in time–development.

2. Stochastic Model

We include temporal stochastic fluctuations to the relevant dynamical parameters of the injected laser system. We shall consider amplitude and frequency fluctuations in the injected coherent field and the frequency of the laser itself.

The detuning and the amplitude of the injected field have well defined temporal mean values; we call these $\langle \Delta_\nu(t)\rangle = \Delta\nu$, and $\langle E_I(t)\rangle = E_I$. For non–zero detuning, a convenient time–scaling $\tau = \Delta\nu t$, and fluctuations are expressed as $\Delta_\nu(\tau)/\Delta\nu = 1 + u_3(\tau)$ and $\Delta_c\sqrt{T}E_I(\tau)/\Delta\nu = F + u_4(\tau)$ leads to the following stochastic differential equation[10],

$$\partial_\tau \mathbf{E} = (i + g(E))\mathbf{E} + F + iu_3\mathbf{E} + u_4, \tag{1}$$

where $g(E) = -\kappa + \gamma/(1+E^2)$ is the adimensional net gain saturation, and F an adimensional parameter related with the amplitude of injected signal. In the case of white noise, the self correlations functions for the aleatory variables $u_j(\tau)$ are :

$$\langle u_j(\tau)u_j(\tau')\rangle = D_j\, \delta(\tau - \tau'), \tag{2}$$

for $j = 3, 4$, $D_3 = \epsilon_3/\Delta\nu^2$ and $D_4 = \Delta_c^2 T\epsilon_4/\Delta\nu^2$, being ϵ_3 and ϵ_4 the fluctuations intensities in frequency and amplitude, respectively. If the aleatory variables are representatives of Ornstein–Uhlenbeck processes with zero mean value, the correlations functions are:

$$\langle u_i(\tau)u_j(\tau')\rangle = \delta_{ij}\frac{D_i\alpha_i}{2}\exp(-\alpha_i|\tau - \tau'|), \tag{3}$$

where α_i^{-1} is the correlation time of the process i^{th}, and D_i its diffusion constant. This type of stochastic processes is solved using Langevin's differential equations of the type

$$\partial_\tau u_j(\tau) = -\alpha_j u_j(\tau) + \zeta_j(\tau), \tag{4}$$

where $\{\zeta_j(\tau)\}$ are uncorrelated white noise whose intensities are $\epsilon_j = D_j\alpha_j^2$. If noise is absent the system presents a Hopf bifurcation at $\mathbf{E} = \mathbf{E}_b\exp(-i\varphi_b)$ $(E_b^2 = \sqrt{\sigma} - 1, \varphi_b = \arctan^{-1}[\kappa(\sqrt{\sigma}-1)], \sigma = \gamma/\kappa)$ with $u_j = 0$, when F reaches the value $F_b = E_b(\kappa^2 E_b^4 + 1)^{1/2}$ [10].

3. Numerical Simulation

Numerical integration of the stochastic differential equations (1) and (4) was done by conventional procedures[11]. After eliminating the transients, the values for the electric field of each one of the 200 simultaneous realizations which simulate the stochastic process, were used to evaluate the Fourier coefficients for each stage in the integration, thus getting the spectral power density through the Wiener–Khinchine theorem relations. All cases were solved with similar numerical parameters: the integration adimensional time–step was 10^{-3}.

For $F > F_b$, the spectra show the normal appearance of an injection locked laser, consisting of a very narrow central peak on a noisy background. In the case of colored noise, this background is larger than the white noise case. Figure 1(a) displays the spectral power density for parameter chosen as to lock all the oscillators in phase. Figures 1(b) and (c), displays the spectral power density which correspond to two cases with $F < F_b$, (b) white noise process and (c) colored noise process. A shift in frequency for the main resonant contribution is apparent, together with a broadening of the resonant peaks when the correlation time for the colored noise is increased. Further increase in this parameter, will wash all higher resonant contributions out of the spectrum.

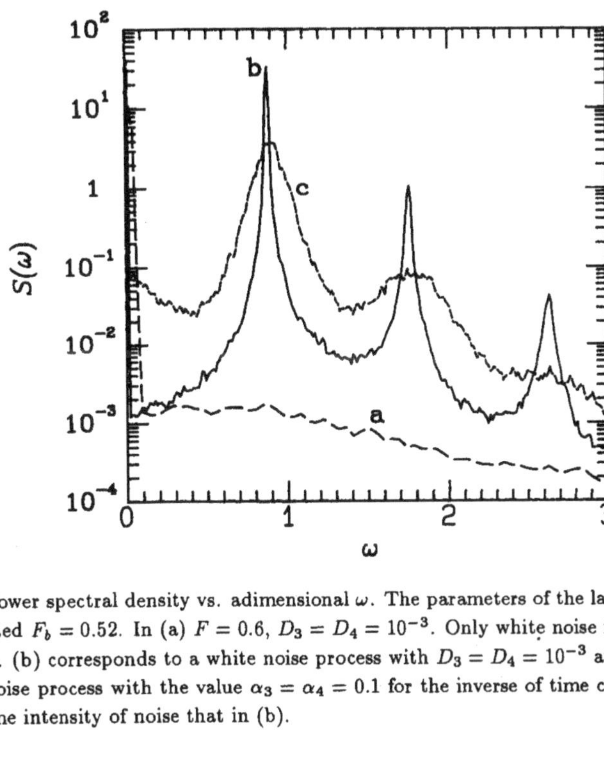

Figure 1: Power spectral density vs. adimensional ω. The parameters of the laser are: $\gamma = 3$, $\kappa = 2$. The evaluated $F_b = 0.52$. In (a) $F = 0.6$, $D_3 = D_4 = 10^{-3}$. Only white noise is present. In (b) and (c) $F = 0.4$. (b) corresponds to a white noise process with $D_3 = D_4 = 10^{-3}$ and (c) corresponds to a colored noise process with the value $\alpha_3 = \alpha_4 = 0.1$ for the inverse of time correlation parameter, and the same intensity of noise that in (b).

Figure 2 shows the spectral power density which correspond to two cases where the dynamical variable F undergoes a time dependent process, represented by $F = F_0(1 + \lambda \cos \omega_f \tau)$. In part (a), the system is above the locking (bifurcation) point, but the oscillation of the driving force loosen the system out of locking, thus appearing a resonance at the detuning between the laser and the external signal. The stochastic resonance at ω_f and its harmonics are apparent. The part (b), shows

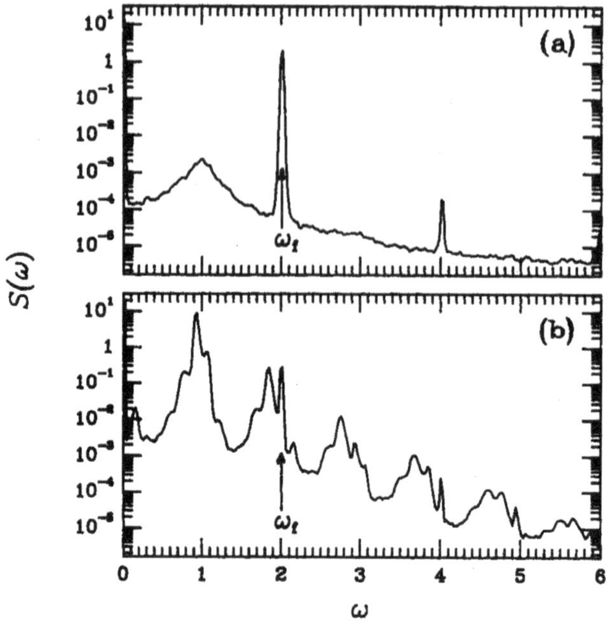

Figure 2: Power spectral density vs. adimensional ω when the dynamical variable F is a time–dependent function. The parameters of the system are: $\gamma = 3$, $\kappa = 2$, $\lambda = 0.5$, and $\omega_f = 2$. In part (a), the system is above the locking point ($F_0 = 0.6$). Note that the oscillation of the driving force loosen the system out of locking, thus appearing a resonance at the detuning between the laser and the external signal. The stochastics resonances at ω_f and its harmonics are also present. In part (b), the parameters are the same except that the system is below the locking point ($F_0 = 0.4$). The frequency ω_f is not natural frequency of the system, when F is constant and equal to F_0. The spectral power density shows a more complex dynamical behavior in which several combination tones are present.

similar spectra but when the system is below the bifurcation point. The resonant peaks shows a much more complex dynamical behavior in which several combination tones are present.

4. Summary

The numerical results presented here agree very well with the results expected by an analysis of the Stochastic Normal Form of this problem which was found to correspond to a generalization of similar problems in laser dynamics.

The introduction of white noise (additive and multiplicative to the parameters) results in the appearance of narrow resonant peaks on a noisy near Lorenzian line shape, while for the colored Ornstein–Uhlenbeck processes, a shift and widening of the resonances can eventually wipe out any feature in the spectrum, which is interpreted as a very long–term memory; thus the signal necessary to lock the system in phase should eventually become very large to be of any practical interest. Such memory processes can for example be introduced by the same system, if a retarded path is provided for the field.

The case of time–changing injection is under study to verify the frequency–sensitive harmonics, which constitutes an interesting selection rule.

Acknowledgements

This work is partially supported by PID 3–155200–88 CONICET (Argentina).

References

1. A. E. Siegman in *Laser* (University Science Books, Mill Valley, California, 1986) and references therein.
2. S. A. Shakir and W. W. Chow, *Opt. Lett.* **9** (1984) 202.
3. H. F. Ranea Sandoval, M. C. Von Reichenbach, R. Rodriguez and F. A. Schaposnik, *Opt.Comm.* **49** (1986) 39.
4. A. Shenzle and H. Brand, *Opt. Commun.* **27** (1978) 485.
5. E. Peacock-Lopez, F. J. de la Rubia, B. J. West and K. Lindenberg, *Phys. Rev.* **A39,** (1989) 4026.
6. F. C.Cheng and L. Mandel, *J. Opt. Soc. Am.* **B8** (1991) 1681.
7. J. O'Gorman, B. J. Hawdon, J. Hegarty and D. M. Heffernan, *J. Appl. Phys.* **66** (1988) 1.
8. M. Aguado and M. San Miguel, *Phys.Rev.* **A37** (1988) 450.
9. M. Aguado, E. Hernández-García and M. San Miguel, *Phys. Rev.* **A38** (1988) 5670.
10. R. C. Buceta, M. S. Torre and H. F. Ranea Sandoval, submited to *Phys. Rev.* A.
11. M. Sancho, M. San Miguel, S. L. Katz, J. D. Gunton, *Phys. Rev.* A **26**, 1589 (1982).

PHASE SEPARATION IN BINARY MIXTURES AT LOW TEMPERATURES

A. M. LACASTA[(a,b)], A. HERNÁNDEZ-MACHADO[(b)], J.M. SANCHO[(b)] and R. TORAL[(c)]

(a) Departament de Física Aplicada, Universitat Politècnica de Catalunya Av. Gregorio Marañón s/n. E-08028 Barcelona. Spain
(b) Departament d'Estructura i Constituents de la Matèria, Facultat de Física, Universitat de Barcelona, Diagonal 647, E-08028 Barcelona, Spain.
(c) Departament de Física, Universitat de les Illes Balears, E-07071 Palma de Mallorca, Spain.

ABSTRACT

Dynamical scaling during the late stages of phase separation in a binary mixture are studied for different values of the volume fraction ϕ of the minority component. We have performed a numerical integration of the deterministic time-dependent Ginzburg-Landau equation with a variable-dependent diffusion coefficient. The forms of the pair-correlation function and the structure function are independent of temperature but dynamics are slower at low temperatures. A crossover between interfacial diffusion and bulk diffusion mechanisms is observed in the behaviour of the characteristic domain size. This effect is explained theoretically in terms of an equation of motion for the interface.

1. Introduction

Pattern formation during phase separation is an interesting nonlinear, non-equilibrium problem[1]. This situation appears, for example, when a binary mixture, initially in a homogeneous equilibrium state, is suddenly quenched into the two-phase coexistence region. Then, the system evolves spontaneously by generating macroscopic domains towards a new two-phase equilibrium state.

The classical theory makes a distinction between nucleation and spinodal decomposition, depending on whether the system reaches equilibrium from a metastable or unstable state. However, it is known that there is no sharp distinction between the two processes. Rather, there is a gradual transition in the dynamic behaviour. The position of the quench modifies the amount of each phase, and therefore the morphology of the patterns formed. By increasing the degree of asymmetry between the volume fraction of the two components, the shape of the domains varies from an interconnected structure, characteristic of spinodal decomposition, to a distribution of circular clusters of the minority phase in a background of the other phase.

From a theoretical point of view, apart from microscopic models, these phase separation processes have usually been described by a time-dependent Ginzburg -

Landau equation for the concentration $\bar{c}$:

$$\frac{\partial}{\partial \tau}\bar{c}(\vec{r},\tau) = \nabla\left(\Gamma\nabla\frac{\delta F(\{\bar{c}\})}{\delta\bar{c}}\right) + \eta(\vec{r},t)\,, \tag{1}$$

where the mobility Γ is a constant and $\eta(\vec{r},t)$ is the thermal noise. $F(\{\bar{c}\})$ is the Ginzburg-Landau free energy

$$F(\{\bar{c}\}) = \frac{1}{2}\int d\vec{r}\{-r\bar{c}^2 + \frac{u}{2}\bar{c}^4 + k(\nabla\bar{c})^2\}\,, \tag{2}$$

where r, u and k are positive phenomenological coefficients.

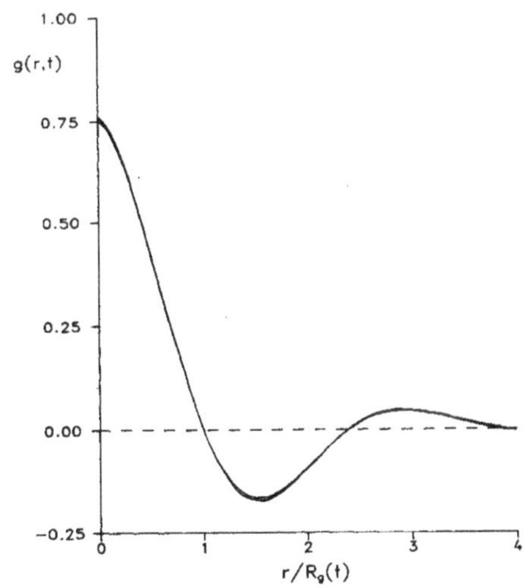

Figure 1: Radial Correlation Function versus the scaled variable $r/R_g(t)$, for $\phi = 50\%$ and five values of a. The curves have been taken at times: $t = 1200$ for $a = 0$; $t = 1400$ for $a = 0.2$; $t = 2000$ for $a = 0.6$; $t = 3000$ for $a = 0.8$ and $t = 5800$ for $a = 1$.

Computer simulations of this model for critical[2-4] and off-critical[5-6] quenches suggest that the late time behavior is well described in terms of a scaling with a characteristic length R(t), which shows a power-law growth in time with an exponent 1/3.

Recently, some authors[7-10] have argued the necessity of modifying the usual Ginzburg-Landau equation by considering a concentration-dependent diffusion coefficient for adequate modelling of a deep quench. This coefficient has been postulated to be

$$\Gamma(\bar{c}) = \Gamma_0(\bar{c}_0^2 - \bar{c}^2) \tag{3}$$

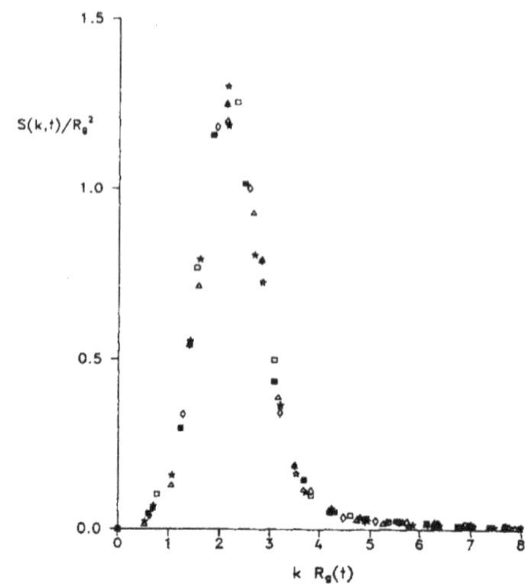

Figure 2: Scaled radial structure function versus the scaled variable $kR_g(t)$, for $\phi = 50\%$ and two values of a: $a = 0$ black symbols and $a = 0.8$ white symbols. The points have been taken at times: 5000(★), 8000(■), 10000(▸), 12000(♦), 9000(⋆), 12000(□), 15000(▹), 18000(◇).

where Γ_0 is a constant, $\bar{c}_0 = \bar{c}_{st}(T = 0)$ and $\bar{c}_{st}(T)$ is the equilibrium value for a given temperature T.

In this contribution we focus on the effects of a concentration-dependent diffusion coefficient on the dynamics of phase separation[10-11]. Since we are interested in low temperatures and late stages of the evolution, the noise term is neglected.

2. Model and Numerical Results

Our macroscopic continuous model can be written in the dimensionless form:

$$\frac{\partial c}{\partial t} = \nabla\left((1 - ac^2)\nabla(-c + c^3 - \nabla^2 c)\right) \tag{4}$$

where $a = (\bar{c}_{st}(T)/\bar{c}_0)^2$ is the only relevant parameter in our study. Its values go from 0 to 1 as temperature is reduced, $a = 1$ for T $= 0$, and we recover the usual G.L. equation for $a = 0$. Now the bulk equilibrium values of the variable c are $c_{st} = \pm 1$.

We have numerically integrated Eq. (4) in a square lattice of size $L = 120$ and periodic boundary conditions, using Euler's method with mesh size $\Delta x = 1$ and time step $\Delta t = 0.025$. We have studied quenches corresponding to three volume fractions of the minority phase: $\phi = 50\%$ (critical case)[10] , $\phi = 30\%$ and $\phi = 5\%$ (both off-critical cases)[11]. We have selected different values for a, and the results have

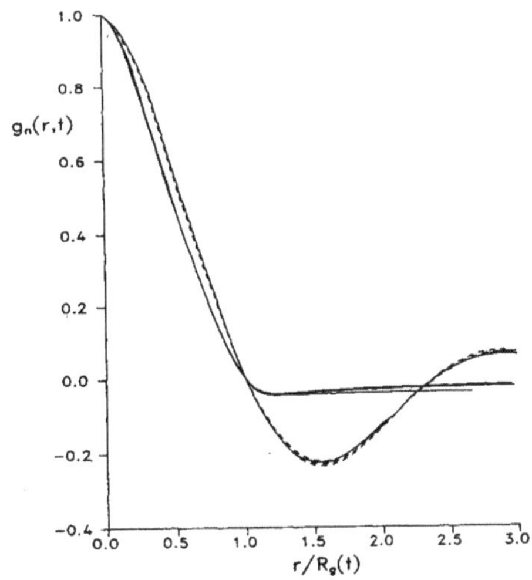

Figure 3: Radial and normalized correlation function, versus the scaled variable $r/R_g(t)$. The dashed lines correspond to the curves obtaines for volume fraction $\phi = 30\%$, and three values of a (0, 0.8 and 1). The solid lines represent the case $\phi = 5\%$ and the same three values of a. We have also represented by a solid line the function obtained for the critical case ($\phi = 50\%$).

been averaged over 10 runs. The system was initially prepared by assigning to each point a concentration $c(\vec{r},t) = c_0 + \tilde{c}(\vec{r},t)$, where $c_0 = 1 - 2\phi$ is the mean concentration and $\tilde{c}$ is a random number uniformly distributed in the interval $(-0.05, 0.05)$.

In order to study the dynamics of the domain growth we introduce the pair-correlation function

$$G(\vec{r},t) = < \frac{1}{L^2} \sum_{x} \left(c(\vec{r}+\vec{x},t)c(\vec{x},t) - c_o^2 \right) > \tag{5}$$

where the brackets mean average over initial conditions.The Fourier transform of $G(\vec{r},t)$ is the structure function $S(\vec{k},t)$.

As our system is isotropic, we introduce the circularly-averaged pair correlation function

$$g(r,t) = \frac{1}{N_r} \sum G(\vec{r},t) \tag{6}$$

and the circularly-averaged structure function

$$s(k,t) = \frac{1}{N_k} \sum S(\vec{k},t) \tag{7}$$

where the sums run over the set of points inside the coronas of radii r and $r + \Delta r$ or k and $k + \Delta k$ respectively. N_r and N_k are the number of points inside such coronas.

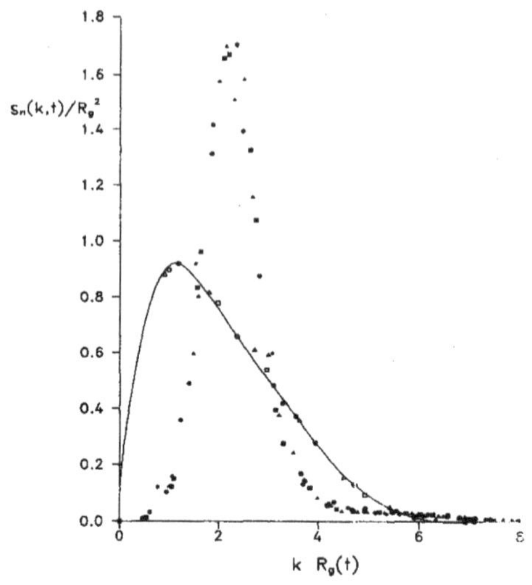

Figure 4: Scaled radial and normalized structure function versus the scaled variable $kR_g(t)$, for $\phi = 5\%$ (white symbols) and $\phi = 30\%$ (black symbols), and three values of a ($a = 0(\triangleright)$; $a = 0.8(\circ)$; $a = 1(\square)$). The stars correspond to the values obtained for $\phi = 50\%$.

The dynamical scaling hypothesis for the pair-correlation function and the structure function states that

$$g(r,t) = g(r/R_g(t)) \tag{8}$$

$$s(k,t) = R_g(t)^2 s(kR_g(t)) \tag{9}$$

where $R_g(t)$ is a relevant length defined as the first zero of the pair-correlation function.

First, we have tested that our numerical scheme reproduces the well-known results of the scaling regime for a=0. At late stages, the scaling pair-correlation and structure functions are time-independent, and their form depends strongly on ϕ.

We find that, for different values of a, the scaling regime is reached for long enough times. The dynamics are slower and so the scaling regime is reached later, as a is increases from 0 to 1. In Fig.1 we present results for the pair-correlation function for the critical case and five values of a. We can see that the form of this scaling function (Eq. (8)) is independent of temperature. In Fig.2, we show the structure function for the critical case, two values of a, and some different times. We find that the scaling function (Eq. (9)) is independent of time and temperature, within our numerical accuracy. In Fig. (3) and Fig. (4), we present the normalized correlation function and the normalized structure function for the three different

a	$n(\phi = 30\%)$	$n(\phi = 50\%)$
0.0	0.32	0.33
0.8	0.26	0.28
1.0	0.20	0.22

Table 1: Exponents n obtained for a fit $R_g(t) = \beta_2 + \beta_1 t^n$, for the volume fractions ϕ=50% and ϕ=30%.

volume fraction values studied. We observe that for ϕ=30% the form of these functions is identical to that obtained for ϕ=50%, but the shape is different for ϕ=5%.

As to the behaviour of the characteristic domain size, the main effect of considering a non-zero value for a is that R_g exhibits a crossover from 1/4 to 1/3 power law as a function of time. This crossover appears at longer times for lower temperatures. In Table 1 we present the effective exponents n obtained from a fit of the form $R_g(t) = \beta_2 + \beta_1 t^n$ for two volume fractions. The crossover behaviour can be understood in terms of the equation describing the motion of the interface[9-10],

$$v(s,t) = \frac{1}{4}(1-a)\sigma \int ds' W(r(s), r(s'))K(s,t) + a\nabla^2 K(s,t) \tag{10}$$

where (u,s) are curvilinear coordinates, $v(s,t)$ is the normal velocity of the interface, σ is the surface tension, $K(s,t)$ is the local curvature and W is defined by the following equation

$$\int ds'' G(r(s), r(s''))W(r(s''), r(s')) = \delta(s - s') \tag{11}$$

where $G(r(s), r(s''))$ is the Green function solution of

$$\nabla^2 G(\vec{r}, \vec{r}\prime) = \delta(\vec{r} - \vec{r}\prime) \tag{12}$$

The first term of Eq. (10) contains the usual bulk diffusion, responsible for the 1/3 power law, and it dominates for long times. The second term contains an interfacial mechanism that is associated with the presence of a tangential flux at the interface, which is the responsible for a 1/4 power law. From Eq. (10) it seems natural to propose the following equation for the characteristic domain size $R(t)$,

$$\frac{dR(t)}{dt} = (1-a)\frac{C_2}{R^2} + a\frac{C_3}{R^3} \tag{13}$$

We obtain 1/4 power law for $a = 1$ and a = 1/3 power law for a =0. For intermediate values of a there is a crossover behaviour between these two values in accordance with the numerical results.

3. Conclusions

We have studied both numerically and analytically the effects of a concentration -dependent diffusion coefficient on the domain growth dynamics of a deterministic time-dependent Ginzburg-Landau model. We find that the form of the scaling functions $g(x)$ and $s(x)$ is independent of temperature. For ϕ=50% and ϕ=30% the form of these functions seems to be identical whereas for ϕ=5% they are clearly different. We obtain that the characteristic domain size presents a crossover from 1/4 to 1/3 power law as a function of time. This effect can be explained theoretically by means of an equation of motion for the interface, and it can be understood in terms of a reduction in the bulk diffusion mechanism that induces a greater role of interfacial diffusion for short time.

Acknowledgements

We acknowledge financial support from the Dirección General de Investigación Científica y Técnica (Spain)(Projects Nos PB90-0030 a nd PB89-0424). A.H.M. and J.M.S. also acknowledge financial support from NATO under Collaborative Research Grant No.900328.

References

1. J. D. Gunton, M. San Miguel and P. S. Shani, *Phase Transitions and Critical Phenomena*, Vol.8. Ed. C. Domb and J.L. Lebowitz (Academic Press, 1983).
2. T. M. Rogers, K. D. Elder and R. C. Desai, *Phys. Rev.* **B37** (1988) 9638.
3. R. Toral, A. Chakrabarti and J. D. Gunton, *Phys. Rev. Lett.* **60** (1988) 2311.
4. E. T. Gawlinski, J. Viñals and J. D. Gunton, *Phys. Rev.* **B39** (1989) 7266.
5. R. Toral, A. Chakrabarti and J. D. Gunton, *Phys. Rev.* **B39** (1989) 901.
6. T. M. Rogers and R. C. Desai, *Phys. Rev.* **B39** (1989) 11956.
7. J. S. Langer, M. Bar-on and H. D. Miller, *Phys. Rev.* **A11** (1975) 1417.
8. K. Kitahara and M. Imada, *Prog.Theor.Phys.Suppl.* **64** (1978) 65.
9. T. Ohta, *J. Phys.* **C21** (1988) L361.
10. A. M. Lacasta, A. Hernández-Machado, J. M. Sancho and R. Toral *Phys. Rev.* **B45** (1992) 5276.
11. A. M. Lacasta, J. M. Sancho, A. Hernández-Machado and R. Toral (submit *Phys. Rev.* B).

SIMULATION OF THE TIME EVOLUTION OF A SYSTEM EXHIBITING A VAPOR-LIQUID TRANSITION

J. GÜÉMEZ
Departamento de Física Aplicada
Facultad de Ciencias, Universidad de Cantabria
E-39005, Santander, Spain

and

S. VELASCO and J.A. WHITE
Departamento de Física Aplicada
Facultad de Ciencias, Universidad de Salamanca
E-37008, Salamanca, Spain

ABSTRACT

We present a computer simulation describing the time evolution of the number of particles N_i in a subvolume V_i of a fluid system (with N particles and volume V) exhibiting a liquid-vapor phase transition. The simulation makes use of the transition probabilities $\omega(N_i \to N_i \pm 1; V_i)$ obtained from the equilibrium probability distribution of particles over a subvolume. The method is applied to a lattice gas model with binary attractive forces in the Van der Waals limit.

1. Introduction

A very important feature of many-body systems exhibiting phase transitions is their kinetic behaviour towards the equilibrium stationary state. Of special interest is the investigation of the dynamics of a homogeneous nonequilibrium state evolving towards an equilibrium state characterized by the coexistence of two phases. In this context the usual Monte Carlo approach consists of either a Glauber or a Kawasaki dynamics applied to an Ising-like model[1].

From a theoretical viewpoint this issue can be addressed by considering a Markovian master equation describing the time evolution of the probability distribution of the system. However, the determination of the transition probabilities appearing in the master equation is usually a rather difficult task. Recently[2] we have related these transition probabilities to the equilibrium probability distribution (known in many model systems). Here we present a simple simulation procedure valid for describing the approach to equilibrium phase coexistence in a model exhibiting a liquid-vapor phase transition. The method is based on a generalized Ehrenfest urn model describing the evolution of the number of particles in a given

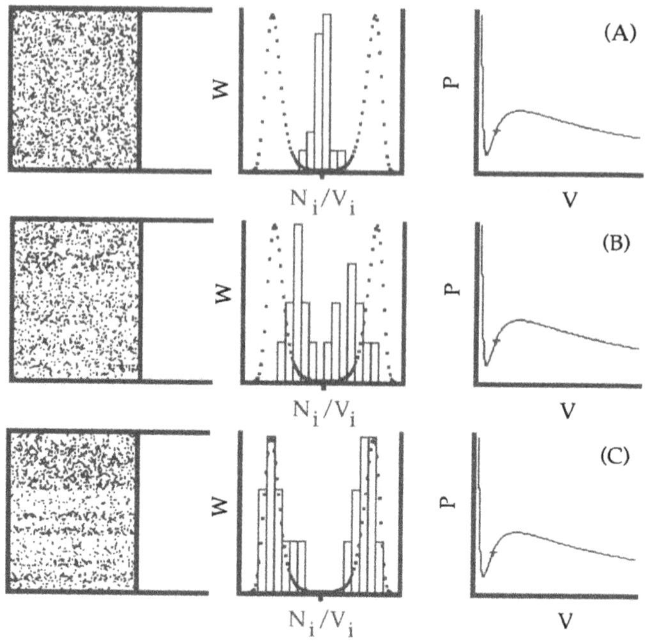

Figure 1: Time evolution of the particle distribution from the initial state (A) towards the equilibrium one (C), at $V = 80$ and $T = 21$. Left: spatial distribution. Center: simulated (bar chart) and theoretical (dots) probability distribution. Right: P–V diagram indicating the thermodinamic state (+).

subvolume of the system.

2. Theoretical Basis

We consider a system of N particles confined to a box of volume V in contact with a thermal bath at temperature T. The volume is divided into a number of subvolumes V_i ($V = \sum_i V_i$). The numbers $\{N_i\}$ of particles in subvolumes $\{V_i\}$ characterize the state of the system. We have previously shown that in the Ehrenfest urn model framework, the time evolution of the $\{N_i\}$ is given by the following rate equation[2]

$$\frac{dN_i(t)}{dt} = D\left[\omega(N_i \rightarrow N_i + 1; V_i) - \omega(N_i \rightarrow N_i - 1; V_i)\right] \tag{1}$$

where D is a parameter playing the role of a diffusion coefficient and the transition probabilities $\omega(N_i \rightarrow N_i \pm 1; V_i) \equiv \omega_i(\pm)$ are given by

$$\omega_i(\pm) = \frac{W_0(N_i \pm 1; V_i)}{W_0(N_i; V_i) + W_0(N_i \pm 1; V_i)}, \tag{2}$$

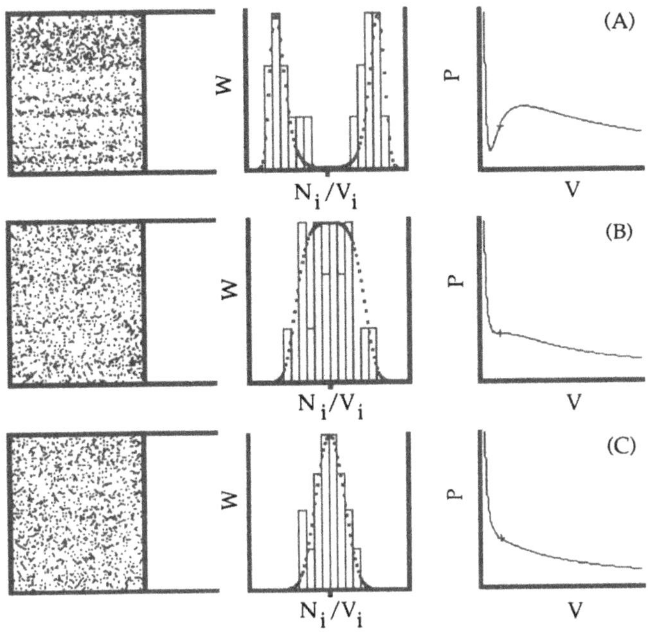

Figure 2: Three different equilibrium states at $V = 80$ and $T = 21$ (A), 25 (B), and 30 (C). Left, center, and right items represent the same as in Fig.1

W_0 being the equilibrium probability distribution of particles over subvolume V_i.

3. Details of the Simulation

In order to perform the computer simulation the total volume V is divided in n equal subvolumes $V_i = V/n$. Starting from an initial distribution of particles in the vessel (usually one chooses an uniform initial distribution), the simulation is carried out according to the following steps:

i) The normalized probability of losing one particle $\omega_i^*(-) = \omega_i(-)/\sum_i \omega_i(-)$ is calculated for every subvolume V_i.

ii) A random number s between 0 and 1 is generated. One particle is removed from the subvolume V_j that fulfils $\sum_{i=1}^{j-1} \omega_i^*(-) < s < \sum_{i=1}^{j} \omega_i^*(-)$.

iii) The normalized probability of gaining one particle $\omega_i^*(+) = \omega_i(+)/\sum_{i \neq j} \omega_i(+)$ is calculated for every remaining subvolume V_i with $i \neq j$.

iv) A random number r between 0 and 1 is generated. One particle is added to the subvolume V_k that fulfils $\sum_{i=1,i\neq j}^{k-1} \omega_i^*(+) < r < \sum_{i=1,i\neq j}^{k} \omega_i^*(+)$. Notice that the chosen subvolume V_k cannot be the one V_j from which the particle was removed.

We have applied the method to a lattice gas model with attractive binary

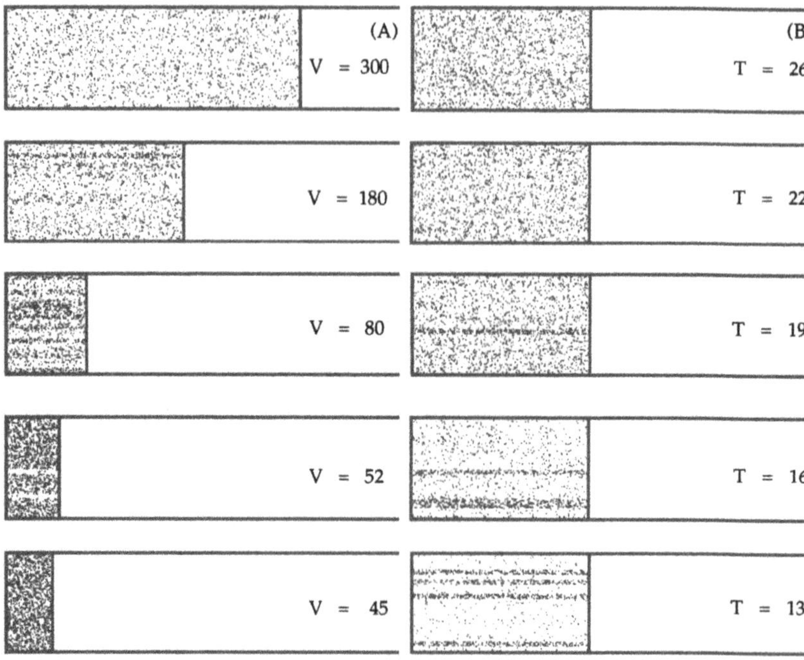

Figure 3: Spatial distribution (A) for an isothermal process at $T = 19$ and (B) for an isochoric process at $V = 180$.

interactions characterized by the following equilibrium distribution[3]

$$W_0(N_i;V_i) = \binom{N}{N_i}\binom{V^* - N}{V_i^* - N_i}\binom{V^*}{N}^{-1} \exp\left\{\frac{a}{kT}\frac{(N-\bar{N}_i)^2}{V_i\,(1-\frac{V}{V_i})}\right\} \tag{3}$$

where $V^* = V/\sigma$, $V_i^* = V_i/\sigma$ and $\bar{N}_i = NV_i/V$, σ and a being parameters measuring the finite volume of particles and the strength of the interaction respectively. The distribution P_0 in Eq. (3) was proven to describe qualitatively the vapor-liquid phase transition in a lattice gas[3]. From Eq. (3), we can derive most of the thermodynamic properties of this lattice gas model. In particular, the pressure is given by[3]

$$\frac{P}{kT} = -\frac{1}{\sigma}\ln\left(1-\frac{N\sigma}{V}\right) - \frac{aN^2}{kTV^2} \tag{4}$$

This is the well-known thermal equation of state for a lattice gas with a mean attractive interaction $-2aN/V$ between each pair of particles[4]. One can easily check that Eq. (4) presents a critical point at

$$P_c = \frac{a\sigma^2}{4}\,(2ln2-1), \quad V_c = 2N\sigma, \quad T_c = \frac{a}{2k\sigma} \tag{5}$$

For temperatures $T > T_c$, the equilibrium probability distribution (3) only exhibits one peak, while for temperatures $T < T_c$ this distribution presents one hump or two humps depending on the density value.

4. Results

We have elaborated a pedagogical program in order to perform the above simulation. This program follows the time evolution of the spatial distribution of N particles over n equal (horizontal) subvolumes of a vessel. The characteristics of the system are chosen from a menu. The initial state corresponds to a random distribution of particles over the subvolumes. In all cases, $N = 2000$, $n = 20$, $\sigma = 0.02$ (arbitrary units), $a = 1$ (arbitrary units). For these values, one has: $V_c = 80$ and $kT_c = 25$. From the initial distribution to the equilibrium distribution, a typical run on a Macintosh IIci spends a time between a few seconds and some 3 minutes depending on the considered termodynamic point.

Figure 1 shows the time evolution from the initial state towards a bimodal equilibrium distribution characteristic of phase coexistence. In Fig. 1C we can see the equilibrium state formed by a mixture of rarified and dense subvolumes. This feature is also shown in the distribution diagram, where the result of the simulation (bar chart) agrees with the theoretical one (dots) given by Eq. (3).

Figure 2 shows the equilibrium state for a temperature below, equal to and above the critical one. These states correspond, respectively, to a two-hump, to a plateau, and to a one-hump probability distribution.

An isothermal process showing the phase transition is presented in Fig. 3A. Analogously, Fig. 3B shows an isochoric process where the phase transition takes place.

References

1. J. Marro and R. Toral, *Am. J. Phys.* **54** (1986) 1114.
2. J. Güémez, S. Velasco and A. Calvo Hernández, *Am. J. Phys.* **57** (1989) 828.
3. J. Güémez and S. Velasco, *Am. J. Phys.* **59** (1991) 335.
4. P. T. Landsberg (Ed), *Problems in Thermodynamics and Statistical Mechanics* (Pion Limited, London, 1971).

ISING MODEL WITH TWO TEMPERATURES: SIMULATION AND THEORY

M. A. MUÑOZ
Departamento de Física Aplicada. Universidad de Granada
Campus Fuentenueva, 18071 Granada, Spain.

and

P. L. GARRIDO
Departamento de Física Moderna. Universidad de Granada.
Campus Fuentenueva, 18071 Granada, Spain.

ABSTRACT

We present here both a Monte Carlo (MC) simulation and a Fokker-Planck equation (FPE) approach to the kinetic Ising model with two competing temperatures[1]. The MC simulation of this non-equilibrium model shows that its macroscopic stationary properties are very sensitive to the analytical form of the microscopic dynamical mechanism. Taking into account the MC results we derive different FPE by using standard techniques. The deficiences of each of these FPE are discussed and one of them is found to be the most suitable in order to treat the model in a field theoretical context.

1. Introduction

Field theoretical methods have proved very useful in the study of equilibrium phase transitions; in particular, they have allowed to clarify the concepts of scaling and universality, and to obtain the values of critical exponents both in static and dynamical problems. An extension of these ideas to deal with nonequilibrium problems is presently of great interest. The model that we study here is the kinetic Ising model with two temperatures [1]. It simulates the presence of two thermal baths acting on a magnetic system. This model is a non-equilibrium one because of the presence of *dynamical frustration*, i. e., due to the competition between different dynamical mechanisms (each one characterized by a different temperature) the detailed balance condition does not hold and it does not seem possible to express the stationary solution as a Gibbs distribution. Since no exact method is available to solve this model a computer simulation has been carried out.

The simulation shows that contrarily to what happens in equilibrium problems, in this model there is a dependence of the macroscopic stationary properties on the analytical form of the microscopic dynamical mechanism (i.e. on the concrete realization of the master equation that defines the model)[2]. Therefore, any description of this model should take into account this dependence. In particular, the starting point for a field theoretical approach to a kinetic model is the so-called

Fokker-Planck Equation (FPE). Then, the latter equation has to contain sufficient microscopic kinetic information to properly describe the macroscopic behavior of the system.

The paper is divided in two parts. In the first one the model is defined and the MC simulation is described. In the second one we introduce different FPE which have been derived by using standard techniques. We discuss their shortcomings and select the most suitable to achieve our goals.

2. The Model and the MC simulation

Let us consider a dynamical Ising model in two dimensions with a total of ν spin variables $s_x = \pm 1, x \in \Lambda$ and Λ being a square lattice. The evolution of the system is governed by a markovian master equation in which only spin flip processes are considered:

$$\partial_t P(\mathbf{s},t) = \sum_x (w(\mathbf{s}^{\mathbf{x}},x)P(\mathbf{s}^{\mathbf{x}},t) - w(\mathbf{s},x)P(\mathbf{s},t)) \tag{1}$$

where $\mathbf{s}$ is a spin configuration, $P(\mathbf{s},t)$ is the probability of finding the system in $\mathbf{s}$ at a given time t, $\mathbf{s}^{\mathbf{x}}$ is the configuration obtained by flipping the spin at x in $\mathbf{s}$, and $w(\mathbf{s},x)$ is the probability per unit time for the transition from the configuration $\mathbf{s}$ to $\mathbf{s}^{\mathbf{x}}$.

For systems whose stationary state is an equilibrium one defined by a Hamiltonian $H(\mathbf{s})$, the rates, $w(\mathbf{s},x)$, that fulfil the detailed balance condition:

$$w(\mathbf{s},x)\exp(-\beta H(\mathbf{s})) = w(\mathbf{s}^{\mathbf{x}},x)\exp(-\beta H(\mathbf{s}^{\mathbf{x}})) , \tag{2}$$

guarantee that the system will evolve from almost all initial conditions to its final equilibrium state. Besides, it is usual to consider $w(\mathbf{s},x)$ depending only upon the variation of $H(\mathbf{s})$ in the transition. The rates verifying the latter two conditions can be written as: $w(\mathbf{s},x) = \varphi(\beta\delta H(\mathbf{s}))$, where $\delta H = H(\mathbf{s}^{\mathbf{x}}) - H(\mathbf{s})$ and $\varphi(z) = e^{-z}\varphi(-z)$. For example, two usual choices are: (i) $\varphi(z) = exp(-z/2)$ and (ii) $\varphi(z) = 1 - tanh(z/2)$. For a fixed Hamiltonian, the stationary properties of these models does not depend on the concrete form of $\varphi(z)$.

The kinetic Ising model with two competing temperatures consists on having a rate which is the superposition of two different mechanisms: $w(\mathbf{s},x) = pw_1(\mathbf{s},x) + (1-p)w_2(\mathbf{s},x)$. Each $w_j, j = 1,2$, when acting apart, drives the system to a different equilibrium distribution, $P_{st,j}(\mathbf{s}) = exp(-\beta_j H_{Ising}(\mathbf{s}))/\mathbf{Z}$ where β_j is the inverse of the jth temperature, $H_{Ising} = -J\sum_{|\mathbf{x}-\mathbf{y}|=1} s_{\mathbf{x}}s_{\mathbf{y}}$ and $0 \leq p \leq 1$. In dimension greater than one, the competition between both mechanisms prevents the system to fulfil a detailed balance condition with respect to any hamiltonian. As a consecuence, the stationary solution of the master equation will be a non equilibrium one.

In order to gain insight about the behavior of this system we have performed a MC simulation. We have taken a 96*96 lattice and considered two kinds of transition rates commonly used in the literature, namely, the ones given by (i) and (ii) above. The values of p and T_1 are maintained fixed in each simulation and T_2

is varied. We will not present here a detailed analysis of the simulations, but only expose the main qualitative results:

1- The curves of the net magnetization m versus T_2 have different shapes for different rates. In particular $m(T_2 = 0) = \pm 1$ and $|m(T_2 = 0)| < 1$ for the defined above rates (i) and (ii) respectively. Different behaviors are also found for the energies, specific heats, susceptibilities, and other magnitudes when the transition rates are modified. The differences are, as in the net magnetization case, both quantitative and qualitative.

2- When T_1 is close to the Onsager temperature, T_O, the system can be described quite well like an equilibrium one, with an effective temperature.

3- Although our results in this respect are not conclusive, it seems that the slopes of the magnetization curves at the critical point differ from the equilibrium one, and they also depend on the parameters as p or T_1. Thus, critical exponents may change. Futher simulations are required in order to better elucidate this point.

The main conclusion is that the analytical form of the transition rates influences dramatically the macroscopic behavior of the system.

3. Fokker Planck Equations (FPE)

As we have pointed out before, a field theoretical approach for this model, requires a FPE describing it as reliably as possible. There are different ways to approximate a given master equation by a FPE; none of them is rigorous, but it is possible to impose several conditions to the FPE that can make it much more reliable.

A usual way to get a FPE in equilibrium dynamical problems consists of replacing the original microcopic dynamical model by a continuous simpler one in such a way that the global properties of the stationary state, i.e. symmetries, phase transitions and criticallity.., are maintained. This is the case of the model A used to study dynamical properties of the non-conservative Ising model [3]. The success of this substitution in the study of equilibrium critical phenomena is guaranteed because of universality. In our nonequilibrium problem we can define a FPE in a similar way: the simplest one we can construct consists of the superposition of two models A, each of them with a different temperature. In this same level of aproximation we can also consider some approaches for driven diffusive systems[4], that are also non-equilibrium problems, but where the system is approximated by a modified model B[3]. This is in principle an oversimplification because the dependence on the microscopic dynamics is played down. Universality in non-equilibrium critical phenomena has not been proved in general. So, the construction of a FPE in the way described above is at least dangerous. We cannot guarantee that the critical behavior of our original model is the same that the one generated in this approximation. In fact, as it can be easily shown, this way to proceed reduces the problem to an effective equilibrium one, that is, the system behaves as if it were at equilibrium with an effective temperature defined by $T_{eff.} = pT_1 + (1-p)T_2$, and any other nonequilibrium effect, such as the dependence on the dynamics, is eliminated. This is in disagreement with what we observe in the MC simulation. Therefore the conclusion is that, as this kind of simplifications is not safe enough, in order to represent the system we need a more elaborated FPE that should describe the original model as closely as possible.

A second way to derive a continuous FPE is provided by truncating a Kramers Moyal expansion of the original master equation [5]. This standard approach respects a dependence on the microscopic dynamics, but it presents important deficiences. In particular, a disadvantage of this approach is that the detailed balance condition does not hold even in the equilibrium case, and so this form to do the continuum limit does not maintain the property that the stationary solution is an equilibrium one. This is a very dissapointing fact because even in the simplest case of systems at equilibrium spurious phenomena may appear. So we consider that this approach is not good enough for our purposes.

In order to overcome these drawbacks, we have derived a new FPE following the scheme introduced by Hanggi et al [6] which guarantees that, in the continuum limit, the stationary solutions for systems with equilibrium states is preserved. The FPE derived is:

$$\frac{\partial P(\phi;t)}{\partial t} = -\int d\mathbf{x} \left[\frac{\delta}{\delta\phi(\mathbf{x})} a[\phi(\mathbf{x})] + \frac{1}{\nu} \frac{\delta}{\delta\phi(\mathbf{x})} L[\phi(\mathbf{x})] \frac{\delta}{\delta\phi(\mathbf{x})} \right] P(\phi, t) \tag{5}$$

where $\phi(\mathbf{x})$ is the continuous magnetization at point $\mathbf{x}$. The coefficients $a_1(\phi, \mathbf{x})$ and $L(\phi, \mathbf{x})$, whose definition will be given elsewhere [7], depend on the concrete form of the transition rates. This FPE reproduces in the equilibrium limit the exact stationary solution and depends on the microscopic dynamics in nonequilibrium cases. Therefore it constitutes a convenient starting point to apply field theoretical techniques [8] to our model. In particular, starting from this FPE it is possible to derive the solution in a path integral form with an asociated lagrangean density which depends on the dynamics. This is the basis to renormalize the theory, to consider renormalization group techniques and to study, for example, the critical exponents of the system.

In conclusion, we have pointed out that the role of microscopic dynamics is essential in a non-equilibrium system and we have derived a suitable FPE to include properly this fact in a field theoretical treatment.

References

1. P.L. Garrido, A. Labarta and J. Marro. *J. Stat.Phys.* **49**, 551, (1987).
2. M. A. Muñoz and P. L. Garrido. Submitted to *Journal of Phys. A*
3. P. C. Hohenberg and B. J. Halperin. *Rev. Mod. Phys.* **49**, 435, (1977).
4. K. t. Leung, B. Schmittmann, and R. K. P. Zia. *Phys. Rev. Let.* **62**, 1772, (1989).
5. N. G. van Kampen, *'Stochastic methods in Physics and Chemistry'* North Holland, 1981.
6. P. Hanggi, H. Grabert, P. Talkner, and H. Thomas. *Phys. Rev.* **A29**, 371, (1984).
7. M. A. Muñoz and P. L. Garrido. In preparation.
8. H. K. Janssen *Z. Physik* **B23**, 377 (1976). R. Bausch, H. K. Janssen and H. Wagner. *Z. Physik* **B24**, 113, (1976).

METROPOLIS, LANGEVIN AND HYBRID SIMULATIONS OF A 1-D SPINLESS FERMION MODEL

GLORIA BUENDÍA
Departamento de Física, Universidad Simón Bolivar
Apartado 89000, Caracas 1080, Venezuela

ABSTRACT

We compare the performance of the Metropolis (exact updating), the Langevin and the hybrid algorithms in the simulation of a pure fermion system; the 1-D spinless fermion model with a half filled band. A Trotter expansion and a Hubbard-Stratonovich transformation are apply in order to eliminate the interacting term of the Hamiltonian. We calculate the density-density correlation function, the order parameter and the autocorrelation. The relation between the accuracy and the speed of the algorithms is analyzed.

1. Introduction

Since a complete understanding of lattice gauge theories, and also of electrons in crystals, can not be achieved without the inclusion of fermions, there has been an enormous amount of effort directed to the development of efficient algorithms to perform numerical simulations of systems with fermionic degrees of freedom. In the path integral formulation of field theories, fermions are represented by anticommuting, Grassmann's type variables, not accessible to direct numerical simulation, thus, they must be replaced by some appropriate boson functional in the action. Unfortunately, the price paid for the elimination of the fermionic variables is the presence of a non-local action that includes a determinant. The computer time needed to update one single link of the lattice, with the Metropolis algorithm, grows with the cube of the lattice volume. This makes highly costly the study of even small systems.

The main purpose of the so-called dynamical simulation algorithms is to develop a global updating procedure where the determinant does not need to be calculated as often as in the Metropolis method. In this category, fall the Langevin and the hybrid methods and the most recent: *the Hybrid Monte Carlo algorithm.* The basic idea behind these techniques is to introduce an extra degree of freedom, a fake time, and describe the evolution of the system by a set of differential equations, in such a way that for large times, the system will reach equilibrium. The computer time required to perform a complete sweep (with a small step size) through the lattice grows linearly with volume, which represents a big improvement over the Metropolis algorithm. But, the numerical solution of the differential equations introduces systematic errors in the physical results. Then, it is not obvious which technique will give the best compromise between accuracy and speed.

We are going to simulate the 1-D spinless fermion model with a half filled band, a pure fermion system. This is a reasonably well-known model, the study of the performance of the algorithms will not be obscured by uncertainties about the correct behaviour of the system. A difference with more complicated models is that this system can be simulated with an exact algorithm at a reasonably cost, providing us a way to perform a systematic study of the errors for the different methods.

2. The Model

The Hamiltonian of the model is

$$H = -t\sum_i (c_i{}^\dagger c_{i+1} + c_{i+1}{}^\dagger c_i) + u\sum_i (n_i - \frac{1}{2})(n_{i+1} - \frac{1}{2}) \tag{1}$$

where $n_i = c_i{}^\dagger c_i$ is the number operator of fermions at site i, t is the single fermion transfer integral between sites, and u is the strength of the nearest-neighbour repulsion. This system has a phase transition when the temperature is zero, then, by decreasing the temperature we will able to study the behaviour of the algorithms near critical points. In order to eliminate the interacting term and perform the integration over the fermionic degrees of freedom we perform a Trotter expansion; we divide the imaginary time interval into L equal subintervals of width $\Delta\tau$, such that $L\Delta\tau = \beta$, where β is the inverse temperature, then we apply a continuous Hubbard-Stratonovich transformation[6]. After these steps we obtain a partition function of the form

$$Z = (2\pi\Delta\tau)^{\frac{NL}{2}} \int_{-\infty}^{+\infty} \prod_i dx_{i,i+1} \exp[-S_B(x)] det[1 + T\exp(-\int_0^\beta d\tau h(\tau))] \tag{2}$$

where

$$S_B(x) = \int_0^\beta d\tau \frac{1}{2} \sum_i x_{i,i+1}{}^2 \tag{3}$$

is the bosonic part of the action, T means time ordering and $h(t)$ is an $N \times N$ matrix, where N is the number of spatial sites of the system. The elements of h for the time slice l are,

$$h(\tau_l)_{ij} = -t(\delta_{i,j-1} + \delta_{i,j+1}) + \sqrt{u}\,[x_{i,i+1}(\tau_l) + x_{i,i-1}(\tau_l)]\delta_{i,j} \tag{4}$$

Introducing an $N \times N$ matrix at each time interval

$$B(l) = \exp[-\Delta\tau h(\tau_l)] \tag{5}$$

we can write

$$Z = (2\pi\Delta\tau)^{\frac{NL}{2}} \int_{-\infty}^{+\infty} \prod_i dx_{i,i+1} \exp[-S_B(x)] det[1 + B_L B_{L-1} .... B_1] \tag{6}$$

Also, there is a discrete version[8] of the Hubbard-Stratonovich transformation which has proven to be more efficient than the continuous one[9], but in order to perform the Langevin and the hybrid updatings that involve differential equations, we are limited to the standard transformation.

We applied an algorithm developed by Blankenbecler, Scalapino and Sugar[7], that consist in writing the change in the effective action in terms of Green's function with a considerable reduction of computer time. This technique was created with the Metropolis algorithm in mind, however we were able to prove that it can be extended successfully to the Langevin and the Hybrid algorithms. The equal-time Green's function at the time slice l, defined as

$$g_l(i,j) = < c_l{}^{\dagger}(i) c_l(j) > \tag{7}$$

can be written in term of the B's matrices as

$$g_l(i,j) = (1 + B_{l-1} B_{l-2} \ldots\ldots B_1 B_L \ldots\ldots B_l)^{-1}(i,j) \tag{8}$$

3. Summary and Results

We work in a lattice of eight spatial sites and select a Trotter expansion parameter, $\Delta\tau = 0.1$, and perform simulations at inverse temperatures, $\beta = 1,2,3,4$. We calculate the density-density correlation function,

$$d(j) = < n(i) n(i+j) > \; ; \; n(i) = 1 - g(i,j) \tag{9}$$

the order parameter,

$$O = \frac{1}{N} \sum_{i=1}^{N} (-1)^i (n_i - \frac{1}{2}) \tag{10}$$

and the autocorrelation function after M sweeps,

$$C(M) = \frac{< O'(m) O'(M+m) - < O'(m) >^2}{< O'^2(m) > - < O'(m) >^2} \tag{11}$$

where $O'(m) = abs[O(m)]$, for sequences of measurements obtained with the Metropolis, the hybrid and the Langevin algorithms. From these results we perform a systematic study of the errors and the relative efficiency of the techniques (for the detailed results see reference 10).

We found that at high temperatures ($\beta = 1,2$), i.e., closer to the phase transition, the hybrid method is definitively the best choice, is faster and gives more accurate results. The Langevin algorithm is faster than the Metropolis but slower than the hybrid. By decreasing the temperature ($\beta = 3,4$), we found that in order to keep the stability of the Langevin and the hybrid updating, we need to use very small step sizes when solving numerically the differential equations, with the consequently slowing down. For the size chosen, the system evolves so slowly that these techniques become even more computationally expensive than the exact updating.

We think that the reason for this abrupt decrease in the allowed step size for small temperatures is related to the fact that at low temperatures, the Green's function becomes poorly conditioned and some of its eigenvalues are very big, then we must proportionately reduce the step size to maintain the stability. A clear example of these problems is the free field theory, where the step size is determined by the biggest frequency of the system[11].

We conclude than the Langevin and the hybrid algorithms in particular the last one, are fast alternatives to the simulation of fermionic systems not very close to a transition point. Closer to phase transitions it is not clear they can be considered a real alternative to the exact updating, probably for big systems they are faster, but at a considerable lost in accuracy.

Acknowledgements

I want to thank Doug Toussaint for his help throughout the course of this work.

References

1. N. Metropolis, A. Rosenbluth, M. Rosenbluth, A. H. Teller, and E.Teller, *J. Chem. Phys.* **21** (1963) 1987.
2. D. Callaway and R. Rahman, *Phys. Rev. Lett.* **49** (1982) 613.
3. G. Parisi and Wu Yongshi, *Sci. Sin.* **24** (1981) 483 ; G. G. Batrouni, G. R. Katz, A. S. Kronfeld, G. P. Lepage, B. Svetitsky and K. G. Wilson, *Phys. Rev. D***32** (1985) 2736.
4. S. Duane, *Nucl. Phys. B* **257** (1985) 652.
5. S. Duane, A. D. Kennedy, B. J. Pendleton and R. Roweth, *Phys. Lett. B* **195** (1987) 216.
6. J. Hubbard, *Phys. Rev. Lett.* **3**, (1959) 77.
7. R. Blankenblecler, D. J. Scalapino and R. L. Sugar, *Phys. Rev. D* **24** (1981) 2278.
8. J. E. Hirsch, *Phys. Rev. B* **23** (1983) 4049.
9. G. M. Buendía, *Phys. Rev. B* **33** (1986) 3519.
10. G. M. Buendía, *Phys. Rev. B* **42** (1990) 8163.
11. G. M. Buendía, *J. Phys. A* **22** (1985) 5065.

COMPUTER CALCULATIONS IN QUANTUM CHROMODYNAMICS

BARTOLOMÉ ALLÉS
Departamento de Física Teórica y del Cosmos
Facultad de Ciencias. Universidad de Granada
18071-Granada, España

ABSTRACT

A Monte Carlo method is used to evaluate vacuum expectation values of dimensionfull operators. The calculations of the gluon condensate and the topological susceptibility in quantum chromodynamics are reviewed.

1. Introduction

Quantum chromodynamics (QCD) is the theory of quarks and gluons [1]. However this theory is intended to explain eventually the physics of strong interactions and hadron properties. The QCD lagrangian is

$$\mathcal{L}_{QCD} = -\frac{1}{4} G^b_{\mu\nu}G^b_{\mu\nu} + \bar{q}_f(i\not{D} - m_f)q_f. \tag{1}$$

In this expression, summation on colour indices b, flavours f, and Lorentz indices $\mu\nu$ is understood. This lagrangian is invariant under the gauge group $SU(3)$.

The vacuum expectation value (VEV) of some operators can help in understanding hadron properties such as their masses. For example, condensates are essential tools for the SVZ sum rules [2], and useful for vacuum modelling. Also, the knowledge of the topological susceptibility is necessary to solve the so-called $U_A(1)$ problem and to determine the mass of the η' particle [3]. In this talk we review the methods used to determine the gluon condensate G_2,

$$G_2 \equiv \langle 0 \mid : \frac{g^2}{4\pi^2} G^b_{\mu\nu}G^b_{\mu\nu} : \mid 0 \rangle, \tag{2}$$

and the topological susceptibility χ,

$$\chi \equiv \int d^4x \, \langle 0 \mid \ T(Q(x)Q(0)) \ \mid 0 \rangle, \tag{3}$$

where $Q(x)$ is the topological charge density,

$$Q(x) \equiv \frac{g^2}{64\pi^2} \epsilon^{\mu\nu\rho\sigma} G^b_{\mu\nu}G^b_{\rho\sigma}. \tag{4}$$

In these equations, g is the coupling constant. In the language of statistical mechanics, G_2 is proportional to the internal energy. On the other hand, the topological susceptibility is the zero frequency vacuum response to the topological charge density [4].

2. The Method

Perturbation theory cannot be trusted when computing quantities like those in Eqs. (2) and (3) because they have nontrivial dimensions in mass. Indeed, if an operator $\mathcal{A}$ has mass dimensions equal to d, the renormalization group tell us that $\langle 0 \mid \mathcal{A} \mid 0 \rangle \sim \exp(-8\pi^2 d/11g^2)$ and this function cannot be expanded in positive powers of the coupling constant g.

The VEV of any operator $\mathcal{A}$ is equal to [5]

$$\langle 0 \mid \ \mathcal{A} \ \mid 0 \rangle \ = \ \mathcal{N} \ \int \mathcal{D}A_\mu^b(x) \ \exp\left(-\int \mathcal{L}_{QCD}\right) \ \mathcal{A}, \tag{5}$$

where $\mathcal{N}$ is a normalization constant and $\mathcal{D}A_\mu^b(x)$ is the measure of the functional integral. Perturbation theory consists in Taylor expanding the interaction terms in the exponential of Eq. (5) in order to get only gaussian integrals which can be performed analytically. A direct integration of Eq. (5) is possible by Monte Carlo methods [6].

However, the integral in Eq. (5) has an infinite number of integration variables: the field $A_\mu^b(x)$ at every spacetime point x and every colour and Lorentz index. Therefore, to perform this integral, firstly we have to discretize the space-time in order to have a finite number of integration variables. The resulting reticulated space-time is a lattice. Allowed physical space-time points lie only on the intersections of the lattice. The distance between neighbour sites is the lattice spacing. Lattices have a finite number of sites and consequently a finite volume. Typical lattices used to study the VEV's in Eqs. (2) and (3) are hypercubes of 16^4 sites with a lattice spacing a. The fact that finite lattices imply a finite volume does not prevent us from obtaining sensible physical results for G_2 and χ. In the next section we see the consequences of having a finite lattice spacing.

Then, a lattice version $\mathcal{A}^L$ of the operator $\mathcal{A}$ can be written. This lattice version has to satisfy the relation $\mathcal{A}^L \stackrel{a \to 0}{\longrightarrow} a^d \mathcal{A} + \mathcal{O}(a^{d+1})$ where d is the mass dimension of $\mathcal{A}$. Usual lattice versions for the operators in Eqs. (2) and (3) are

$$\langle 0 \mid (1 \ - \ \Pi_{\mu\nu}) \mid 0 \rangle \tag{6}$$

for the gluon condensate, and

$$\sum_x \langle 0 \mid \ Q^L(x) Q^L(0) \ \mid 0 \rangle \tag{7}$$

for the topological susceptibility. $Q^L(x)$ is [7]

$$Q^L(x) \ = \ - \ \frac{1}{2^4 32\pi^2} \ \sum_{\mu,\nu,\rho,\sigma=\pm 1}^{\pm 4} \epsilon^{\mu\nu\rho\sigma} \mathrm{tr}(\Pi_{\mu\nu}(x)\Pi_{\rho\sigma}(x)). \tag{8}$$

In these equations $\Pi_{\mu\nu}$ is the plaquette in the $\mu - \nu$ direction,

$$\Pi_{\mu\nu}(x) \equiv U_{x,\mu} U_{x+\hat{\mu},\nu} U^{\dagger}_{x+\hat{\nu},\mu} U^{\dagger}_{x,\nu},$$

$$U_{x,\mu} \equiv e^{igaT^b A^b_\mu(x)}, \tag{9}$$

where g and a are respectively the coupling constant and lattice spacing and T^b are the generators of the gauge group in the fundamental representation. The correct continuum limit of these operators is satisfied as it can be easily seen by expanding Eq. (9) for small values of a.

3. Monte Carlo Data

Once the lattice is defined and the lattice version A^L of the original operator is given, we can perform the Monte Carlo calculation of the integral Eq. (5). We will use the so-called quenched approximation which essentially consists in removing fermion loops. To this end we can take as the QCD lagrangian only the first term in Eq. (1).

The lattice spacing a acts as a cutoff of the theory and therefore the VEV $\langle 0 \mid A^L \mid 0\rangle$ calculated on the lattice will include mixings with operators of equal or lower dimensions having the same quantum numbers of $\mathcal{A}$. As a consequence, Eqs. (6) and (7) are not equal respectively to G_2 and χ but to the following a-dependent expressions [8,9]

$$\langle 0 \mid (1 - \Pi_{\mu\nu}) \mid 0 \rangle = \frac{\pi^2}{36} Z_G\, a^4 G_2 + \sum_{n\geq 1} c_n g^{2n} + \mathcal{O}(a^6), \tag{10}$$

$$\sum_x \langle 0 \mid Q^L(x) Q^L(0) \mid 0 \rangle = Z_\chi^2\, a^4 \chi + a^4 G_2 \sum_{n\geq 2} b_n g^{2n} + \sum_{n\geq 3} d_n g^{2n} + \mathcal{O}(a^6). \tag{11}$$

The last terms in these expressions (those proportional to c_n and d_n) are called perturbative tails. They represent mixings with the unity operator. The second term in Eq. (11) is a mixing with the gluon condensate [10]. Finally, Z_G and Z_χ are multiplicative finite renormalizations which relate the lattice and continuum definitions of the respective operators [11].

All of these coefficients can be computed in perturbation theory. In fact, their knowledge is very important for the correct evaluation of G_2 and χ. For example, the perturbative tail in Eq. (10) represents almost the 90% of the Monte Carlo signal while in Eq. (11) it amounts to the 50% of the whole signal. On the other hand, in the scaling window, $Z_\chi^2 \sim 0.1$.

Unfortunately, the evaluation of these coefficients is by far not easy. For instance, c_1 and c_2 in Eq. (10) can be calculated with moderate effort, but the next coefficient, c_3, needs the computation of 30 three-loop Feynman diagrams! For this reason, an algebraic computer program has been developed [12] using the software *MATHEMATICA* [13].

This program works in three stages. Firstly it writes down the vertices of the lagrangian and of the operators. Due to the fact that on the lattice the rotational symmetry is lost and that we work with a finite lattice spacing a, the vertices have longer expressions and there are more kinds of them. For instance in the continuum there are no more than 3-legged vertices for the topological charge, instead, on the lattice, there are vertices with an arbitrarily large number of legs. Secondly the diagram is calculated and we arrive to an expression that has to be integrated. Performing this integral is the last step. For three or more loops, these integrals must be done on small lattices and the result can be extrapolated to larger lattices.

The expressions for the renormalization constants in Eqs. (10) and (11) are [7,8,10,11,12]

$$Z_G = 1 + \mathcal{O}(g^4), \tag{12}$$

$$Z_\chi = 1 - 0.91\ g^2 + \mathcal{O}(g^4), \tag{13}$$

$$\sum_{n\geq 1} c_n g^{2n} = 0.333\ g^2 + 0.034\ g^4 + 0.014\ g^6 + \mathcal{O}(g^8), \tag{14}$$

$$\sum_{n\geq 3} d_n g^{2n} = 1.66\ 10^{-5}\ g^6 + 6.50\ 10^{-7}\ g^8 + \mathcal{O}(g^{10}), \tag{15}$$

$$\sum_{n\geq 2} b_n g^{2n} = 1.76\ 10^{-4}\ g^4 + \mathcal{O}(g^6). \tag{16}$$

The lattice spacing a and the bare coupling g are related through the renormalization group equation

$$a\ \Lambda_L = \left(\frac{16\pi^2}{11g^2}\right)^{51/121} \exp\left(-\frac{8\pi^2}{11g^2}\right). \tag{17}$$

In this formula, Λ_L is a renormalization group invariant mass parameter and its value is approximately $\Lambda_L \simeq 4.4$ MeV [14].

Now, with these expressions we can perform a best fit of all Monte Carlo data to extract G_2 and χ from Eqs. (10) and (11), respectively. The results are [8,15]

$$G_2 = 2.79(33)\ 10^8\ \Lambda_L^4, \qquad \chi = 2.77(32)\ 10^6\ \Lambda_L^4, \tag{18}$$

in good agreement with the expected experimental values.

Acknowledgements

I wish to thank, for their hospitality, the Department of Physics of the University of Pisa where most of this work was done. This work has been partially supported by the Spanish-Italian "Accion Integrada" number $A17$ and a CICYT contract.

References

1. T. Muta, *Foundations of Quantum Chromodynamics* (World Scientific, Singapore, 1987).
2. M. A. Shifman, A. I. Vainshtein and V. I. Zakharov, *Nucl. Phys.* **B147** (1979) 385, 448, 519.
3. E. Witten, *Nucl. Phys.* **B156** (1979) 269; G. Veneziano, *Nucl. Phys.* **159** (1979) 213.
4. R. J. Crewther, *Rivista del Nuovo Cimento* **2** (1979) 63.
5. C. Itzykson and J. B. Zuber, *Quantum Field Theory* (McGraw-Hill, New York, 1980).
6. M. J. Creutz, *Quarks, gluons and lattices* (Cambridge University Press, Cambridge, 1983); C. Rebbi *Lattice Gauge Theories and Monte Carlo Simulations* (World Scientific, Singapore, 1983).
7. P. Di Vecchia, K. Fabricius, G. C. Rossi and G. Veneziano, *Nucl. Phys.* **B192** (1981) 392; *Phys. Lett.***B108** (1982) 323.
8. M. Campostrini, G. Curci, A. Di Giacomo and G. Paffuti, *Z. Phys.* **C32** (1986) 377; M. Campostrini, A. Di Giacomo and Y. Gündüç, *Phys. Lett.* **B225** (1989) 393.
9. M. Campostrini, A. Di Giacomo, H. Panagopoulos and E. Vicari, *Nucl. Phys.* **B329** (1990) 683.
10. B. Allés and M. Giannetti, *Phys. Rev.* **D44** (1991) 513.
11. M. Campostrini, A. Di Giacomo and H. Panagopoulos, *Phys. Lett.* **B212** (1988) 206; B. Allés and E. Vicari, *Phys. Lett.* **B268** (1991) 241.
12. B. Allés, M. Campostrini, A. Feo and H. Panagopoulos, *Lattice Perturbation Theory Done Symbolically: A Three-Loop Result for the Topological Susceptibility*, Pisa preprint IFUP-TH 31/92; *The Three-Loop Lattice Free Energy*, Pisa preprint IFUP-TH 32/92.
13. S. Wolfram *Mathematica* (Addison-Wesley Publishing Company, 1988).
14. C. Michael, *Nucl. Phys. B (Proc. Suppl.)* **4** (1988) 209.
15. M. Campostrini, A. Di Giacomo, Y. Gündüç, M. P. Lombardo, H. Panagopoulos and R. Tripiccione, *Phys. Lett.* **B252** (1990) 436.

SIMULATION OF A SET OF SUPERCONDUCTING SPHERES IN METASTABLE STATE IN A BIDIMENSIONAL LATTICE

S. LAIN and J. G. ESTEVE
Departamento de Física Teórica, Facultad de Ciencias
Universidad de Zaragoza, 50.009 Zaragoza, Spain

ABSTRACT

We propose a model to study the diamagnetic interactions among small superconductor spheres. The ensuing results are analyzed, and it is proposed a qualitatively explanation, but the problem remains open.

1. Introduction

The particle physics tendencies in recent years need a new non- conventional detection methods. Among the new generation of detectors those most studied are those of solid state and in particular those that make use of superheated superconducting spheres (SSS)[1].

These small grains have a few microns of diameter and they are superconducting metals of type I, which undergo a first order phase transition between the normal and superconducting states. As in any other first order phase transition, metastable states can be reached if for a fixed temperature (below the critical one) the applied magnetic field B varies between B_{sc} and B_{sh}. Starting with $B > B_{sc}$ and decreasing the strength of B, a normal conducting sample remains in a metastable state until certain value $B = B_{sc} < B_c$ is reached; at this point the transition to the superconducting state occurs. In the same way, starting from a superconducting sample, the field strength can be increased above B_c without phase transition and the sample remains in a metastable superconducting state until certain value $B = B_{sh} > B_c$ is reached, at this point the sample becomes normal conducting.

In order to discriminate between interesting and non interesting events taking place in the detector, it is very important that one can simulate the physics of the detector which in the present case is limited by the existence of diamagnetic interactions between the superconducting spheres. There is a little theoretical and experimental information about this kind of interactions so that we propose a simplified model. The system we work over is the following:

(1) We shall work over a bidimensional lattice made by small superconducting spheres with the same radius a and lattice constant R (distance between centers of consecutive spheres). This lattice is in the influence of a uniform perpendicular magnetic field.

(2) We neglect the penetration of the magnetic field into the superconducting spheres. This is justified when the separation between spheres' surfaces is greater

than 5 per cent of their radii.

(3) We shall always work at constant temperature.

(4) We assume that a superconducting grain completely reaches the normal state as soon as the magnetic field value at any point on its surface is greater than $\vec{B}_{sh}$. This assumption is certainly correct if the external field strength $\vec{B}_{ex}$ is larger than $\vec{B}_c$ since then the total volume of the sphere is in a metastable state as it will be our case.

2. Mathematical Development

Let us assume that in the space between the spheres there is no density current and magnetization. Then we can use Maxwell equations

$$\nabla \times \vec{B}(\vec{r}) = 0 \tag{1}$$

$$\nabla \cdot \vec{B}(\vec{r}) = 0 \tag{2}$$

so that we can write $\vec{B}(\vec{r}) = -\nabla U(\vec{r})$ and $U(\vec{r})$ is solution of the Laplace equation

$$\Delta U(\vec{r}) = 0 \tag{3}$$

The boundary conditions are the following:

(1) The normal component of $\vec{B}(\vec{r})$ on the spheres' surface is null, due to the Meissner-Oschenfeld effect and the continuity of the normal component of magnetic field in the interphase.

(2) If we separate the spheres sufficiently far the magnetic field has to be the applied one, i.e.

$$-\nabla U(\vec{r}) \rightarrow \vec{B}_{ex} \qquad r \rightarrow \infty$$

Solving the Laplace equation for the potential with these two boundary conditions we get a general solution including every interaction among the spheres.

In order to make the problem easier we shall suppose that the interactions are only between pairs of spheres (we make this assumption because we hope it will be the dominant contribution in dilute lattices). That means that the field at the surface of a grain is evaluated as a sum of two-body interactions, neglecting interactions of higher order. This approximation was studied by Mazur[2].

The contribution to the magnetic field on a sphere's surface by a second one will be

$$\frac{B^2_{2\rightarrow 1}(\theta,\phi)}{B^2_{ex}} = (\cos\phi\sin\theta)^2 \cdot \left\{ \left[\sum_{q=1}^{\infty} (q-1)! P''_q(\cos\theta) d_q \right]^2 - \right.$$
$$\left. \left[\sum_{q=1}^{\infty} (q-1)!(P'_q(\cos\theta) + P''_q(\cos\theta)\cos\theta) d_q \right]^2 \right\} + \left(\sum_{q=1}^{\infty} (q-1)! P'_q(\cos\theta) d_q \right)^2 \tag{4}$$

where θ and ϕ are the polar angles on sphere 1, $P'(\cos\theta)$ and $P''(\cos\theta)$ are the derivates 1st and 2nd of Lagrange polinomials on the surface and d_q are geometrical coefficients depending on the ratio between the spheres' radius and the distance between the centres of the spheres, which follows some kind of recurrence.

$$d'_q = (-1)^{q+1}\frac{(q+1)!}{2q+1}d_q \tag{5}$$

$$d'_q = \delta_{1,q} + \sum_{p=1}^{\infty}\frac{(q+p)!}{q!p!}\frac{qp}{(q+1)(p+1)}\left(\frac{a}{R}\right)^{q+p+1} d'_p \tag{6}$$

To evaluate the last expression we calculate coefficients until the infinite sum in Eq. (4) converges with a given accuracy (10^{-4}). This convergence is obtained with fifteen terms in the sum.

3. Simulation and Results

Limited by the memory capacity and CPU time of our computer, we work in a 100 × 100 lattice which is large enough to avoid the boundary effects being important.

We have divided the neighbouring spheres to a given one in consecutive layers. For a fixed sphere in the lattice, the nearest eight grains constitute the first layer, the next sixteen ones the second layer, the next twenty four ones the third layer and so on.

In the first place we choose the number of layers which contribute to the magnetic field on the surface of each sphere. Then we evaluate Eq. (4) for all the neighbouring spheres of a given one and we sum all these contributions with the magnetic field of an isolated sphere; this is made for each grain of the lattice and we perform the following steps:

1.- We fix a certain external magnetic field $\vec{B}_{ex}$, and we evaluate the maximum field strength on the surface of each sphere, taking in account all two-body diamagnetic interactions.

2.- We count the number of superconducting spheres which have a field strength larger or equal than $\vec{B}_{sh}$ in at least a point of their surface.

3.- We choose at random a grain of those specified before. This grain undergoes the phase transition, and consequently, the magnetic field in the spheres in its neighbourhood decreases.

4.- The previous step involves a recalculation of the magnetic field on its neighbouring spheres' surfaces.

5.- We return to the second step until the number of spheres able to undergo the phase transition, for a given value of external field, is null. In this case we increase the external field value.

6.- We repeat the steps from two to five until all spheres are in a normal state.

As result we get a graphic with the fraction of superconducting spheres versus the external field strength.

The plot shows the results of the simulation considering 2,3 and 4 layers for a ratio $R/a = 2.5$, (see Fig.1)

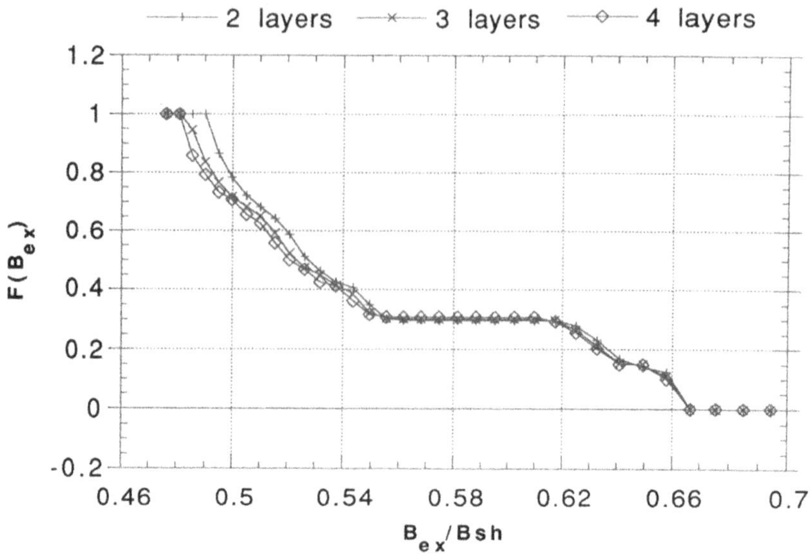

Figure 1: Fraction of superconducting spheres versus applied field strength for a ratio $R/a = 2.5$ and different number of layers.

We realize that the three curves are similar, so that from now on we shall deal with only three layers. In the following graph we show the results for different ratios R/a (Fig.2)

The main two points to take into account are

(1) Instead of showing a monotonously decreasing curve, for the shorter lattice constants R we find three sections: two decreasing ones and a plain zone or "plateau". However for higher R's we get a continuous fall.

(2) The "plateau" evolves with R until its disappearance but it is always located at $F(\vec{B}_{ex}) = 0.3$ as it is shown in the plots.

We can explain the "plateau", at least qualitatively, in the following way: For higher values of R, we are working in a dilute regime (one sphere is far away from the others) so the diamagnetic interactions between them are not very important and each time a sphere becomes normal conducting it does not greatly affect the spheres which are located in its neighbourhood; consequently, the plot shows a rapidly decreasing function with increasing field strength (for $R \to \infty$ the plot should approach a θ function).

At lower values of R, we are working in the concentrate regime (each sphere is near its neighbours) so the diamagnetic interactions among them become crucial and when a sphere becomes normal conducting it strongly decreases the maximun field strength on the surface of its neighbours. In this way the falling of the fraction

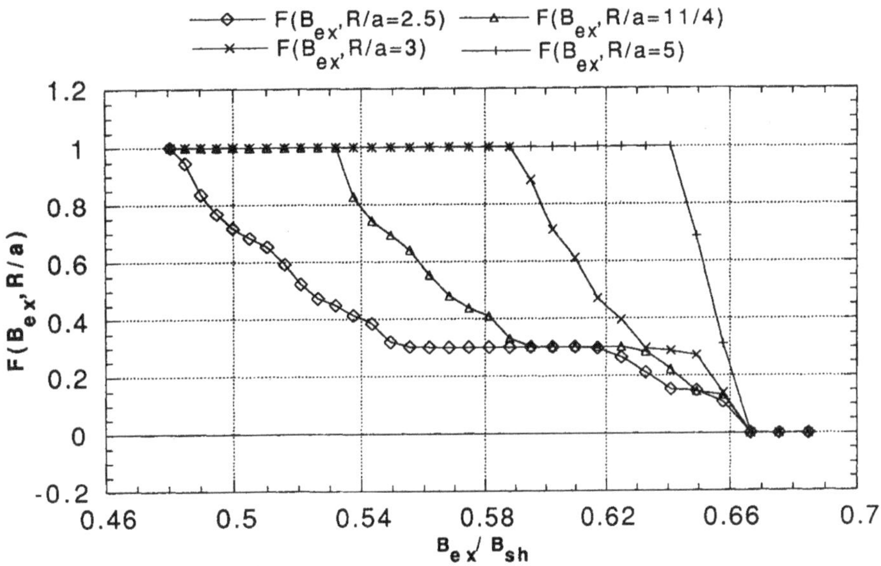

Figure 2: Fraction of superconducting spheres versus applied field strength for different R/a ratios and three layers.

of superconducting spheres with increasing field strength is lower than in the dilute regime, and it forces an effective dilution of the system (a given superconducting grain sets further and further from its neighbouring superconducting spheres).

Between both regímes and interpolating between them, there is the "plateau"; its existence is a consequence of the evolution of the system during the concentrate regime and the fact that the diamagnetic interactions between the spheres are now less important as a consequence of the effective dilution of the system. However, we have not found any theoretical explanation for the fact that the plain zone is always located at $F(B_{ex}) = 0.3$ independently of the initial value of R $(2 < R/a \leq 3.5)$; this fact could be related to some "conformal" property.

The existence of the "plateau" for small values of R has to be proved experimentally and nowadays it is an open question.

4. Conclusions

We have performed the simulation of a superconducting lattice of micrograins in metastable state in two dimensions put into a perpendicular external magnetic field. In this system we have studied the diamagnetic interactions among the microspheres and we have obtained the ratio of superconducting spheres versus the applied magnetic field.

These graphics show, for the lower ratios R/a studied (R is the lattice constant and a the spheres' radius), the appearance of a "plateau" while for higher ratios R/a it disappears. We have interpreted the "plateau" as an interpolated state between two qualitatively different dilution regimes: initially homogenization and, at the end, regular dilution.

References

1. H. Bernas, J. P. Burger, G. Dutscher, C. Valette and S. Williamson, *Phys. Lett.* **24 A** (1967) 721. G. Waysand, in *Rencontre de Moriond on Masive Neutrinos*, ed. J. Than Van (Edit. Frontieres, Gif-sur-Yvette, 1984). L. Stodolsky, in *Proceedings of Low Temperature Detectors for Neutrinos and Dark Matter*, ed. K. Pretzl, N. Schmitz, L. Stodolsky (Springer Verlag, 1987), p. 1. S. Lain and J. G. Esteve, DFTUZ92.40, Preprint.
2. P. Mazur and U. Geigenmuller, *Physica* **136 A** (1986) 316. U. Geigenmuller, *On electric and magnetic properties of dispersions of spheres*, PhD. thesis (University of Leyden, 1987).

DYNAMICAL ASPECTS OF ^{3}He DROP COLLISIONS

M. GUILLEUMAS, M. BARRANCO, M. PI

Departament d'Estructura i Constituents de la Materia
Facultat de Física, Universitat de Barcelona,
Diagonal 647, 08028 Barcelona, Spain

and

E. SURAUD
GANIL, BP5027, 14021 Caen Cedex, France

ABSTRACT

We studied the dynamics of two colliding ^{3}He drops by solving the Vlasov equation in phase space. We have used a set of coherent states as an overcomplete basis for the decomposition of the phase-space distributions.

In a recent work[1] we have studied different aspects of the fission and fusion of ^{3}He drops from a static point of view, using a density functional arising from an effective ^{3}He-^{3}He interaction[2]. It is the aim of this contribution to present some preliminary results obtained applying that density functional to the dynamical description of ^{3}He drop reactions.

The ^{3}He dynamics is described by the Vlasov equation [3,4] which gives the time evolution of the one-body phase-space distribution under the influence of the mean field of the drop:

$$\left[\frac{\partial}{\partial t} + \frac{\mathbf{p}}{m}\nabla_{\mathbf{r}} - (\nabla_{\mathbf{r}}V)\nabla_{\mathbf{p}}\right] f(\mathbf{r},\mathbf{p},t) = 0\,. \tag{1}$$

In this equation $f(\mathbf{r},\mathbf{p},t)$ is the phase-space distribution function and $V(\mathbf{r})$ is the self-consistent single-particle potential derived from the effective interaction.

Equation (1) is a semiclassical approximation to more sophisticated self-consistent quantum evolution equations. In particular, it can be derived as the $\hbar \to 0$ limit of the Wigner transform of the time dependent Hartree-Fock equation.

The Vlasov equation, which is formally equivalent to the Liouville equation [5], is difficult to solve numerically, especially for finite systems. Let us briefly outline the pseudo-particle method [3], which we have used in the present application. It consists in projecting the distribution function $f(\mathbf{r},\mathbf{p},t)$ over a complete basis of the Hilbert space relative to the phase-space metric $d\tau = d\mathbf{r}\,d\mathbf{p}$:

$$f(\mathbf{r},\mathbf{p},t) = \int d\mathbf{r}_0\, d\mathbf{p}_0\, w(\mathbf{r}_0,\mathbf{p}_0)\, d(\mathbf{r}-\mathbf{r}_0,\mathbf{p}-\mathbf{p}_0,t)\,, \tag{2}$$

where $w(\mathbf{r_0}, \mathbf{p_0})$ is a static phase-space distribution and $d(\mathbf{r}-\mathbf{r_0}, \mathbf{p}-\mathbf{p_0}, t)$ is a set of dynamical functions which plays the role of a moving basis for the distribution function.

Inserting the distribution function (2) into Eq. (1), one obtains a suplementary condition: the ansatz (2) provides us with a solution of the Vlasov equation if the basis functions are themselves solutions. The Generalized Coherent States (GCS) [6] fulfil this condition, and when the local harmonic approximation is valid the GCS are gaussian wave packets:

$$d(\mathbf{r}-\mathbf{r_0}, \mathbf{p}-\mathbf{p_0}, t) = \frac{1}{(2\pi\sigma_r\sigma_k)^3} \exp\left[-\frac{(\mathbf{r}-\mathbf{r_0})^2}{2\sigma_r^2} - \frac{(\mathbf{p}-\mathbf{p_0})^2}{2\sigma_p^2}\right] . \tag{3}$$

In numerical calculations one discretizes Eq. (2) as

$$f(\mathbf{r}, \mathbf{p}, t) = \frac{A}{N} \sum_{i=1}^{N} w(\mathbf{r_i}, \mathbf{p_i})\, d(\mathbf{r}-\mathbf{r_i}, \mathbf{p}-\mathbf{p_i}, t)\,, \tag{4}$$

where A is the number of atoms and N is the number of coherent states that are randomly selected in a phase-space volume defined by $w(\mathbf{r}, \mathbf{p})$. The number N is chosen by requiring numerical stability of the solutions against variations of N. We have taken N equal to 30 times the number of atoms in the drop.

Direct substitution of Eq. (3) into Eq. (1) shows that $d(\mathbf{r}-\mathbf{r_0}, \mathbf{p}-\mathbf{p_0}, t)$ is a solution if the gaussian centers $\mathbf{r_0}$ and $\mathbf{p_0}$ are solutions of the Hamilton equations taken in the Ehrenfest sense, i.e.,

$$\frac{d\mathbf{r_0}}{dt} = \frac{\mathbf{p_0}}{m} \tag{5.a}$$

$$\frac{d\mathbf{p_0}}{dt} = -\int d\mathbf{r}\, d\mathbf{p} \nabla V(\mathbf{r})\, d(\mathbf{r}-\mathbf{r_0}, \mathbf{p}-\mathbf{p_0}, t)\,. \tag{5.b}$$

It means that the explicit time dependence of the solutions (4) of the Vlasov equation arises from the time dependence of $(\mathbf{r_0}, \mathbf{p_0})$ according to Eq. (5). Once the initial conditions are fixed i.e., the value of $w(\mathbf{r_0}, \mathbf{p_0})$, to solve the Vlasov equation one has to follow the trajectory of the phase-space gaussian cells which play the role of a moving basis. To do so, we have discretized Eq. (5) using the 'leap-frog' method.

When the potential is local, the force $\nabla V(r)$ is analytical and can be easily implemented. The problem for ^{3}He is more complex, as the presence of a large effective mass makes the potential non-local.

As an application, we have studied the head-on collision of two $^3\mathrm{He}_{250}$ drops at $E_{\mathrm{Lab}} = 250\,\mathrm{K}$. This corresponds to a relative wave vector $k_r = 0.35\,\mathrm{Å}^{-1}$ and $E_{\mathrm{CM}} = 125\,\mathrm{K}$. Since the binding energy of $^3\mathrm{He}_{250}$ in our model is around 313 K and the Fermi wave vector of liquid ^{3}He at saturation is $0.78\,\mathrm{Å}^{-1}$, the reaction is rather gentle and will eventually lead to the fusion of the drops [1] after evaporating a few atoms that carry away a sizeable part of the excitation energy.

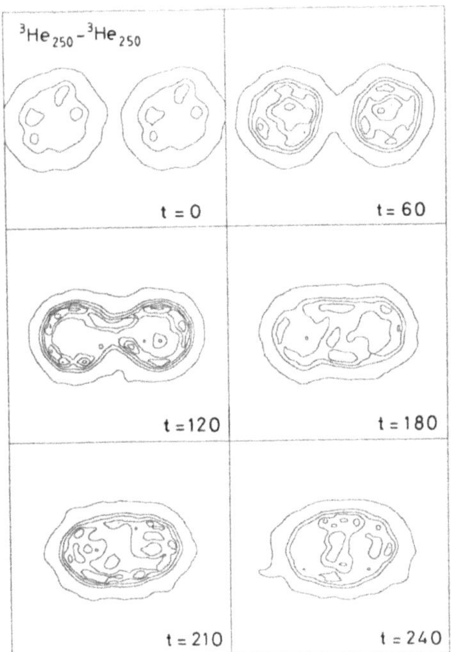

Figure 1

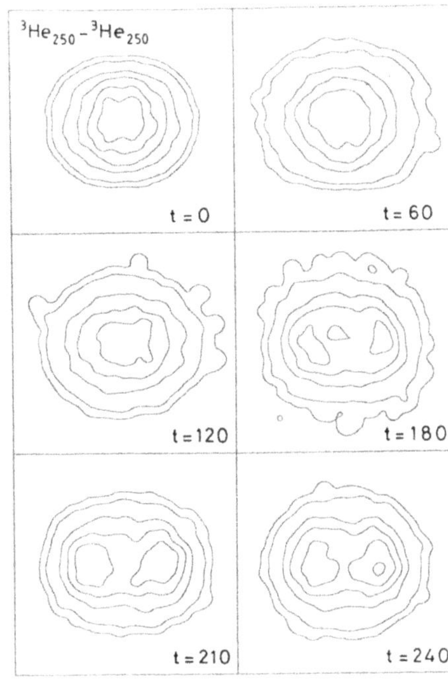

Figure 2

A convenient way of visualizing the reaction consists in representing as a function of time the particle density in the $y = 0$ plane in coordinate space, and in the $k_y = 0$ plane in momentum space:

$$\rho(x, z, t) = \int d\mathbf{p}\, f(x, y = 0, z, \mathbf{p}, t)$$

$$\rho(k_x, k_z, t) = \int d\mathbf{r}\, f(\mathbf{r}, k_x, k_y = 0, k_z, t)\,.$$

The Figure displays several snapshots of $\rho(x, y = 0, z, t)$ taken at intervals of $60 \cdot 10^6$ Å/c sec, and Figure 2 shows the corresponding $\rho(k_x, k_y = 0, k_z, t)$. The outermost equidensity lines of $\rho(x, y = 0, z, t)$ and of $\rho(k_x, k_y = 0, k_z, t)$ correspond to $10^{-3}\,\text{Å}^{-3}$ and $1\,\text{Å}^{-3}$, respectively. Since $f(\mathbf{r}, \mathbf{p}, t)$ is solution of the Vlasov equation, where no individual collisions are taken into account, the direction of motion remains a privileged direction and the system can only develop quadrupole deformations. At present, a systematic study of ^{3}He-^{3}He reactions for different energies and impact parameters is under way. The inclusion of atomic collisions in the Landau-Vlasov approximation is also being developed.

Acknowledgements

This work has been supported in part by the DGICYT, Spain, grant PB89-0332, and by the IN2P3-CICYT exchange program. M.G. gratefully acknowledges the financial support of the Departament d'Ensenyament de la Generalitat de Catalunya.

References

1. M. Guilleumas, F. Garcias, M. Barranco and E. Suraud, *Z. Phys. D*, to be published.
2. S. Stringari, *Phys. Lett.* **A107** (1985) 36.
3. Ch. Grégoire, B. Remaud, F. Sébille, L. Vinet and Y. Raffray, *Nucl. Phys.* **A465** (1987) 317.
4. P. Ring and P. Schuck, *The Nuclear Many-Body Problem* (Springer-Verlag, New York, 1980).
5. H. Goldstein, *Classical Mechanics* (Addison-Wesley, London, 1959).
6. C. Cohen-Tannoudji, B. Diu, F. Laloë, *Mécanique Quantique* (Hermann, Paris, 1977).

IMPLEMENTATION OF THE BETTIS ALGORITHMS FOR THE NUMERICAL INTEGRATION OF PERTURBED OSCILLATING SYSTEMS

J. VIGO and J. M. FERRÁNDIZ

Departamento de Matemática Aplicada a la Ingeniería,
E.T.S. de Ingenieros Industriales, 47011- Valladolid, Spain

ABSTRACT

The multistep Bettis methods possess very good properties for integrating problems with periodic or quasi-periodic perturbed solutions. Their initialization requires the computation of a certain number of coefficients, which can be fairly inconvenient when considering a superposition of oscillations with various frequencies, especially for high orders, so only the simplest versions corresponding to the case of a single frequency are usually used. In this paper we present a simpler alternative procedure independent of the order. The explicit formulas corresponding to two different or confluent frequencies are shown and some illustrative numerical examples are offered.

1. Introduction

We do not believe it necessary to insist on the importance and frequency with which the problems that can be considered perturbations of others with periodic, or more generally, quasi-periodic solutions appear in Physics. In the computational field, this fact has given rise to several attempts to design numerical integration methods especially adapted to oscillatory phenomena. Gautschi[1] developed a basic theory and algorithms capable of integrating oscillations with a given frequency as well as a finite number of the corresponding harmonics without truncation error.

In 1969 Stiefel and Bettis[2] introduced modifications to the Cowell method which permitted the exact integration of Fourier polynomials at various frequencies, as well as products of these by ordinary polynomials. In the said article explicit expressions of the coefficients for methods of orders four and six were presented and a procedure for the calculation of coefficients was established, based on solving an interpolation problem with points given on a circle, which would be used by Bettis[3] to modify the methods of Störmer, Adams-Bashforth and Adams-Moulton in the same way. Two years latter, Lyche[4] published some theoretical results which allowed us to contemplate the above methods from a more general perspective.

Although the Bettis methods behave excellently for the numerical integration of perturbed oscillator, their use cannot be considered as extensive, possibly due to the difficulties in their implementation. The said integrators work like the classic multistep ones with fixed stepsize, but require the modification of a number of coefficients equal to two times the number of frequencies when the exact integration of a quasi-periodic problem is in question. Additionally, when, for same of the frequencies, we wish to integrate functions where the coefficients of the circular functions can be polynomials of a certain degree, the number of coefficients increases proportional to the double of the degree of the polynomial. For the simplest case of a harmonic oscillator at one frequency only, Bettis proposed two different algorithms for calculating the two required coefficients, which should be used alternately, depending on the magnitude of the product of the frequency by the step-longitude. When the latter is large, it is sufficient to solve a couple of recurrences, whereas when it is small, it is better to use expressions in series previously obtained.

The problems are much greater in the case of various frequencies or when the oscillations amplitudes are non-constant polynomials, what we may call, using the usual terminology in Celestial Mechanics, mixed secular terms, so that in practice the potential user usually decides not to programme them if he does not have much time.

In this paper we propose alternative algorithms, to calculate the coefficients which are considerably simpler, at least in the usual cases. Independently of the chosen order, it is sufficient to compute the last two coefficients as in the case of one frequency, and then to solve a system of equal dimension to the number of required coefficients, and whose unknowns are the differences between the sought coefficients and those corresponding to a single frequency. The solution can be achieved numerically or analytically and the help of a general purpose symbolic manipulation system, although not strictly speaking necessary, significantly contributes to the convenience of the process and the reliability of the result.

In the following section, after the description of the Bettis methods, explicit formulas are offered for the case of two frequencies different or confluent, obtained using MATHEMATICA, running on a PC.

Finally, a few numerical examples are presented in section 3 that illustrate the behaviour of the methods under consideration.

2. General Properties and Implementation of Bettis Methods

To simplify the description, let us first of all refer to the Bettis predictor obtained by modifying the corresponding Adams-Bashforth one.

Let us consider a initial value problem

$$y' = f(t,y), \quad y(t_0) = y_0, \tag{1}$$

where f is sufficiently regular and let us denote $t_m = t_0 + mh$, $f_m = f(t_m, y_m)$.

The Adams-Bashforth method of order n with a fixed step h is given by the formula

$$y_m = y_{m-1} + h \sum_{k=0}^{n-1} \alpha_k \nabla^k f_m \,, \tag{2}$$

applied from a set $y_0, \cdots, y_n$ of starting points and where $\nabla^k f_m$ are the backward differences, that is, $\nabla^{k+1} f_m = \nabla^k f_m - \nabla^k f_{m-1}$, $\nabla^0 f_m = f_m$ and the coefficients α_k

coincide with those of the McLaurin expansion of the function $\xi/(1-\xi)\log(1-\xi)$, which converges in $|\xi|<1$.

The Bettis predictor of n steps corresponding to a given frequency ω and associated to Eq. (2), can be deduced from this by simply substituting the α_k's with new coefficients β_k's, of which the first $n-2$ coincide with those of Adams, and the remaining two, β_{n-2} and β_{n-1} are defined with the condition that the equation

$$\frac{\xi}{(1-\xi)\log(1-\xi)} = \sum_{k=0}^{n-1} \beta_k \xi^k \tag{3}$$

identically holds along the whole path $\xi = 1 - e^{-2i\sigma}$, with $\sigma = \omega h/2$, so that these two variables both depend on ω and h. Let us note that the condition (3) is equivalent to the local truncation error being ruled out when the function $f(t) = z^{t/h}$ is integrated, with $z = 1/(1-\xi) = e^{i\omega h}$, so that the Bettis algorithm

$$y_m = y_{m-1} + h \sum_{k=0}^{n-1} \beta_k \nabla^k f_m \tag{4}$$

exactly integrates the functions $\cos\omega t$ and $\sin\omega t$, if the starting values are assumed to be exact and round-off error is ignored.

The condition (3) can be reduced to the form

$$\beta_{n-2}(\xi) + \beta_{n-1}(\xi)\xi = \sum_{j=0}^{\infty} \alpha_{j+n-2}\xi^j\,, \tag{5}$$

which is adequate for the calculation (analytical or numerical) of the two coefficients where σ, or its equivalent ξ, is sufficiently close to zero, since the convergence is rapid and, besides, there is no problem in obtaining the necessary α's.

The real existence of the algorithm can be deduced from the following proposition, Ferrándiz and Vigo[5], whose first paragraph could be deduced from Theorem 4.1 of Lyche, and which can be proved without difficulty by the usual methods of complex variable:

Proposition 1: The last two Bettis coefficients can be extended to analytical functions $\beta_{n-1}(\xi)$, $\beta_n(\xi)$, in $\xi = 0$ satisfying Eq. (5) and in addition:

i) They are real on the path $\xi = 1 - e^{-2i\sigma}$.

ii)$\beta_{n-1}(0) = \alpha_{n-1}$, $\beta_n(0) = \alpha_n$.

iii)$\beta'_{n-2}(0) = 0$, $\beta'_{n-1}(0) = 0$.

The property ii) indicates that when the frequency vanishes, the Bettis predictor is reduced to that of Adams.

With help of Proposition 1, some algebra allows us to obtain the following:

Proposition 2: The Bettis method, Eq. (5), has order n, and its truncation error is of the form:

$$\mathcal{L}_\beta(y,h) = h^{n+1}C_{n+1}(y^{n+1)} + \omega^2 y^{n-1)}) + O(h^{n+2}) \tag{6}$$

where C_{n+1} is the error constant of the Adams method, Eq. (2). Besides, the method is stable and convergent to the same order.

This follows from well-known results by Grigorieff[6] and Stetter[7].

Remark 1: It is important to note that when the solution $y(t)$ is a perturbation of a harmonic oscillation of frequency ω, the components of Eq. (6) in the space generated by the functions $\cos\omega t$ and $\sin\omega t$ are cancelled in view of Eq. (5), so that the local error is in general of the order of magnitude of the perturbation.

Remark 2: Let us note that the above results are still valid with slight modifications when the predictor, Eq. (2), is replaced by another scheme in differences, such as those of Adams-Moulton, Störmer, etc., whose coefficients can be deduced by the development in series of a generating function.

For the computation of coefficients in the case of a single frequency, Bettis[3] establishes a procedure that can be outlined as follows. First we define a variable $u = 4\sin^2\sigma$, and two sequences of functions $R_m(u)$, and $S_m(u)$ through the recurrences:

$$\begin{aligned} R_{m+1} &= u(R_m - R_{m-1}), \quad R_0 = 2, \;\; R_1 = u, \\ S_{m+1} &= u(S_m - S_{m-1}), \quad S_0 = 0, \;\; S_1 = u, \end{aligned} \tag{7}$$

and starting from these two functions $F_r(u)$, whose expression depends on the integrator under consideration, and which, for that of Adams-Bashforth are

$$F_{n-1} = \frac{\mu^\star(u)R_{n-1}}{u}, \quad F_{n-2} = \frac{\mu^\star(u)R_n}{u}, \tag{8}$$

with

$$\mu^\star(u) = \frac{\sin\sigma}{2\sigma\cos\sigma}. \tag{9}$$

The two coefficients needed are then calculated using the relation

$$\beta_p = \frac{(-1)^{n-1-p}}{u^{n-2}}\{F_p^\star + \sum_{k=0}^{n-3}\alpha_k u^{k-1}S_{q-k}\}, \tag{10}$$

where $p = n-1$ when $q = n-2$, and $p = n-2$ when $q = n-1$

Let us note that this algorithm is not adequate when n is small, due to the appearance of powers of u as divisors. For this case we recommend the direct use of the relation (5) or similar ones.

In the case of various frequencies the procedure is considerably more complicated. Without going into details, which can be seen in Bettis[3] let us say that it is possible to consider functions $F(u)$ as being defined by relations similar to those in Eq. (8), but with different indices for the functions R or S, different powers of u in the denominator and functions which play the role of $\mu^\star$ with greater numerical problems, such as

$$\mu(u) = \frac{1}{4}(\frac{1}{\sin^2\sigma} - \frac{1}{\sigma\sin\sigma\cos\sigma}), \tag{11}$$

corresponding to the integrator of Adams-Bashsfoth. The last two coefficients can be explicitly calculated using relations that, in the case of ν different frequencies, are of the type

$$\beta_p = K_p + \sum_{k=1}^{\nu} F_p(u_k)\Pi^\star, \quad p = n-1, n-2 \tag{12}$$

with $u_k = 4\sin^2\sigma_k$, $\sigma_k = \omega_k h/2$ and

$$\Pi^\star = \prod_{m=1;m\neq k}^{\nu} \frac{1}{u_k - u_m}, \tag{13}$$

and K_{n-1}, K_{n-2} being certain constants.

To find the lowest coefficients it is necessary to know those of the method of order $n-2$ and to identify terms in a certain expression, so, in short, it is necessary to calculate the lower coefficients for orders $4,6,8,\cdots n$.

We shall now explain an alternative procedure for calculating the coefficients in the case of several frequencies, which is more direct an easier to use in cases which are common in practice.

Let us begin by considering a multistep numerical scheme similar to, although not necessarily equal to that of Eq. (2), as there is no problem in substituting the first derivative in Eq. (1) for another of superior order, (and with this h for h^r in Eq. (2)). It is sufficient to replace the Equation (3) for its analogue

$$\sum_{k=0}^{\infty} \alpha_k \xi^k = \sum_{k=0}^{n-1} \beta_k \xi^k \tag{14}$$

to conclude that from this, a Bettis method can be deduced for a given frequency ω_i, making α_j equal to β_j for $j = 0,\cdots,n-3$ and applying the Proposition 1.

Let us introduce the notations:

$$\sigma_j = \omega_j h/2, \; u_j = 4\sin^2\sigma_j, \; \xi_j = 1 - e^{2i\sigma_j}, \; \beta_k[j] = \beta_k(\xi_j). \tag{15}$$

In the case of two frequencies, in order to obtain the corresponding Bettis method, it is sufficient to consider a set γ_k of coefficients such that the equalities

$$\sum_{k=0}^{n-1} \beta_k(\xi)\xi^k = \sum_{k=0}^{n-1} \gamma_k(\xi)\xi^k \tag{16}$$

are verified when ξ takes the values ξ_1 and ξ_2, where, additionally $\gamma_k(\overline{\xi}_i) = \gamma_k(\xi_i)$, where the bar represents the complex conjugation. A solution of Eq. (12) can be obtained when all the coefficients γ's, except the last four, coincide with the corresponding β's, and thus with the original α's. Imposing this condition, and evaluating Eq. (12) in ξ_1 and ξ_2, we have the following two equations:

$$\sum_{k=n-4}^{n-1} \{\gamma_k - \beta_k[1]\}\xi_1^k = 0,$$

$$\sum_{k=n-4}^{n-1} \{\gamma_k - \beta_k[1]\}\xi_2^k = (\beta_{n-2}[2] - \beta_{n-2}[1])\xi_2^{n-1} + (\beta_{n-1}[2] - \beta_{n-1}[1])\xi_2^n \tag{17}$$

Adding the two conjugated equations, we obtain a system in the four unknowns $\gamma_k - \beta_k[1]$ whose solution leads to the expressions:

$$\begin{aligned}
\gamma_{n-1} &= \beta_{n-1}[1] - \frac{(\beta_{n-1}[1] - \beta_{n-1}[2])u_2}{u_2 - u_1} \\
\gamma_{n-2} &= \beta_{n-2}[1] - \frac{u_2(\beta_{n-2}[1] - \beta_{n-2}[2] - u_1(\beta_{n-1}[1] - \beta_{n-1}[2]))}{u_2 - u_1} \\
\gamma_{n-3} &= \beta_{n-3}[1] + \frac{u_1u_2(\beta_{n-2}[1] - \beta_{n-2}[2] - \beta_{n-1}[1] + \beta_{n-1}[2])}{u_2 - u_1} \\
\gamma_{n-4} &= \beta_{n-4}[1] - \frac{u_1u_2(\beta_{n-2}[1] - \beta_{n-2}[2])}{u_2 - u_1}
\end{aligned} \tag{18}$$

These formulas generalize those (25) of Stiefel and Bettis[2], in that they are valid for any order and numerical scheme of the type under consideration. The calculation of the $\beta_k[i]$ is performed as in the case of a single frequency. Naturally, when ω_1 and ω_2 are very close, the expressions Eq. (14) are not really adequate due to the presence of the denominator $u_2 - u_1$. In such a case it is preferable to use the coefficients for confluent frequencies, which can be deduced from the previous ones just by taking limits for $\omega_2 \longrightarrow \omega_1 = \omega$ which gives us

$$\begin{aligned}
\gamma_{n-1} &= \beta_{n-1} + u\frac{d\beta_{n-1}}{du}, \\
\gamma_{n-2} &= \beta_{n-2} + u(\frac{d\beta_{n-2}}{du} - \frac{ud\beta_{n-1}}{du}), \\
\gamma_{n-3} &= \beta_{n-3} - u^2(\frac{d\beta_{n-2}}{du} - \frac{d\beta_{n-1}}{du}), \\
\gamma_{n-4} &= \beta_{n-4} + u^2\frac{d\beta_{n-2}}{du}.
\end{aligned} \tag{19}$$

An analogous process to that described can be followed when the number of frequencies is greater than two, using a system similar to Eq. (13).

The possibility of using symbolic manipulators allows us to solve such systems on computers, so it seems superfluous to set out a list of formulas analogous to Eq. (14) for three or more frequencies. In any case as the order of the method can not be less than double the number of frequencies, it is not realistic to consider and arbitrarily high number of unknowns; so the algorithm describe, even though it could be improve, is sufficient in practice.

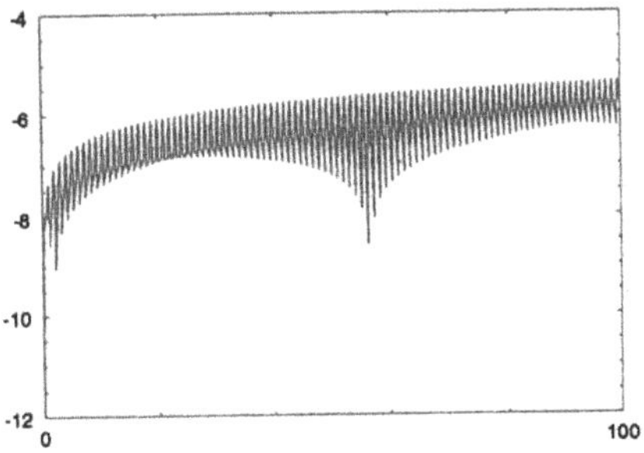

Figure 1: Stomer-Cowell method. Step π/40. Order 4

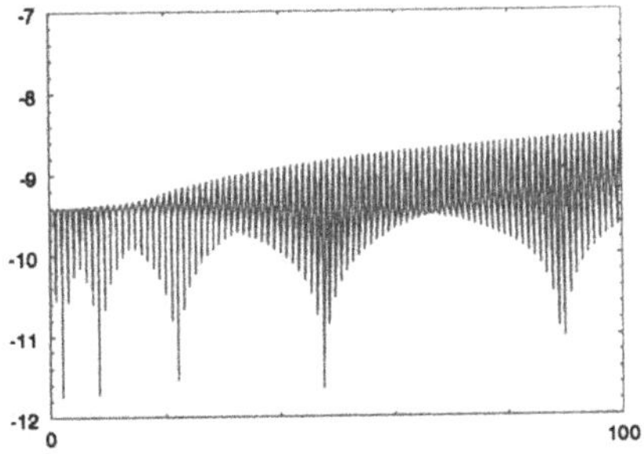

Figure 2: Bettis single frequency method. Step π/40. Order 4

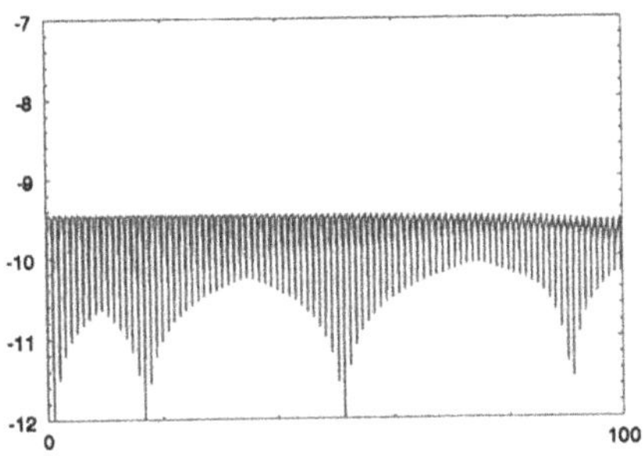

Figure 3: Bettis double frequency method. Step π/40. Order 4

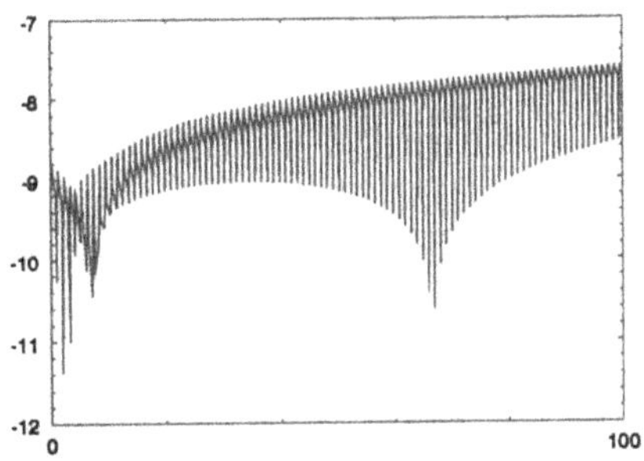

Figure 4: Stormer method. Step π/20. Order 10

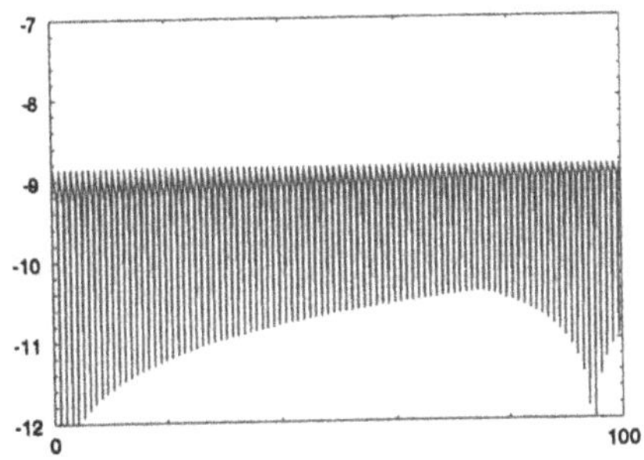

Figure 5: Bettis single frequency method. Step π/20. Order 10

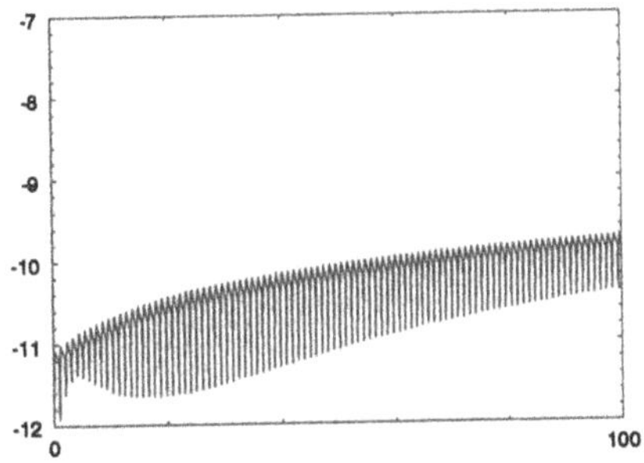

Figure 6: Bettis two frequencies method. Step π/20. Order 10

3. Numerical Examples

In this section we shall present some examples to illustrate the behaviour of the Bettis methods, and to give some idea of the advantage that can be obtained by using algorithms adapted to more than one frequency.

Let us consider a perturbed oscillator which satisfies the Duffing equation

$$\begin{aligned} y'' + y &= \epsilon y^3, \\ y(0) &= 1, \\ y'(0) &= 0, \end{aligned} \tag{20}$$

where ϵ is a small parameter. In Kirchgraber[8] an asymptotic solution can be seen with an error to the order of ϵ^7, which will be taken as exact solution for estimating the error in the integrations that follow.

Firstly, let us present figures corresponding to $\epsilon = 10^{-3}$, using a predictor-corrector pair of forth order with a step-size of $\pi/40$.

In the figures, the vertical axis is the decimal logarithm of the absolute value of the error, and the horizontal axis shows the number of revolutions (understood as an increase of 2π of the independent variable). The comparison is made several times per revolution.

In Figure 1, the ordinary Störmer-Cowell pair is used, and in Figure 2 both the predictor and corrector are the corresponding Bettis algorithms, with a single frequency $\omega = 1$. We can see that the decrease in the error is to the order of the small parameter. In Figure 3 the Störmer-Cowell-Bettis pair has been used with a double frequency equal to one. An improvement of an order of magnitude is obtained after 100 revolutions, and we can also see that the error growth is slower. In connection with this last aspect we could refer the reader to the paper of Ferrándiz and Novo[9] and Ferrandiz et al[10]

In Figures 4, 5 and 6 we have taken a relatively high order with a larger step-size, $\pi/20$, and using only a predictor, that of Störmer in first of the figures, that of Störmer-Bettis with frequency one in Figure 5 and with frequencies one and three in the last.

The graphs show that the Bettis methods have good stability properties, apart from the accuracy they provide. In addition, the use of a method with two frequencies improves the accuracy with a small increase in the initialization cost: computing an additional pair of last Bettis coefficients of a single frequency and using the formulas Eq. (14).

Acknowledgements

The authors acknowledge the financial support of the Spanish CYCIT under grant ESP88-541.

References

1. W. Gautschi, *Numer. Math.* **3** (1961) 381.
2. E.L. Stiefel and D.G. Bettis, *Numer. Math.* **13** (1969) 154.
3. D.G. Bettis, *Numer. Math.* **14** (1970) 421.
4. T. Lyche, *Numer. Math.* **19** (1972) 65.
5. J.M. Ferrándiz and J. Vigo in *proceedings of the XII CEDYA*, In Press.
6. R.D. Grigorieff, *Numer. Math.* **42**,(1983) 359.
7. H.J. Stetter *Analysis of Discretization Methods for Ordinary Differential Equations* (Springer, Berlin, 1973).
8. U. Kirchgraber, *Numer. Math.* **53** (1988) 621.
9. J.M. Ferrándiz and S. Novo, in *Predictability, Stability and Chaos in N-Body Dynamical Systems*, A.E. Roy ed., Plenum, (1990) 515.
10. J.M. Ferrándiz, J. Vigo and P. Martín in *Proceedings of the 3rd International Symposium on Spacecraft Flight Dynamics*, ed J.J. Hunt, ESA, SP-**326** (1991) 49.

THE LOGISTIC MAP DRIVEN BY CORRELATED NOISE

J. M. GUTIÉRREZ , A. IGLESIAS
Departamento de Matemáticas
Universidad de Cantabria
Avda. de los Castros s/n 39005, Santander, Spain

and

M. A. RODRÍGUEZ
Departamento de Física Moderna
Universidad de Cantabria
Avda. de los Castros s/n 39005, Santander, Spain

ABSTRACT

The interplay between chaos and noise yields complex and rich phenomena even in simple cases. In this paper we deal with one of such cases, a Logistic Map driven by a dichotomic noise, showing and analysing several noise induced phenomena.

1. Introduction

Most of the works devoted to noise in dynamical systems consider low intensity white noise. With this kind of noise only small changes of the original system without noise are possible. For instance, in a Logistic Map (LM) this noise smears the details of the high order bifurcation structure[1]. When one uses noise with great intensity and color, strong changes of the deterministic behavior are expected. This is an unexplored question and, as far as we know, there are not many results in the relevant literature. We have investigated this problem on a simple model, a LM driven by a dichotomic noise. Linear maps with this kind of noise can appear as Poincare maps of physical systems and have been recently analysed in the context of stochastic resonance[2]. The LM has a more complex evolution and then is difficult to identify resonances due to the noise but as we can see other interesting noise induced phenomena occur. Our aim is twofold, on the one hand, we try to identify the mechanisms which are responsible for such changes, on the other hand we study the meaning of characteristic parameters such as Lyapunov exponents. As it is well known the meaning of such parameters in the presence of noise is not clear[3].

2. The Model

We consider a LM with a multiplicative noise in the form:

$$x_{n+1} = \mu(1+\xi_n)x_n(1-x_n). \tag{1}$$

ξ is a noise taking two possible values $+\Delta$ and $-\Delta$ with a probability $\alpha < 1$ of keeping

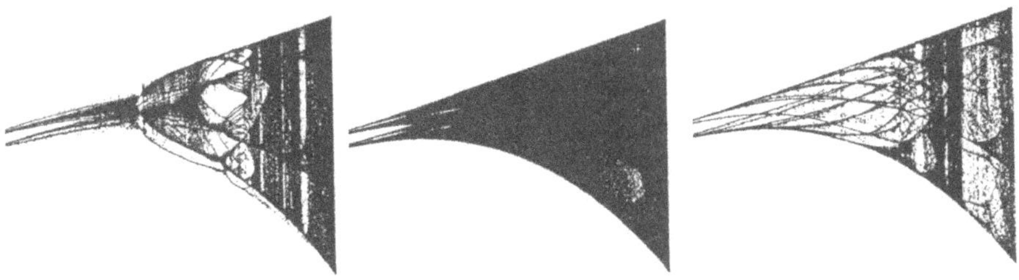

(A) (B) (C)

Figure 1: Bifurcation diagrams of the noisy LM. Values of parameters are $\Delta = 0.05$ and $\alpha = 0.99(A), 0.5(B), 0.01(C)$ respectively.

the same value as in the previous step. Since the noise is bounded, the iteration is well defined for $\mu(1+\Delta) < 4$. As a measure of the correlation time τ we take the average duration of the most probable sequence of the noise. For $\alpha > \frac{1}{2}$ this sequence is formed by repetitive values $+\Delta$ or $-\Delta$ with τ given by $\frac{\alpha}{1-\alpha}$. For $\alpha = \frac{1}{2}$ we have a white noise, i.e. all sequences are equally weighted. For $\alpha < \frac{1}{2}$ the most probable sequence is formed by alternated values of $+\Delta$ and $-\Delta$. For this sequence τ reads $\frac{1-\alpha}{\alpha}$. In the following analysis we distinguish three cases, the limit of repetitive sequence, $\alpha \to 1$, the white noise case, $\alpha = \frac{1}{2}$ and the limit of alternated sequence, $\alpha \to 0$. Bifurcations diagrams are shown for these cases with the same intensity $\Delta = 0.05$ in Figure 1.

3. Numerical Analysis

Limit $\alpha \to 1$. The exact limit yields a superposition of two logistic maps f_+ and f_-, with $f_\pm = \mu(1 \pm \Delta)x(1-x)$. As a consequence the invariant measure and the Lyapunov exponent are given by the arithmetic mean: $\rho(x) = \frac{1}{2}\rho_+ + \frac{1}{2}\rho_-$ and $\lambda = \frac{1}{2}\lambda_+ + \frac{1}{2}\lambda_-$ respectively. This simple fact is very instructive for understanding the meaning of the Lyapunov exponent in noisy maps. Clearly, the sign of the Lyapunov exponents now is not evidence of chaotic or regular motion. It denotes only average behaviors. So probably it would be necessary to compute some other parameter related to the dispersion of the Lyapunov exponent but such a question goes beyond the scope of this work. In Figure 1a we show the bifurcation diagram for α=0.99. It can be easily analysed. It is mainly composed by the superposition of the bifurcation diagrams of f_+ and f_- and, as we analyse in a following section, in a lower proportion by the transient to an attractor of $f\pm$ when starting from an attractor of $f\mp$.

White noise $\alpha = \frac{1}{2}$. All sequences of the noise now have the same probability. Deterministic components disappear in almost all the region of parameters except near superstable points (see Fig. 1b).

Limit $\alpha \to 0$. The exact limit corresponds to a LM with a periodic modulation. There are two possible attractors corresponding to the composed maps

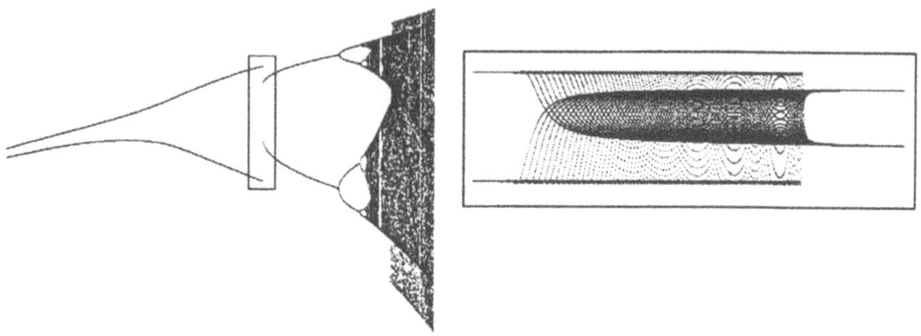

Figure 2: Bifurcation diagram of the LM with a periodic modulation ($\alpha = 0$) and the same initial conditions. We show the discontinuity on an enhanced scale.

f_-f_+ and f_+f_-. The basin of attraction is a fractal set with a Cantor structure. Bifurcation diagrams with fixed initial conditions present discontinuities in the values of μ where the initial condition is marginal (Fig. 2). Near these discontinuity points there are metastable attractors with a very intrincate form that are structurally stable (Fig. 2). In Figure 1c we present a bifurcation diagram for $\alpha = 0.01$. The noise, even with a very large correlation time, leads to the mixing of the above mentioned attractors. When increasing the intensity Δ the contribution of transients from both attractors is more evident.

4. Analytical Results

Several of the above mentioned numerical observations can be analysed by perturbative methods. We start by defining the joint probability, $P_n(x, \pm\Delta)$, of having a value x in the LM with $\pm\Delta$ in the noise at time n. Then the following master equation for this joint probability holds:

$$P_{n+1}(x, \pm\Delta) = \alpha \int \delta(x - f_\pm(y)) P_n(y, \pm\Delta) dy + (1-\alpha) \int \delta(x - f_\pm(y)) P_n(y, \mp\Delta) dy \qquad (2)$$

and, as consequence, in the limit $n \to \infty$, the equation for the stationary state $P(x, \pm\Delta)$ is:

$$P(x, \pm\Delta) = \alpha \int \delta(x - f_\pm(y)) P(y, \pm\Delta) dy + (1 - \alpha) \int \delta(x - f_\pm(y)) P(y, \mp\Delta) dy \tag{3}$$

The stationary probability density for the noisy map $\rho(x)$ is obviously given by:

$$\rho(x) = P(x, +\Delta) + P(x, -\Delta) \tag{4}$$

With these equations in mind let us now consider the three cases defined in the last section.

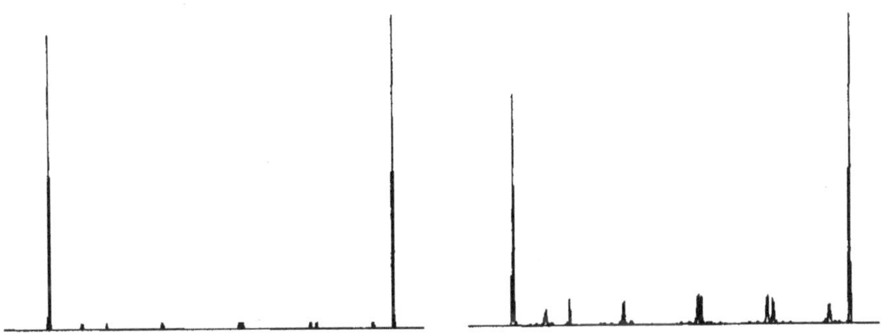

(a) (b)

Figure 3: Invariant densities of the noisy map with $\Delta = 0.05$ and two values of $\alpha(0.01(a), 0.1(b))$. In both figures it is possible to see first order contributions. In (b) we observe also second order contributions.

1.- Limit $\alpha \to 1$. We take $\alpha = 1 - \epsilon$, being ϵ a small quantity, and consider perturbations in the form:

$$P(x, \pm\Delta) = P_0(x, \pm\Delta) + \epsilon P_1(x, \pm\Delta) \tag{5}$$

Substituting (5) in (3) we have for the zero order in ϵ:

$$P_0(x, \pm\Delta) = \int \delta(x - f_\pm(y)) P_0(y, \pm\Delta) dy \tag{6}$$

and for the first order:

$$P_1(x, \pm\Delta) = \int \delta(x - f_\pm(y))[P_0(y, \mp\Delta) - P_0(y, \pm\Delta)] dy + \int \delta(x - f_\pm(y)) P_1(y, \pm\Delta) dy \tag{7}$$

The equation (6) has a normalized solution given by:

$$P_0(x,\pm\Delta) = \frac{1}{2}\rho_\pm(x) \tag{8}$$

where $\rho_\pm(x)$ is the invariant density of the logistic map $f_\pm(x)$ respectively. The integral equation (7) can be solved. From (4),(8) and an iterative n solution of (7) we finally obtain:

$$\begin{aligned}\rho(x) = \frac{1}{2}(1-n\epsilon)\rho_+(x) + \frac{1}{2}(1-n\epsilon)\rho_-(x) + \frac{\epsilon}{2}\int \delta(x - f_+(y_1))\rho_-(y_1)dy_1 + \ldots +\\ +\frac{\epsilon}{2}\int\cdots\int \delta(x-f_+(y_1))\ldots\delta(y_{n-1}-f_+(y_n))\rho_-(y_n)dy_1\ldots dy_n +\\ +\frac{\epsilon}{2}\int \delta(x-f_-(y_1))\rho_+(y_1)dy_1 + \ldots +\\ +\frac{\epsilon}{2}\int\cdots\int \delta(x-f_-(y_1))\ldots\delta(y_{n-1}-f_-(y_n))\rho_+(y_n)dy_1\ldots dy_n + o(\epsilon^2).\end{aligned} \tag{9}$$

This solution gives, as we obtained in the numerical analysis, the superposition of ρ_+ and ρ_- and in order ϵ the transient from an atractor of $f_\pm$ when starting from an atractor of $f_\mp$ respectively.

2.- White noise, $\alpha = \frac{1}{2}$. In this case the equation for the invariant density takes a simple form:

$$\rho(x) = \int\{\frac{1}{2}\delta(x-f_+(y)) + \frac{1}{2}\delta(x-f_-(y))\}\rho(y)dy \tag{10}$$

A perturbative analysis for small intensity Δ is possible. Taking, in general, $f_\pm(y) = f(y) \pm \Delta g(y)$ and being Δ the small parameter we obtain from (10):

$$\rho(x) = \int \delta(x-f(y))\rho(y)dy + \Delta^2 \int \delta''(x-f(y))P(y)dy \tag{11}$$

A perturbative solution of this equation can be obtained as:

$$\rho(x) = \rho_0(x) + \Delta^2 \frac{\partial^2}{\partial x^2}[g(f^{-1}(x))\rho_0(f^{-1}(x))] \tag{12}$$

where $\rho_0(x)$ is the invariant density for the map $f(x)$ and f^{-1} the inverse function of $f(x)$. A better approach could be used with path integral methods as in ref.(4) but in any case it only gives a smoothing of the unperturbed solution of $\rho_0(x)$. The great intensity case, as concerns analytical results, is an open problem.

3.- Limit $\alpha \to 0$. Taking $\alpha = \epsilon$ in (3) and making an identical perturbative analysis as in the case with $\alpha \to 1$, we obtain for the zero order equation:

$$P_0(x,\pm\Delta) = \int \delta(x - f_\pm(f_\mp(y)))P_0(y,\pm\Delta)dy, \tag{13}$$

and for the first order:

$$P_1(x,\pm\Delta) = \int \delta(x-f_\pm(y))\{P_0(y,\pm\Delta) - P_0(y,\mp\Delta)\}dy + \int \delta(x-f_\pm(y))P_1(y,\mp\Delta)dy. \tag{14}$$

Now $P_0(x, \pm\Delta)$ is given by:

$$P_0(x, \pm\Delta) = \frac{1}{2}\rho_{\pm}(x) \tag{15}$$

where $\rho_{\pm}(x)$ is the invariant density of the composed map $f_{\pm}(f_{\mp}(x))$. Also the equation (14) can be iteratively solved. We obtain for the invariant density:

$$\rho(x) = \frac{1}{2}(1-n\epsilon)\rho_+(x) + \frac{1}{2}(1-n\epsilon)\rho_-(x) + \frac{\epsilon}{2}[\int \delta(x - f_+(y_1))\rho_+(y_1)dy_1 +$$

$$+ \ldots + \int \ldots \int \delta(x - f_+(y_1))\delta(y_1 - f_+(y_2)) \ldots \delta(y_{n-1} - f_+(y_n))\rho_+(y_n)dy_1 \ldots dy_n] +$$

$$+ \frac{\epsilon}{2}[\int \delta(x - f_-(y_1))\rho_-(y_1)dy_1 + \ldots +$$

$$+ \int \ldots \int \delta(x - f_-(y_1))\delta(y_1 - f_-(y_2)) \ldots \delta(y_{n-1} - f_-(y_n))\rho_-(y_n)dy_1 \ldots dy_n] \tag{16}$$

in agreement with numerical results. In Figure 3 we show the invariant density for two values of ϵ. As our analysis indicates it is easy to see that small peaks correspond to transients from both atractors.

5. Conclusions

We have studied numerically and analytically a simple example of interplay between chaos and noise: a Logistic Map driven by dichotomic noise. Analytical results are possible in two extreme cases, namely, when the value of noise either does not change or changes with high probability. In the first case a superposition of two maps is obtained. The invariant density and Lyapunov exponent are the mean. In the second case a composition of maps is obtained. These results are illustrative for understanding the loss of meaning of the Lyapunov exponent. Now a negative value of the Lyapunov exponent clearly does not indicate regular motion.

Perturbative analysis around these limit cases also is possible. We obtain the first correction as a transient of a map when the initial conditions are the attractor of the other map. It is in agreement with numerical results. In the same manner higher order corrections could also be calculated.

To conclude, it is worth to emphasising the main point of this work; this simple example could be a good candidate for carrying out a study, including an analytical one, of some complex phenomena involving chaos and noise.

Acknowledgements

Financial support from Universidad de Cantabria and DGICYT (Spain) Project No. PS90-0098 is acknowledged.

References

1. J. P. Crutchfield, J. D. Farmer and B. A. Huberman, *Phys. Rep.* **92** (1982) 45.

2. S. J. Frazer and R. Kapral, *Phys. Rev.* **A45** (1992) 3412. A. J. Irwin, S. J. Frazer and R. Kapral, *Phys. Rev. Letters* **64** (1990) 2343.
3. J. Bene and P. Szepfalusy, *Phys. Rev.* **A37** (1988) 371.
4. P. Reimann and P. Talkner, *Phys. Rev.* **A44** (1991) 6348.

ATMOSPHERE COOLING RATE CALCULATION USING A NARROW-BAND MODEL

A. SERRANO and J. M. MONTANERO
Departamento de Física, Facultad de Ciencias
Universidad de Extremadura, E-06071 Badajoz, Spain

ABSTRACT

Cooling rate calculations based on the transfer equations require several integrals. The integral over all frequencies needs a hard computational work due to the spectrum shape of the different absorbing gases in the atmosphere. Therefore, some approximations and a band model instead of a line-by-line integral are used.

Our aim is to obtain the cooling rate due to infrared radiation of an atmosphere with standard distribution of pressure, temperature, and concentration by means of a narrow-band model. The infrared heating rate due to the presence of absorbing-emiting substances is given by[1]

$$\rho C_p \frac{\partial \theta}{\partial t} = \frac{\partial F}{\partial z} \tag{1}$$

where ρ is air density, C_p is the specific heat of air at constant pressure, θ is the air temperature, and F is the net infrared radiation flux.

We divide the frequency spectrum into twenty bands and the atmosphere into thirty three levels to obtain the radiation flux crossing each level. Our task is to solve the transfer equation. For a non-scattering atmosphere in local thermodynamic equilibrium (LTE) conditions, the spectral flux may be written [2] as a function of B_ν, the Planck function for the radiation of a black body, and $T_\nu^f(z',z)$, the flux transmission function between the levels z and z'. This is for a stratified atmosphere with clear sky conditions.

Several integrations are needed to get the net flux. The frequency integration of the transfer equation is a very hard numerical task because of the irregular shapes of the absortion spectrums. Therefore, band models have been developped. These models make use of the averages over many lines, embracing spectral ranges, and them allow us to make the frequency integration in a numerical way with not much resolution. We use the random model of Malkmus[2]. It is also supposed that all absortion lines can be described by a Lorentz line shape. This assumption and the approximation of LTE reduce the data validity to a height less than $25Km$. because it is there where the two approximations may be meaningful.

As band-model theory has been developped for homogeneous absortions paths, an important step is to reduce the calculations for the nonhomogeneous atmosphere to calculations through a homogeneous path with fixed pressure p^*, the amount of absorbing material, m^*, and the temperature, θ^*. This is done by the Hulst, Curtis and Godson (HCG) approximation and p^* and m^* are given by

$$m^* = \int_{path} dm \frac{\phi(\theta)}{\phi(\theta^*)} \quad p^* m^* = \int_{path} dmp \frac{\psi(\theta)}{\psi(\theta^*)} \tag{2}$$

where $\phi(\theta)$ and $\psi(\theta)$ are functions evaluated in Ref. 2.

Under this approximation the spectral flux is[2]:

$$F_i(z) = F_i^+(z) - F_i^-(z) \tag{3}$$

$$F_i^+(z) = \pi B_i(\theta_g) T_i^f(0,z) + \pi \int_0^z dz' B_i(z') \frac{\partial T_i^f(z',z)}{\partial z'} \tag{4}$$

$$F_i^-(z) = -\pi \int_z^\infty dz' B_i(z') \frac{\partial T_i^f(z',z)}{\partial z'} \tag{5}$$

$$F(z) = \sum_i F_i(z) \Delta k_i \tag{6}$$

where the transmission function depends now on the scaling pressure p^*, scaling temperature θ^* and scaling amount of material m^* for each path and Δk is the spectral width.

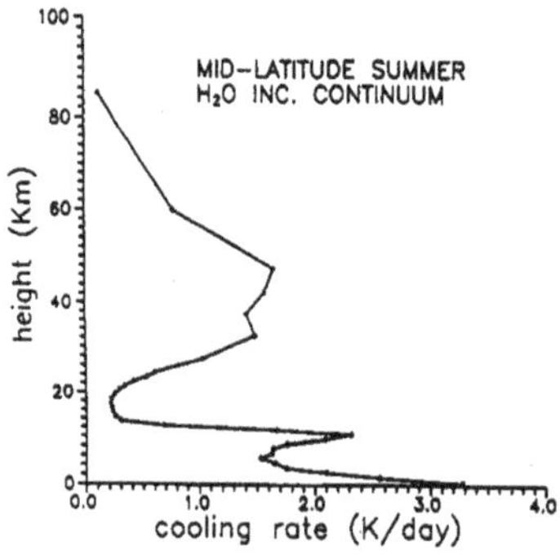

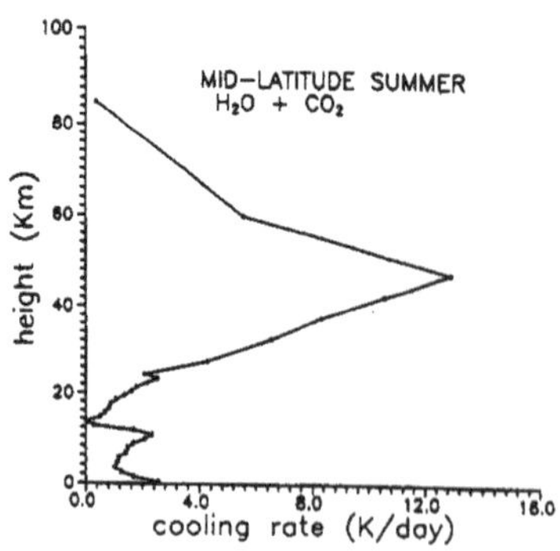

Figure: Cooling rate profiles for the middle latitude summer model atmosphere in two cases, as described in the text.

Diffusivity approximation[2] allow us to perform the angular integration using the Elssaser value 1.66. The results are obtained via a numerical integration by the trapezoidal method. The pictures show the cooling rate profiles for the middle latitude summer model atmosphere in two cases.

References

1. L. Garand, *Journal of the Atmospheric Sciences* **40** (1983) 230-244.
2. R.M. Goody and Y.L. Yung in *Atmospheric Radiation* (Oxford University Press, New York, 1989).

PARTICIPANTS

L. Acedo	*Departamento de Física, Facultad de Ciencias, Universidad de Extemadura, 06071-Badajoz.*
A. Achahbar	*Departamento de Física Aplicada, Facultad de Ciencias, Universidad de Granada, E-18071-Granada.*
J. J. Alonso	*Departamento de Física Aplicada, Facultad de Ciencias, Universidad de Granada, E-18071-Granada.*
A. I. Alonso de Mena	*Departamento de Matemática Aplicada a la Ingeniería, E.T.S. de Ingenieros Industriales, P° del Cauce s/n, Universidad de Valladolid, E-47011-Valladolid.*
B. Allés	*Departamento de Física Teórica y del Cosmos, Facultad de Ciencias, Universidad de Granada, E-18071-Granada.*
E. Battaner	*Departamento de Física Teórica y del Cosmos, Facultad de Ciencias, Universidad de Granada, E-18071-Granada.*
G. M. Buendía	*Departamento de Física, Universidad Simón Bolívar, Apartado 89000, Caracas, Venezuela.*
J. Cabrera	*Department of Statistics, Rutgers University, Hill Center, Busch Campus, New Brunswick, New Jersey 08904, U.S.A.*
J. L. Cabrera	*Departamento de Física Fundametal, U.N.E.D., Apartado 60141, E-28080-Madrid.*
J. A. Calzada	*Departamento de Física Teórica, Facultad de Ciencias, P° del Prado de la Magdalena s/n, Universidad de Valladolid, E-47011-Valladolid.*
E. Cobos	*Universidad de Granada.*
A. Egea	*Departamento de Física, Facultad de Ciencias, Universidad de Murcia, E-30071-Murcia.*
J. F. Fernández	*Centro de Física, Instituto Venezolano de Investigaciones Científicas, Apartado 21827, Caracas, Venezuela.*
P. L. Garrido	*Departamento de Física Moderna, Facultad de Ciencias, Universidad de Granada, E-18071-Granada.*
A. Gros	*E.U. Formación de Profesorado, c/ Greco 10, Departamento de Física Teórica y del Cosmos, Universidad de Granada, E-11702-Ceuta.*
J. Güémez	*Departamento de Física Aplicada, Facultad de Ciencias, Universidad de Salamanca, E-37008-Salamanca.*
M. Guilleumas	*Departament d'Estructura i Constituents de la Matèria, Facultat de Física, Diagonal 647, Universitat de Barcelona, E-08028-Barcelona.*
J. M. Gutiérrez	*Departamento de Matemáticas, Estadísitica y Computación, Facultad de Ciencias, Universidad de Cantabria, E-39071-Santander.*
A. Iglesias	*Departamento de Matemáticas, Estadísitica y Computación, Facultad de*

	Ciencias, Universidad de Cantabria, E-39071-Santander.
M. Iñarrea	*E.U. Formación de Profesorado, c/ Luis de Ulloa s/n, Universidad de Zaragoza, E-26004-Logroño.*
H. R. Jauslin	*Départment de Physique, Université de Bourgogne, 6 Bd. Gabriel, 21100-Dijon, France.*
A. M. Lacasta	*Departament de Física Aplicada, Universitat Politècnica de Catalunya, Av. Gregorio Marañón s/n, E-08028-Barcelona.*
S. Laín	*Departamento de Física Teórica, Facultad de Ciencias, Universidad de Zaragoza, E-50009-Zaragoza.*
M. A. Lerma	*Facultad de Informática, Universidad Politécnica de Madrid, E-28028-Madrid.*
J. R. Linares	*Departamento de Física Aplicada, Facultad de Ciencias, Universidad de Granada, E-18071-Granada.*
J. M. López	*Departamento de Física Moderna, Av. de los Castros s/n, Universidad de Cantabria, E-39005-Santander.*
R. López	*Dapartamento de Ciencias e Ingeniería de la Tierra, Terreno y Materiales, Escuela Politécnica Superior de Ingeniería, c/ Sevilla 1, Universidad de Cantabria, E-39001-Santander.*
R. de la Llave	*Department of Mathematics, University of Texas, Austin, Texas 78712, U.S.A.*
J. Mach	*Departamento de Química Física, Facultad de Química, c/ Martí i Franquès 1, Universidad de Barcelona, E-08028-Barcelona.*
J. Marro	*Departamento de Física Aplicada, Facultad de Ciencias, Universidad de Granada, E-18071-Granada.*
F. Martínez	*Departamento de Física Aplicada, Facultad de Ciencias, Universidad de Granada, E-18071-Granada.*
L. V. Martínez	*A. M. de San Martín, c/ Sto. Cristo 12, E-47162-Valladolid.*
M. A. Muñoz	*Departamento de Física Aplicada, Facultad de Ciencias, Universidad de Granada, E-18071-Granada.*
E. Pérez	*Instituto de Astrofísica de Canarias, E-38200-La Laguna.*
M. Pérez-Victoria	*Departamento de Física Teórica y del Cosmos, Facultad de Ciencias, Universidad de Granada, E-18071-Granada.*
I. M. Polo	*Departamento de Física Aplicada, Facultad de Ciencias, Universidad de Granada, E-18071-Granada.*
M. C. Romero	*Departamento de Física Aplicada, Facultad de Ciencias, Universidad de Granada, E-18071-Granada.*
L. F. Rull	*Departamento de Física Atómica y Nuclear, Facultad de Física, Apartado 1065, Universidad de Sevilla, E-41080-Sevilla.*
J. P. Salas	*Departamento de Física Aplicada, Colegio Universitario de la Rioja, c/ Obispo Bustamante 3, Universidad de Zaragoza, E-26001-Logroño.*
J. J. Sandonís	*Departamento de Ciencias e Ingeniería de la Tierra, y Materiales, Facultad de Ciencias, Universidad de Cantabria, E-39071-Santander.*
E. Santos	*Departamento de Ciencias e Ingeniería de la Tierra, y Materiales,*

	Facultad de Ciencias, Universidad de Cantabria, E-39071-Santander.
A. Serrano	*Departamento de Física, Área de Física Teórica, Facultad de Ciencias, Av. Elvas s/n, Universidad de Extremadura, E-06071-Badajoz.*
A. Sherar	*Instituto de Física "Arroyo Seco" (IFAS), Universidad Nacional de la Provincia de Buenos Aires, Pinto 399, 7000-Tandil, Buenos Aires, Argentina.*
J. D. Sierra	*Departamento de Física Aplicada, E.U. Politécnica de Logroño, c/ Luis de Ulloa 20, Universidad de Zaragoza, E-26004-Logroño.*
T. M. Sintes	*Departamento de Física, Facultad de Ciencias, Cta. de Valldemosa km. 7'5, Universidad de las Islas Baleares, E-07071-Palma de Mallorca.*
J. R. Solana	*Departamento de Física Aplicada, Facultad de Ciencias, Av. de los Castros s/n, Universidad de Cantabria, E-39071-Santander.*
N. Stollenwerk	*Theoretical Ecology Group, ATO Research Centre, Germany.*
H. Szichman	*Department of Applied Mathematis, SOREQ Nuclear Research Center, Yavnc 70600, Israel.*
M. S. Torre	*Universidad Nacional de la Provincia de Buenos Aires, Pinto 399, 7000-Tandil, Buenos Aires, Argentina.*
J. A. Vacas	*Departamento de Física Aplicada, Facultad de Ciencias, Universidad de Granada, E-18071-Granada.*
J. L. Vallés	*Institute for Advanced Materials, European Joint Research Center, P.O.B. 2, 1755ZG Petten, Netherlands.*
J. Vigo	*Departamento de Matemática Aplicada a la Ingeniería, E.T.S. de Ingenieros Industriales, P° del Cauce s/n, Universidad de Valladolid, E-47011-Valladolid.*

www.ingramcontent.com/pod-product-compliance
Lightning Source LLC
Chambersburg PA
CBHW041304210926
51851CB00002B/4
9789810211639